Oral Medicine - A Clinical Guide

Ramesh Balasubramaniam
Sue-Ching Yeoh • Tami Yap • S. R. Prabhu
Editors

Oral Medicine - A Clinical Guide

Springer

Editors
Ramesh Balasubramaniam
UWA Dental School
The University of Western Australia
Nedlands, WA, Australia

Sue-Ching Yeoh
Sydney Dental School
The University of Sydney
Camperdown, NSW, Australia

Tami Yap
Melbourne Dental School
The University of Melbourne
Carlton, VIC, Australia

S. R. Prabhu
School of Dentistry
The University of Queensland
Herston, QLD, Australia

ISBN 978-3-031-36796-0 ISBN 978-3-031-36797-7 (eBook)
https://doi.org/10.1007/978-3-031-36797-7

This Springer imprint is published by the registered company Springer Nature Switzerland AG
The registered company address is: Gewerbestrasse 11, 6330 Cham, Switzerland

Paper in this product is recyclable.

In memory of our dear colleagues,
Dr Elishka Marvan and Dr Samah
Abdel Hafeth

Foreword

I am honored to write the foreword to this excellent handbook *Oral Medicine: A Clinical Guide*, a guide for clinicians and students to use in their daily practice.

The Editors have called upon authors, mostly oral medicine specialists from Australia and New Zealand and, while the handbook is of international relevance, its Australasian flavor makes it particularly valuable and interesting to those of us working in this part of the world. The production of this book demonstrates the maturing of the discipline of oral medicine in Australasia. From my perspective, it has been wonderful to watch the development of the specialty in this region. When I was doing my postgraduate study at the University of Melbourne in the early 1980s, it was in the fields of both oral medicine and oral pathology. These disciplines are now separate dental specialties, albeit with some overlap. There are still significant difficulties to be resolved or worked around for the specialty to ponder. Can it find a comfortable home between dentistry and medicine? What qualifications are required to be registered as an oral medicine specialist that reflects clinical experience, didactic training and research? What training programs are needed to best educate oral medicine trainees in an efficient yet robust manner; should they be university based, as is usual for dental specialties in Australasia, or hospital/college based as for most medical specialties?

But while these issues are being worked through, it is clear that the discipline is thriving. Bringing together such an excellent group of authors, both specialists and trainees, to produce this book provides evidence of the success of oral medicine in Australasia. In 2022 there were 50 Oral Medicine Specialists recognised by their relevant registering body in Australasia, 44 in Australia and six in New Zealand. I am sure Professor Peter Reade, the "Father of Oral Medicine in Australasia" would be looking down proudly. Peter was the Foundation President of the Oral Medicine Society of Australia and New Zealand (1994–1997). While based in Melbourne, he played an important role promoting scientific investigation in oral medicine and oral pathology throughout all of Australia and New Zealand and internationally. I was the Founding President of the Oral Medicine Society of Australia and New Zealand (Victorian Branch) from 1995 to 1998 at which time I returned to New Zealand and concentrated solely on the practice of oral pathology. These societies have now become the Oral Medicine Academy of Australasia, which acts as the peak body to represent the interests of the profession of oral medicine.

I commend the Editors for their efforts to compile this clinical guide and I am sure readers will value it tremendously.

March 2023

Alison M. Rich
Department of Dentistry
University of Otago
Otago, New Zealand

Preface

The Oral Medicine Academy of Australasia defines Oral Medicine as "the specialist branch of dentistry concerned with the diagnosis, prevention, and predominantly nonsurgical management of medically related disorders and conditions affecting the oral and maxillofacial region, in particular oral mucosal disease and orofacial pain, as well as the oral health care of medically complex patients". Thus, Oral Medicine serves as the important bridge that connects oral health with general health, with a biopsychosocial approach to patient care.

Oral Medicine—A Clinical Guide is a concise, quick reference and study aid for health practitioners in all stages of training and clinical practice. The authorship represents Oral Medicine in Australia and New Zealand with contributions from almost every Oral Medicine Specialist and trainee in the region, along with collaborative academics and specialist medical practitioners.

The handbook encompasses 111 concise chapters on distinct topics in the field of Oral Medicine. The early chapters focus on history taking, clinical examination and diagnostic tests. These are followed by chapters that focus on orofacial diseases and disorders, including mucosal, salivary gland and relevant jawbone pathology, as well as oral manifestations of systemic diseases. Orofacial pain and temporomandibular disorders are broadly covered, as is the field of dental sleep medicine. Each chapter can be read as a standalone topic with cross-referencing to chapters on connected topics. To facilitate access to more comprehensive information, each chapter has a reading list.

It is our vision that this handbook serves as a clinical guide in Oral Medicine for health practitioners at various stages of training and clinical practice and our desire for this "guide" to remain a reliable resource along the clinical journey.

This handbook is the product of the immense support of many individuals. First and foremost, we thank our partners and children for their unwavering support of all our professional endeavours. Recognisably, this project could only be executed given the immense contribution and support of our expert authors. We are also grateful for the administrative assistance provided by Ms Mysia Dumlao and her colleagues at Perth Oral Medicine and Dental Sleep Centre. Finally, we wish to thank Professor Alison Rich for her gracious foreword and Springer Nature, particularly Alison Wolf and her team, for the excellent production quality of this book.

Perth, WA, Australia — Ramesh Balasubramaniam
Sydney, NSW, Australia — Sue-Ching Yeoh
Melbourne, VIC, Australia — Tami Yap
Brisbane, QLD, Australia — S. R. Prabhu
January 2024

Contents

Part III Oral Considerations in Systemic Disease

Part I

The Oral Medicine Patient

History Taking

1

S. R. Prabhu

1.1 Introduction

The key to diagnosis is in the history. Eliciting a full patient history through open-ended questioning and active listening offers critical clues to the diagnosis. It is reported that between 70% and 90% of medical diagnoses can be determined by the history alone.

The following is an outline for history taking.

1.2 The Presenting Complaint

It is important to identify and/or confirm the reason that the patient has been referred for specialist assessment. The concerns of the referring clinician may or may not completely align with the primary concern of the patient. However, both must be addressed. Commencement with open questioning such as 'How can I help you today?' is appropriate to initiate the interaction. Allowing the patient enough time to answer and trying not to interrupt or direct the conversation are important factors.

1.3 History of the Presenting Complaint

Key components of the history of the presenting complaint for mucosal lesions include location, duration and onset of symptoms, aggravating and alleviating factors, any attributable causes, previous episodes, nature/character/severity of

S. R. Prabhu (✉)
School of Dentistry, The University of Queensland, Herston, Qld, Australia
e-mail: r.prabhu@uq.edu.au

R. Balasubramaniam et al. (eds.), *Oral Medicine - A Clinical Guide*,
https://doi.org/10.1007/978-3-031-36797-7_1

symptoms, any treatment used to relieve symptoms and associated symptoms such as fever, and swelling. It is also important to collate the existence of accessible previous investigations and correspondence to aid in understanding the evolution of the patient's concern.

In chronic pain conditions, the complaint may be a protracted experience for the patient. The SOCRATES acronym is a useful tool for exploring orofacial pain. This includes **S**ite, **O**nset, **C**haracter, **R**adiation, **A**ssociations, **T**ime course, **E**xacerbating or relieving factors and **S**everity of pain (on a pain scale of 0–10).

1.4 Medical History

Any significant present and past medical history should be recorded in the patient's notes as these may impact planned oral care. If the patient has a medical condition, information on how well it is controlled and what treatments the patient is receiving should be obtained. Complications and past hospitalization history should also be recorded. History of head and neck cancer in the past is important particularly when dealing with a patient suspected of potentially malignant or malignant disorders. History of allergies is key to the safe administration of medications and other treatments. It is important to know if a patient is pregnant and if so what gestation, or if the patient is currently breastfeeding, as these details may significantly impact the management of oral care issues. Medication history is important as it relates to the presenting complaint, side effects of current medications, contraindications of treatments or identification of current medical conditions. Brief review of systems is essential since this involves a systematic enquiry on a variety of major body systems and symptoms. If positive answers are revealed, these need to be investigated in greater detail.

1.5 Dental History

Past dental history is helpful in assessing the patient's current state of oral health. History of dental treatment received by the patient, including patient's experience before, during and after the treatment are important.

1.6 Family and Social History

With a prevalence of genetic inheritance in certain oral disorders, the consideration of family history is important. A patient's ethnicity may also impact certain oral disorders. Social history is useful in assessing the impact of lifestyle on the general and oral health. Information on occupation, smoking, alcohol consumption, and recreational drug habits provide clues to the aetiology of a host of oral diseases.

At the conclusion of history taking exercise, clinician provides a brief summary of what has been discussed and what follows next.

Recommended Reading

Erian D, Quek SYP, Subramanian G. The importance of the history and clinical examination. J Am Dent Assoc. 2018;149(9):807–14.

Gandhi JS. William Osler: a life in medicine. Br Med J. 2000;321:1087.

Greenwood M. Essentials of medical history-taking in dental patients. Dent Update. 2015;42(4):308–10, 313–5.

Prabhu SR. Patient interview and history taking. In: Prabhu SR, editor. Textbook of oral diagnosis. Oxford University Press; 2007. p. 12–20.

Clinical Examination

2

S. R. Prabhu

2.1 Introduction

Clinical signs and symptoms established during clinical (physical) examination offer important clues to diagnosis and to outline an appropriate management plan. Each patient should be clinically evaluated with both an extra-oral and intra-oral examination.

2.2 General Examination

As soon as the clinician meets the patient, patient's general appearance is passively observed. This observation may give clues to underlying medical conditions and contribute to the diagnosis and treatment plan for the oral complaints.

2.3 Extra-oral Head and Neck Examination

Examination starts with inspection of facial symmetry, and abnormalities if any including those of the vermilion. The lymph nodes in the head and neck area should be palpated gently for tenderness, texture or enlargements. Lymph nodes are generally not palpable. However, small palpable mobile and non-tender lymph nodes are normal. Abnormal lymph nodes are generally larger, fixed and may be tender. Tender lymph nodes usually indicate inflammation. Oral cancer metastasis to lymph nodes often results in fixed, non-tender, firm node enlargement.

S. R. Prabhu (✉)
School of Dentistry, The University of Queensland, Herston, Qld, Australia
e-mail: r.prabhu@uq.edu.au

R. Balasubramaniam et al. (eds.), *Oral Medicine - A Clinical Guide*,
https://doi.org/10.1007/978-3-031-36797-7_2

Midline structure of the neck including the thyroid cartilage should be observed and palpated for asymmetry and movement during swallowing. Examination for orofacial pain disorders includes palpation of muscles of mastication, cervical muscles and cervical vertebrae. Mandibular opening, excursions and protrusion should be observed. The path of mandibular opening should be noted for deflection or deviation. TMJ sounds such as clicking and crepitation should be noted. Auscultation and palpation for joint noises in all movements provide important diagnostic clues. Neurological examination for sensory neurological deficit over the distribution of trigeminal nerve and assessment of facial nerve functions are components of a comprehensive extra-oral examination.

2.4 Intra-oral Examination

Detailed intra-oral examination should include systematic inspection and palpation of all regions of the oral cavity. If there is any soft tissue abnormality, the type of abnormality (e.g., exophytic mass, white or pigmented patch, or ulcer) should be noted including its location, size, colour, surface texture (e.g., smooth, papillary lobulated) and consistency (e.g., soft, firm, fluctuant). Although the presenting complaint is unrelated to teeth, appraisal of the condition of the dentition, occlusion and evidence of oral hygiene is of relevance. Major salivary glands should be palpated and milked to assess salivary gland duct orifices patency and salivary flow and to evaluate the quality of saliva (e.g., frothy versus serous).

2.5 Variations of Normal Anatomical Oral Structures and Orofacial Anatomy

Variations of normal soft tissues are common in the mouth. Often these may be misinterpreted as pathology. Images shown below (Figs. 2.1a–i) are meant for quick reference. Oral health clinicians should have a good knowledge of the variation of normal orofacial structures to be able to differentiate normal from abnormal.

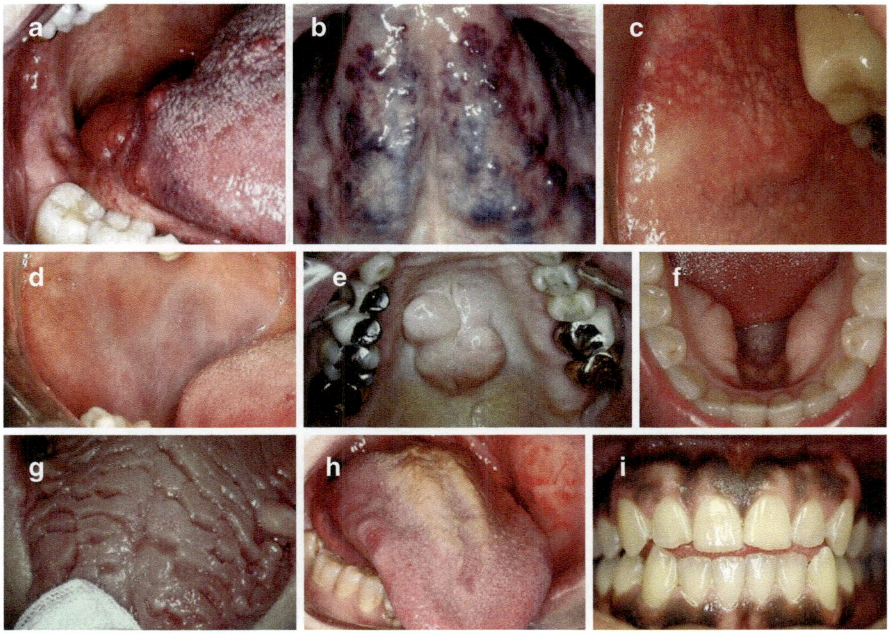

Fig. 2.1 (**a**–**i**) Oral anatomical variants (**a**) Lymphoid aggregates (**b**) lingual varicosities (**c**) fordyce granules (**d**) leukoedema (**e**) palatal torus (**f**) lingual tori (**g**) fissured tongue (**h**) hairy tongue (**i**) physiological pigmentation. (Source: Panagakos FS, Migliorati CA, editors. Diagnosis and Management of Oral Lesions and Conditions: A Resource Handbook for the Clinician [Internet]. 2014. Available from: https://doi.org/10.5772/57597)

2.6 Differential Diagnosis

Integration of information gathered from the history and clinical examination forms the basis for formulating a differential diagnosis. A well-structured differential diagnosis lists possible diagnoses in order of probability, the most probable being listed first. One of the strategies that can be used in formulating differential diagnosis is to put the facts gathered into a 'surgical sieve' based on aetiology and pathogenesis. Categories include multiple causes such as traumatic, developmental, inflammatory/immunologic, infectious, neoplastic, iatrogenic, idiopathic or systemic disease. Another method employed in differential diagnosis of lesions is to categorise them according to their predominant presenting clinical features such as blisters, ulcers, patches and swellings or lumps. With regard to orofacial pain diagnosis, a similar approach broadly classifying pain as either dental, musculoskeletal, neurovascular, vascular or neuropathic is the first step. Once the broad category has been established, specific diagnosis may be determined. Final diagnosis is made on the basis of all available clinical observations, imaging and laboratory investigations.

Recommended Reading

Hardiman R, Kujan O, Kochaji N. Normal variation in the anatomy, biology, and histology of the maxillofacial region. In: Farah C, Balasubramanian R, McCullough M, editors. Contemporary oral medicine. Springer, Cham; 2019. p. 20–78.

Prabhu SR. Diagnostic approaches to common oral symptoms. In: Prabhu SR, editor. Textbook of oral medicine. Oxford University Press; 2007. p. 1–11.

Prabhu SR. Anatomical variants and normal structures often mistaken as pathological lesions. In: Prabhu SR, editor. Handbook of oral diseases for medical practice. Oxford University Press; 2016. p. 12–16.

Blood Tests

3

S. R. Prabhu

3.1 Introduction

Blood tests in oral medicine practice may be ordered to reach a diagnosis, monitor disease, provide baseline screening preceding commencement of therapy, therapy monitoring, and when clinician suspects an underlying haematologic or other related systemic disorder associated with or contributing to oral presenting signs and symptoms.

3.2 Haematology: Full Blood Count

Full Blood Count (FBC), also known as complete blood count (CBC) is the most common blood test ordered by clinicians. A FBC essentially includes red blood cell count, mean red cell volume, white blood cell, differential count, platelet count, and haemoglobin concentration. In oral medicine practice, a full blood count test might be requested if patients present with persistent or recurrent oral ulcerations, opportunistic infections, symptoms of oral burning, mucosal petechiae or gingival bleeding, or other oral signs and symptoms suggestive of haematological abnormalities or unexplained immunosuppression.

S. R. Prabhu (✉)
School of Dentistry, The University of Queensland, Herston, Qld, Australia
e-mail: r.prabhu@uq.edu.au

R. Balasubramaniam et al. (eds.), *Oral Medicine - A Clinical Guide*,
https://doi.org/10.1007/978-3-031-36797-7_3

3.3 Blood Chemistry: Blood Sugar Estimation

The oral medicine practitioner is often required to consider assessment of glycaemic control of patients who require long-term corticosteroid treatment for oral lesions of immunobullous disorders, chronic oral candidosis, and diabetes-related hyposalivation. This biochemical test includes random glucose, fasting glucose, post glucose challenge, and glycosylated haemoglobin (HbA1c). The fasting glucose and post glucose challenge are commonly used to help diagnosis of diabetes mellitus, while the HbA1c gives a long-term information of glycaemic control.

3.4 Haematinics

In oral medicine practice, Vitamin B_{12}, folate estimation and iron studies are requested for patients presenting with recurrent and persistent oral ulceration, symptoms of oral burning, taste disturbances, oral candidosis, anaemia-associated atrophic glossitis, or oral manifestations associated with oral inflammatory bowel disease. Iron studies include serum iron, serum ferritin, total iron-binding capacity (TIBC), and transferrin saturation.

3.5 Inflammatory Markers, Serology, Human Leukocyte Antigen (HLA), and Antinuclear Cytoplasmic Antibody (ANCA) Testing

Erythrocyte sedimentation rate (ESR) and cross-reactive protein (CRP) estimations are used as markers of inflammation. Serological tests are requested in patients with suspected giant cell arteritis, Sjögren syndrome, rheumatoid arthritis, and lupus erythematosus. These tests are aimed at confirming the diagnosis and, in some cases to monitor the disease activity. Tests for rheumatoid arthritis include estimation of rheumatoid factor (RF) and anti-CCP *(anti-Cyclic Citrullinated Peptide)* antibody estimation. The anti-CCP also differentiates rheumatoid arthritis from other types of arthritis. Based on the indications, HLA B27 for seronegative arthropathy, coeliac HLA screening, Behcet's HLA screening, pre-carbamazepine screening, and pre-dapsone screening are occasionally requested by oral medicine practitioners. Hepatitis B and C serology, TB and HIV screening tests are also occasionally undertaken. Antinuclear cytoplasmic antibody (ANCA) testing is done for suspected small vessel vasculitides.

3.6 Coagulation Panel

An assessment of clotting function in addition to a full blood count would normally be required for patients presenting with spontaneous gingival bleeding with no apparent cause, oral purpura, history of prolonged bleeding following tooth extraction and those on anticoagulant therapy or liver disease requiring biopsy for oral lesions. A full blood count will detect abnormalities in platelet count and platelet morphology, whereas platelet function tests detect abnormalities in platelet function. Prothrombin time (PT) expressed as a ratio compared to international normalized ratio (INR) assesses the extrinsic pathway of the clotting cascade. An activated partial thromboplastin time (aPTT) assesses the intrinsic pathway of the clotting cascade and thrombin time assesses conversion of fibrinogen into fibrin. Detailed discussion on these is beyond the scope of this chapter.

Recommended Reading

Gomes AOF, Silva Junior A, Noce CW, Ferreira M, Maiolino A, Torres SR. The frequency of oral conditions detected in hematology inpatients. Hematol Transfus Cell Ther. 2018;40(3):240–4.

Hodgson T, Carey B, Hayes E, Ni Riordain R, Thakrar P, Viggor S, et al. Laboratory medicine and diagnostic pathology. In: Farah C, Balasubramaniam R, McCullough M, editors. Contemporary oral medicine. Springer, Cham; 2019. p. 237–292.

Prabhu SR. Haematological investigations. In: Prabhu SR, editor. Textbook of oral diagnosis. Oxford University Press; 2007. p. 202–212.

Imaging

4

S. R. Prabhu

4.1 Introduction

In oral medicine practice, clinicians encounter a diverse group of orofacial hard and soft tissue lesions and disease processes that require the use of imaging techniques to reach a diagnosis.

4.2 Ionising Radiation-Based Imaging Techniques

Ionising radiation-based imaging techniques comprise conventional plain films and specialized and advanced techniques such as sialography and computed tomography (CT) imaging. Only advanced techniques are briefly discussed below.

- **Sialography** is utilised to image the flow within salivary glands most commonly the parotid gland. Sialography usually involves the injection of a small amount of contrast medium into the salivary duct of a single gland, followed by routine X-ray projections. A series of radiographs are then taken to determine the flow of the contrast medium to identify any ductal obstructions and the rate of excretion of the contrast medium from the gland.
- **Computed tomography and Cone Beam Computed Tomography:** Computed tomography (CT) utilizes ionising radiation to produce three-dimensional images. CT can be divided into 2 categories, namely fan beam and cone beam. In the former, a rotating narrow fan shaped X-ray beam is produced by the source and transmitted through the patient and then detected by a ring of sensitive

S. R. Prabhu (✉)
School of Dentistry, The University of Queensland, Herston, Qld, Australia
e-mail: r.prabhu@uq.edu.au

R. Balasubramaniam et al. (eds.), *Oral Medicine - A Clinical Guide*,
https://doi.org/10.1007/978-3-031-36797-7_4

detectors. Cone beam computed tomography (CBCT) is a form of CT which utilises a cone shaped beam aimed at a flat panel detector. The acquired data is reconstructed in multiplanar reformats allowing the visualisation of the maxillofacial complex in multiple planes (axial, sagittal, coronal). CT imaging can be further enhanced by the use of contrast media. CT scan with contrast medium is useful in outlining the extent of ductal system, hollow structures such as cysts or blood vessels.

4.3 Non-ionising Radiation-based Imaging Techniques

- **Ultrasound:** Ultrasound (US) is a non-ionising radiation-based diagnostic imaging technique. In this technique, high frequency sound waves (>20 kHz) are transmitted into the body using a transducer, and the echo produced from tissue interface is detected and displayed on a screen. Other uses of ultrasound imaging include guiding the needle for fine needle aspiration and therapeutic injection.
- **Magnetic resonance imaging (MRI):** MRI is a non-ionising radiation-based diagnostic imaging technique, using strong magnetic fields, magnetic field gradients, and radio waves to generate images. *Gadolinium*-based contrast agent may improve the visualisation of normal and abnormal tissue. MRI is an excellent imaging tool for soft tissue lesions of the head and neck region including space occupying lesions. Magnetic resonance angiography (MRA) is a non-invasive imaging method for the evaluation of the blood vessels used to study arterial anatomy and flow in the maxillofacial region such as in a suspected vascular malformation. Clinicians should be mindful that most cardiac pacemakers and a variety of metallic implants and coils may contraindicate the use of MRI.

4.4 Functional Imaging Techniques

- **Positron emission tomography (PET)** is a nuclear medicine imaging technique that measures physiological function by looking at blood flow, metabolism, neurotransmitters, and radiolabelled drugs. A radiolabelled biological compound such as F-18 fluorodeoxyglucose (FDG) is injected intravenously. Isotopes used are radioactive and decay rapidly. PET is used for detecting and identifying metastatic tumours.
- **Scintigraphy** is an imaging technique in nuclear medicine wherein radioisotopes are administered internally, and the emitted radiation is captured by external detectors to form two-dimensional images. It is a useful imaging technique for salivary gland lesions, chronic bone infections, or neoplasia. It uses a radioactive isotope (technetium 99) to visualise particular types of cells and their functions in salivary gland lesions.

Recommended Reading

Hodgson T, Carey B, Hayes E, Ni Riordain R, Thakrar P, Viggor S, et al. Laboratory medicine and diagnostic pathology. In: Farah C, Balasubramaniam R, McCullough M, editors. Contemporary oral medicine. Springer, Cham; 2019. p. 237–292.

Shah N, Bansal N, Logani A. Recent advances in imaging technologies in dentistry. World J Radiol. 2014;6(10):794–807.

Tissue Pathology and Cytopathology

5

S. R. Prabhu

5.1 Introduction

Histopathology, immunohistopathology, and cytology play an important role in the diagnostic workup in oral medicine. This chapter briefly highlights the procedures involved and applications of these investigations.

5.2 Tissue Pathology

5.2.1 Histopathology

Histopathologic diagnosis of lesions is carried out on the tissue samples obtained by incisional, punch or excisional biopsy methods. Biopsy is indicated for malignant lesions, potentially malignant disorders, vesiculobullous disorders and most mucosal lesions for definitive diagnosis. When incisional biopsy is contemplated, the most representative sample of the lesion must be selected. Inclusion of the periphery of the lesion to include surrounding tissue is appropriate for incisional biopsy for an ulcerated lesion as the ulcer has no epithelial covering. For lesions covered with intact epithelium, the incisional peripheral sampling of healthy tissue may lead to a missed or under-diagnosis of the lesion. Removing a wedge of a minimal depth of 3 mm, minimal length of 3–6 mm, and minimal width of 1–2 mm, is ideal. Often, multiple biopsies may be necessary for more extensive lesions and/or those with varied appearances. Most biopsy tissue samples for histopathology are sent to the laboratory in 10% buffered formalin.

S. R. Prabhu (✉)
School of Dentistry, The University of Queensland, Herston, Qld, Australia
e-mail: r.prabhu@uq.edu.au

R. Balasubramaniam et al. (eds.), *Oral Medicine - A Clinical Guide*,
https://doi.org/10.1007/978-3-031-36797-7_5

5.2.2 Immunofluorescence

Immunobullous lesions of the oral mucosa usually require immunofluorescence (IF) staining. Tissue specimens for these tests should not be placed in formalin, as this alters proteins and significantly diminishes the accuracy of results. Immunofluorescence technique involves frozen microtome sectioning of fresh biopsy samples. The specimen should be placed in saline solution, a special medium (Michel's solution) or frozen in liquid nitrogen and transported to the laboratory as soon as possible. Fresh tissue samples are also required for flow cytometry and microbiological culturing.

5.2.3 Molecular Biological Tests

Molecular biological tests are increasingly used for screening and for identifying genetic abnormalities, surrogate markers of gene expression and for identifying bacteria and viruses. Molecular biology tests include protein-targeted modalities, such as immunohistochemistry (IHC) or nucleic acid targeted modalities such as polymerase chain reaction (PCR) and in situ hybridisation (ISH) techniques.

IHC often selects a surrogate marker of gene expression, such as p16 for the human papilloma virus. Detection of specific in situ nucleic acid sequences can utilize chromogenic in situ hybridisation (CISH) or fluorescent in situ hybridisation (FISH) methods. Example of routine ISH methods includes detection of the Epstein–Barr virus (EBV) in tissue or human herpes virus 8 (HHV8) within cells.

Polymerase chain reaction (PCR) is a highly specific and sensitive technique by which minute quantities of specific DNA and/or RNA segments are enzymatically amplified to detect various genetic and microbial nucleic acids in a variety of disorders. PCR requires extraction of the DNA or RNA from clinical samples and is commonly used in typing specific viral and bacterial infections.

5.3 Cytopathology

Cytopathological diagnosis rests upon the alterations and abnormalities observed in a single cell or a group of cells. Cytological investigations used in oral medicine setting include Tzanck smear preparations for the diagnosis of vesicular immunobullous and herpes virus diseases. A variety of adjunctive techniques are also available to potentially enhance the diagnosis of the cytologic specimens. These include DNA analysis, immunocytochemistry and molecular analysis.

Non-invasive swabs and smears are commonly used to detect the presence of viruses, bacteria and fungi. Swabs can be used for culture as well as PCR-based assessment. Smears are typically fixed and stained before light-microscope assessment.

Recommended Reading

Hodgson T, Carey B, Hayes E, Ni Riordain R, Thakrar P, Viggor S, et al. Laboratory medicine and diagnostic pathology. In: Farah C, Balasubramaniam R, McCullough M, editors. Contemporary oral medicine. Springer, Cham; 2019. p. 237–292.

Lee JYJ, Wan H. Application of immunofluorescence in the diagnosis of oral diseases. J Dent Oral Biol. 2017;2(2):1029.

Logan RM, Goss AN. Biopsy of the oral mucosa and use of histopathology services. Aust Dent J. 2010;55 Suppl 1:9–13.

Part II

Oral Mucosal and Salivary Pathology

Oral Human Herpes Virus Infections

6

Ivy Tan and Michael McCullough

6.1 Introduction

Human herpes viruses (HHVs) generally present stereotypically, unless in immunocompromised host, with most primary infections being self-limiting over 7–10 days, with signs and symptoms of cervical lymphadenopathy, fever, and sore throat. In those immunocompromised, latent viruses can be reactivated, with severe or prolonged recurrence which require prophylactic and medical management with antivirals.

Ninety percent of adults have been infected by one or multiple HHVs. Global seroprevalence estimates (2016) of HSV-1 were 66.6% (age: 0–49 years) and HSV-2 were 13.2% (age: 15–49 years). It should be noted that the burden of HHV-8 disease remains significant for HIV positive individuals.

6.2 Clinical Presentation and Investigations

HSVs are an 8 member DNA virus family, with humans the natural host, with epithelial tropism through contact with oral fluids and/or genital lesions and lifelong latency. The two serotypes, HSV-1 and HSV-2 primarily infect oral and genital areas, respectively, but cross-over can occur. The primary infection of HSV-1, primary herpetic gingivostomatitis, usually occurs in childhood, is less severe compared to adult onset, and presents as small, round, friable vesicles that breakdown and coalesce to large ulcerations. It is transported in a retrograde fashion to the trigeminal sensory ganglion commonly involving the mandibular branch. Latency ensues, and

I. Tan · M. McCullough (✉)
Melbourne Dental School, The University of Melbourne, Carlton, VIC, Australia
e-mail: m.mccullough@unimelb.edu.au

R. Balasubramaniam et al. (eds.), *Oral Medicine - A Clinical Guide*,
https://doi.org/10.1007/978-3-031-36797-7_6

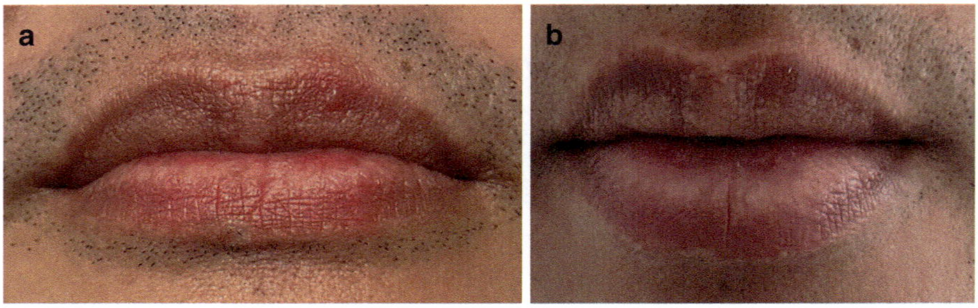

Fig. 6.1 (**a**) Upper lip showing early vesicle eruption on top left lip close to vermillion; (**b**) day 3 after use of topical antiviral day 2 (**b**—photograph—patient supplied)

recurrence presenting as herpes labialis, commonly located on the lip vermilion borders (see Fig. 6.1a, b). This may be heralded by prodromal symptoms of tingling, with subsequent erythematous papules and eruption of vesicles. Secondary reactivation can occur through triggers such as sunlight, wind, friction, and stress. HSV-1 may rarely be associated erythema multiforme, an acute mucocutaneous hypersensitivity reaction typically occurring 1-week post-virus reactivation.

The primary infection of VZV (HHV-3), also known as chicken pox, is well recognized. Transmitted through contact with droplets or active lesions, high vaccination rates have lowered its incidence. VZV remains latent within the spinal or sensory ganglia, its recurrence presenting as herpes zoster/shingles. The prodromal phase of herpes zoster may be confused with odontogenic pain, presenting as a unilateral eruption of vesicles along a dermatome, most commonly the ophthalmic branch of the trigeminal nerve (with an added risk of blindness with corneal involvement). Additionally, the incidence of post herpetic neuralgia (PHN), a chronic neuropathic pain syndrome, is a more common complication amongst the elderly (>70 years).

The primary infection of EBV infection mononucleosis or glandular fever is also well recognized, however the oral manifestations of EBV (HHV-4) and CMV (HHV-5) are more common in the immunocompromised, as EBV associated hairy leukoplakia, non-Hodgkin lymphomas, and CMV-related stellate-like tongue ulcerations.

HHV 6 and 7 are not known to have any oral manifestations, occurring as common cutaneous childhood exanthematous disease (HHV-6) and single rose-coloured, scaly plaques (HHV-7). HHV-8 prominence is primarily associated HIV positive patients presenting as Kaposi's sarcoma.

The clinical picture is usually diagnostic; however, a PCR test may be useful when there are suspecting features. Additional diagnostic tests are presented in the accompanying table (Table 6.1).

Table 6.1 Summary of common oral manifestations, diagnosis, and management of human herpes viral infections

Human herpes virus	Oral disease	Tropism	Clinical presentation	Diagnosis/additional tests	Management
HSV-1	1^0—Primary herpetic gingivostomatitis 2^0—Herpes labialis	Epithelial cells Latent in sensory V3 ganglion	Vesicles, erosions, ulcerations	Clinical picture or PCR. Serological tests not helpful	Supportive measures, palliation e.g. fluids, analgesics, antipyretics Application of 5% topical acyclovir on prodromal stage in reactivated infections every 4 h while awake/ 5 times a day for 4 days. Systemic antivirals in immunocompromised if persistent reactivation occurs, e.g. valaciclovir for HZV: 1 g tds 7 days to commence within 72 h of rash For HSV: 2 g twice daily (12 hourly) for 1 day; initiate at symptom onset; max 2 doses For immunocompromised: Prophylactic dose 500 mg twice daily.
HSV-2	Similar to HSV-1				
VZV HHV 3	1^0—Chicken pox 2^0—Shingles	Dermato-mucosatropic Latent in sensory ganglia of cranial nerves (usually trigeminal) Spinal ganglia	Vesicular eruptions distributed in a classic unilateral fashion according to the respective dermatome or trigeminal branch.	Clinical picture or PCR	Early treatment of systemic antivirals imperative for prevention of post herpetic neuralgia onset, e.g., valaciclovir for HZV: 1 g tds 7 days to commence within 72 h of rash Involvement of V1 requires ophthalmology review

(continued)

Table 6.1 Continued

Human herpes virus	Oral disease	Tropism	Clinical presentation	Diagnosis/additional tests	Management
EBV HHV 4	1^0—Infection mononucleosis 2^0—Oral hairy leukoplakia	Epithelial cells B lymphocytes	Clinical fatigue is common	Clinical biopsy with IHC for EBV derived antigen Detection of anti EBV antibodies in serum (Monospot/Paul Bunnell tests) Atypical lymphocytosis in full blood count	Supportive measures, palliation eg fluids, analgesics, antipyretics No management required for oral hairy leukoplakia
CMV HHV 5	1^0—Glandular fever	Salivary gland cells	In immunocompromised individuals, lateral tongue ulcerations	Differential to EBV- negative Paul Bunnell test	Supportive measures, palliation eg fluids, analgesics, antipyretics
HHV 8	Kaposi's sarcoma	Broad cellular tropism	Solid malignancy	Clinical biopsy	Radiation +/− chemotherapy in a multidisciplinary setting

6.3 Management and Prognosis

Prophylactic doses for the immunocompromised patients and early treatment with antivirals are important to reduce the duration and occurrence of severe symptoms, particularly topical application of antivirals for herpes labialis ideally on prodromal onset, for 7 days.

The ubiquitous nature of the HHV increases the likelihood of patients presenting to dental practitioners for diagnosis and treatment. A good understanding is encouraged, for clinicians to be aware of the different impact of individual HHV infections as almost all human herpes viruses are ubiquitous.

Recommended Reading

Celentano A, Tovaru S, Yap T, Adamo D, Aria M, Mignogna MD. Oral erythema multiforme: trends and clinical findings of a large retrospective European case series. Oral Surg Oral Med Oral Pathol Oral Radiol. 2015;120(6):707–16.

Clarkson E, Mashkoor F, Abdulateef S. Oral Viral Infections: Diagnosis and Management. Dent Clin North Am. 2017;61(2):351–63.

Fatahzadeh M. Oral Manifestations of Viral Infections. Atlas Oral Maxillofac Surg Clin North Am. 2017;25(2):163–70.

Johnson RW, Rice AS. Clinical practice. Postherpetic neuralgia. N Engl J Med. 2014;371(16): 1526–33.

Porter S, Leão JC, Gueiros LA. Oral and maxillofacial viral infections. In: Farah C, Balasubramaniam R, McCullough M, editors. Contemporary oral medicine: a comprehensive approach to clinical practice. 2019. p. 983–1007.

Yap T, Khor S, Kim JS, Kim J, Kim SY, Kern JS, et al. Intraoral human herpes viruses detectable by PCR in majority of patients. Oral Dis. 2021;27(2):378–87.

Coxsackie Viruses

7

Suhaib Alqudah

7.1 Introduction

Hand, foot, and mouth disease (HFMD) and herpangina (HA) are two common paediatric viral infections that are characterised by febrile illness, maculopapular rash, and vesicles affecting the mouth, hands, and feet.

HFMD and HA share most of their epidemiological features. Both conditions affect children under the age of 5 years in more than 90% of cases, typically in the spring to summer seasons.

HFMD and HA are caused by subtypes of coxsackieviruses from the genus enteroviruses. Coxsackievirus A16 (CV-A16) and enterovirus E71 (EV-71) are responsible for the majority of HFMD and HA infections.

7.2 Clinical Presentation and Investigations

Herpangina and HFMD share many clinical symptoms. The two conditions are sometimes considered a continuum. Both conditions are characterised by febrile illness, and vesiculo-ulcerative oral enanthem. The incubation period ranges from 3 to 10 days, and the patient is most infectious during the first week of illness. Active

S. Alqudah (✉)
Melbourne Dental School, The University of Melbourne, Carlton, VIC, Australia

Faculty of Dentistry, Jordan University of Science and Technology, Irbid, Jordan
e-mail: salqudah@student.unimelb.edu.au

R. Balasubramaniam et al. (eds.), *Oral Medicine - A Clinical Guide*,
https://doi.org/10.1007/978-3-031-36797-7_7

viruses may be present in the stool for up to 3 months after resolution of clinical disease.

In HFMD, symptoms might be subclinical or present as a mild fever in a third of cases. Other individuals may present with low grade fever that resolves in 2 days, followed by a sore mouth, throat, and poor appetite. Oral lesions start as erythematous macules affecting the tongue, buccal mucosa, palate, and anterior pillar of fauces. The macules develop into vesicles which later coalesce to form painful oral ulceration measuring 1-10 mm in diameter. Cutaneous lesions follow oral lesions and appear as erythematous macules typically on the hands, palms, and feet progressing to non-itchy vesicles which heal in 7–10 days without scarring. Atypical presentation, severe cutaneous involvement, and serious complications such as aseptic meningitis and acute flaccid paralysis rarely occur.

Herpangina is associated with high grade fever that can reach 40 °C or above and can lead to febrile seizures. It can be associated with respiratory symptoms including cough, rhinorrhoea, and sore throat, headache, dysphagia, and abdominal pain. Oral lesions appear on the soft palate, uvula, and anterior pharyngeal wall as greyish white papulo-vesicular lesions that ulcerate in 1–2 days to leave a small area of erosion surrounded by erythematous halo. Extra-oral manifestations are uncommon, and lesions resolve in 3–7 days. Similar to HFMD, rare complications associated with EV-71 may occur.

HFMD and HA are generally diagnosed clinically upon identification of their classical features. In complicated cases, cell culture or polymerase chain reaction (PCR) may be necessary.

7.3 Management and Prognosis

Management is usually supportive in the form of pain relief, maintaining hydration, and controlling fever using non-steroidal anti-inflammatory drugs or paracetamol. Antiviral therapy is neither available nor effective for enterovirus infections. Therefore, prevention is best and involves measures such as personal hygiene, disinfection of exposed surfaces, social distancing, and isolation of the infected children from schools or day-care centres. Enteroviruses have the ability to block some of the innate immune system pathways and escape immune surveillance. Additionally, HFMD and HA can be caused by multiple virus strains and genotypes. Therefore, vaccinations are not effective, and reinfection can occur in 6% of young children, predominantly in the first 13 months after the first infection.

Recommended Reading

Corsino CB, Ali R, Linklater DR. Herpangina. StatPearls. Treasure Island (FL): StatPearls Publishing; 2021.

Esposito S, Principi N. Hand, foot and mouth disease: current knowledge on clinical manifestations, epidemiology, aetiology and prevention. Eur J Clin Microbiol Infect Dis. 2018;37(3):391–8.

Saguil A, Kane SF, Lauters R, Mercado MG. Hand-Foot-and-Mouth Disease: Rapid Evidence Review. Am Fam Physician. 2019;100(7):408–14.

Yu H, Li XW, Liu QB, Deng HL, Liu G, Jiang RM, et al. Diagnosis and treatment of herpangina: Chinese expert consensus. World J Pediatr. 2020;16(2):129–34.

8 Orofacial Human Papillomavirus Infections

Simone Belobrov, Ivy Tan, and Michael McCullough

8.1 Introduction

HPV is a group of DNA viruses with over 200 subtypes with varied epithelial tropism that are risk stratified by their oncogenic potential. High-risk types, particularly HPV-16 and 18, are now targeted through the national vaccination program. Transmission occurs through contact with active lesions, also conferring lifelong latency. Low-risk types can cause benign lesions presenting clinically as oral papillary lesions. High-risk types are associated with oropharyngeal squamous cell carcinoma (OPSCC) that has a better prognosis than those caused by smoking and alcohol.

HPV infection is extremely common worldwide. It is thought that sexually active adults will have an asymptomatic HPV infection. Benign HPV-related mucosal lesions are generally asymptomatic.

S. Belobrov · I. Tan
Royal Dental Hospital of Melbourne, Carlton, VIC, Australia

Melbourne Dental School, The University of Melbourne, Carlton, VIC, Australia

M. McCullough (✉)
Melbourne Dental School, The University of Melbourne, Carlton, VIC, Australia
e-mail: m.mccullough@unimelb.edu.au

R. Balasubramaniam et al. (eds.), *Oral Medicine - A Clinical Guide*,
https://doi.org/10.1007/978-3-031-36797-7_8

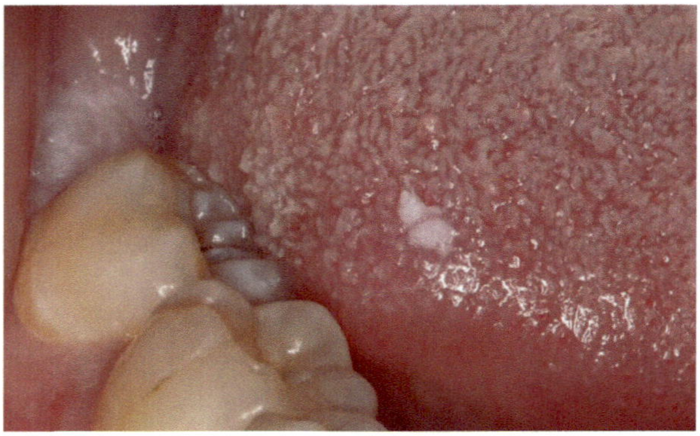

Fig. 8.1 A white exophytic papillary lesion on the tongue that was histopathologically confirmed to be a squamous papilloma

8.2 Clinical Presentation and Investigations

8.2.1 Squamous Papilloma

Squamous papilloma, the most common oral mucosal HPV lesion presents as small finger-like projections causing a rough or cauliflower-like surface, usually solitary. Often located on the palate, tongue, and labial mucosa (Fig. 8.1). HPV types 6 and 11 are the predominant causative genotypes isolated.

8.2.2 Verruca Vulgaris

Verruca vulgaris, often clinically indistinguishable from squamous papilloma and mostly observed on cutaneous surfaces (including the vermilion border), intra-orally is usually located on the labial mucosa and anterior tongue. They are usually solitary lesions with autoinoculation the main mode of transmission. HPV types 1 and 57 are the predominant causative genotypes, however HPV 6 and 11 are sometimes isolated.

8.2.3 Condyloma Acuminatum

Condyloma acuminata are oral warts acquired sexually from oral-genital contact. Maternal transmission is also possible. If these lesions are found in young children it is advisable to screen for sexual abuse; however, additional routes of infection (through fomites) can occur. They have a similar appearance to the squamous papilloma, although multiple in nature, mostly located on the labial mucosa, soft palate, lingual frenum, and tongue. HPV types 2, 6, and 11 are the more commonly isolated genotypes.

8.2.4 Multifocal Epithelial Hyperplasia (Heck's Disease)

Multifocal epithelial hyperplasia presents as multiple soft flat elevated nodules commonly located on the lower lip and buccal mucosa that generally occur in early childhood, are asymptomatic, persist for several years and spontaneously regress with time. These can cluster in particular ethnic groups, may present in multiple family members, and associated with poor living conditions and low socio-economic. HPV 13 and 32 are the predominant isolated genotypes.

Following assessment for typical presentation, the diagnosis of HPV-related lesions can be confirmed by histopathology. Immunohistochemical staining is utilised in some cases to identify viral proteins. Identifying the possible source of the causative virus is rarely helpful.

8.3 Management and Prognosis

Oral HPV-related benign lesions can resolve without treatment, particularly multifocal epithelial hyperplasia. If aesthetics, repeated trauma during mastication or viral transmissibility is a concern, they can be removed surgically. Patients should still be advised to have any other cutaneous warts managed by the appropriate physician. Lesions are biopsied when a definitive diagnosis is necessary.

Although HPV can be sexually transmitted, it is important to recognise that other modes of transmission are also possible. Regression of lesions can occur without intervention; however, biopsy may be necessary for precise diagnosis, decrease viral transmissibility, and/or to remove unsightly or inconvenient lesions.

Recommended Reading

Belobrov S, Cornall AM, Young RJ, Koo K, Angel C, Wiesenfeld D, et al. The role of human papillomavirus in p16-positive oral cancers. J Oral Pathol Med. 2018;47(1):18–24.

Betz SJ. HPV-Related Papillary Lesions of the Oral Mucosa: A Review. Head Neck Pathol. 2019;13(1):80–90.

Meites E, Gee J, Unger E, Markowitz L. Human papillomavirus. In: Hall E, Wodi AP, Hamborsky J, Morelli M, Schillie S, editors. Epidemiology and prevention of vaccine-preventable diseases. Public Health Foundation; 2021. p. 165–178.

Porter S, Leão JC, Gueiros LA. Oral and maxillofacial viral infections. In: Farah C, Balasubramaniam R, McCullough MJ, editors. Contemporary oral medicine. Springer; 2019. p. 983–1107.

Orofacial Viral Infections 9

Katrusha Hull

9.1 Measles

9.1.1 Introduction

Measles (rubeola) is a highly contagious infection caused by a paramyxovirus virus, genius Morbillivirus, a single-stranded RNA virus spread through respiratory droplets or direct contact with the infected nasal or throat secretions.

Prevalence varies widely and is directly correlated to vaccine introduction in 1963 and its use.

9.1.2 Clinical Presentation and Investigations

Initial presentation is usually a high fever 10–12 days after exposure to the virus lasting 4–7 days. The clinical picture occurs over three stages (Table 9.1).

Diagnosis via typical cutaneous features and confirmed by virus isolation and serologic studies.

K. Hull (✉)
Peter MacCallum Cancer Centre, Melbourne, VIC, Australia

Royal Dental Hospital of Melbourne, Carlton, VIC, Australia

Monash Medical Centre, Clayton, VIC, Australia
e-mail: katrusha.hull@dhsv.org.au

R. Balasubramaniam et al. (eds.), *Oral Medicine - A Clinical Guide*,
https://doi.org/10.1007/978-3-031-36797-7_9

Table 9.1 Clinical presentation of measles

	Symptoms and signs	Oral features
Initial stage	Runny nose, cough, high fever, conjunctivitis	Koplik spots: multiple areas of mucosal erythema especially on buccal and labial mucosa with white macules overlying
Second stage	Ongoing fever, maculopapular erythematous rash initially on face with eventual downwards spread	Fading of Koplik spots
Final stage	Resolution of fever and fading of rash	

9.1.3 Management and Prognosis

Management is supportive to ensure good nutrition, adequate fluid, and treatment for dehydration. Significant complications include blindness, encephalitis, and severe respiratory infection with most measles-related deaths as a result of complications in children under 5 or adults over 30 years.

9.2 Mumps

9.2.1 Introduction

Also known as epidemic parotitis, mumps is a highly infectious acute febrile sialadenitis caused by a *Rubulavirus*, a member of the paramyxovirus family. It is a single-stranded RNA virus spread via saliva droplet infection with a 2–3 week incubation.

Mainly presenting in children and young adults, the prevalence of epidemic parotitis has decreased substantially in regions of the world where the MMR vaccine is routinely given.

9.2.2 Clinical Presentation and Investigations

The clinical features of epidemic parotitis are summarised in Table 9.2.

Epidemic parotitis can affect all exocrine glands. Testicular involvement occurs in 25% of post pubertal males leading to significant pain. On resolution, atrophy of the affected testicle can lead to sterility which may be permanent.

Diagnosis is made via clinical presentation combined with demonstration of mumps-specific IgM antibody in serum or fourfold rise in specific IgG titres between acute and convalescent samples. IgM detection is of limited value in unvaccinated individuals.

Table 9.2 Clinical presentation of mumps

	Symptoms and signs	Duration
Prodrome	Headache, fever, and malaise	
Extra-oral salivary gland features	Tender uni-or bilateral enlargement of the major salivary glands. Typically, parotitis.	Salivary gland changes peak within 2–3 days then resolve within 1–2 weeks.
Intra-oral features	Possible erythema and enlargement of Wharton's and Stensen's duct openings	

9.2.3 Management and Prognosis

Possible complications include orchitis, meningitis, and encephalitis. Mumps however is most often self-limiting with management palliative with analgesics and antipyretics. Avoidance of sour foods and drinks may decrease salivary discomfort.

9.3 Rubella

9.3.1 Introduction

Rubella virus is a single-stranded RNA virus of the togaviridae family spread in airborne droplets of respiratory secretions. Viral multiplication occurs in cells of the respiratory system followed by viremic spread to target organs.

Rubella occurs worldwide in children, adolescents, and young adults with peak infection in late winter or early spring in unvaccinated populations.

9.3.2 Clinical Presentation and Investigations

Postnatal infection is often asymptomatic but may present as a mild, self-limiting illness characterised by lymphadenopathy and low-grade fever followed by a diffuse rash. Unlike measles, acute rubella infection is not often associated with pronounced oral features, but red palatal macules are possible. A characteristic maculopapular rash appears on the face-neck which quickly spreads to the trunk, upper extremities, and legs, fading on the face while progressing downwards. Lesions are initially discrete then coalesce causing a flushed appearance. Resolution is within 3–5 days. Congenital disease may cause a number of significant anomalies depending on gestational age. Oral features are outlined in Table 9.3.

Diagnosis is via typical cutaneous features and confirmed by virus isolation and serologic studies.

Table 9.3 Possible oral features of congenital rubella syndrome

	Clinical features
Odontogenic	Missing teeth, enamel hypoplasia in primary or permanent dentition
Non-odontogenic	Narrow maxillary arch, deep palate

9.3.3 Management and Prognosis

No specific treatment is available, and management is supportive. A woman infected with rubella in the first trimester has up to 90% chance of giving birth to a baby with congenital rubella syndrome leading to birth defects including, but not limited to, blindness, deafness, and heart disease.

Recommended Reading

Centers for Disease Control and Prevention. Global Measles. And Rubella [Internet]. Global Immunisation [cited 16 November 2023] Available from: https://www.cdc.gov/globalhealth/measles/index.html.

World Health Organization. Measles and rubella surveillance data [Internet]. Geneva: WHO; 2023 [cited 1 November 2023]. Available at: https://www.who.int/teams/immunization-vaccines-and-biologicals/immunization-analysis-and-insights/surveillance/monitoring/provisional-monthly-measles-and-rubella-data.

Necrotising Periodontal Disease 10

S. R. Prabhu

10.1 Introduction

Necrotising periodontal disease is a spectrum of conditions that affect the periodontium through ulceration and necrosis. This group includes necrotising (ulcerative) gingivitis (NUG), necrotising (ulcerative) periodontitis (NUP), and necrotising stomatitis. NUG is limited to the gingiva. When progression of NUG involves destruction of gingival tissues, loss of alveolar bone, and periodontal ligament, it is classified as NUP. Necrotising stomatitis, also known as Cancrum oris or *Noma*, involves the destruction of structures beyond the mucogingival junction.

Necrotising periodontal disease is a non-communicable infectious disease predominantly involving anaerobic bacteria and spirochaetes. Associations with human herpes viruses and *Candida* species have also been reported. Predisposing factors include poor oral hygiene, malnutrition, psychological stress, smoking, and immunosuppression related to systemic disease (e.g. diabetes and HIV/AIDS). The disease involves decreased neutrophil chemotaxis and phagocytosis, and tissue destruction is mediated by bacterial metabolites such as collagenases, endotoxins, hydrogen sulphide, and fibrinolysin.

S. R. Prabhu (✉)
School of Dentistry, University of Queensland, Herston, Qld, Australia
e-mail: r.prabhu@uq.edu.au

R. Balasubramaniam et al. (eds.), *Oral Medicine - A Clinical Guide*,
https://doi.org/10.1007/978-3-031-36797-7_10

The global prevalence of necrotising periodontal disease ranges from 0.03% to 9.4%. Evidence suggests increased prevalence in two populations: children from low socioeconomic backgrounds in developing countries and in HIV-infected patients, in whom prevalence of necrotising periodontal disease varies between 4.3 and 16%.

10.2 Clinical Presentation and Investigations

In NUG, the interdental papillae are painful, inflamed, and bleed easily with "punched out" crater-like ulcerations often covered by a white pseudomembranous slough (Fig. 10.1). Halitosis is common. Systemic signs and symptoms may include low-grade fever, malaise, and cervical lymphadenopathy. In NUP, erythema also includes the alveolar mucosa and there is severe loss of periodontal attachment without deep pocket formation (Fig. 10.2). Progressive tissue destruction resulting in perforation of the mucosa, muscle, and skin, and exposure of the bone is the hallmark of necrotising stomatitis.

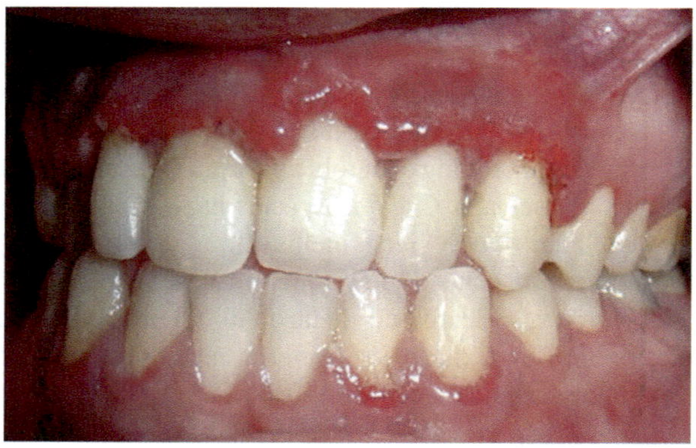

Fig. 10.1 Acute necrotising ulcerative gingivitis. Crater-like ulcerative lesions involving gingival papillae and attached gingiva are characteristic features of ANUG. (Source: Malek R, Gharibi A, Khlil N, Kissa J. Necrotizing Ulcerative Gingivitis. Contemp Clin Dent. 2017;8(3):496–500)

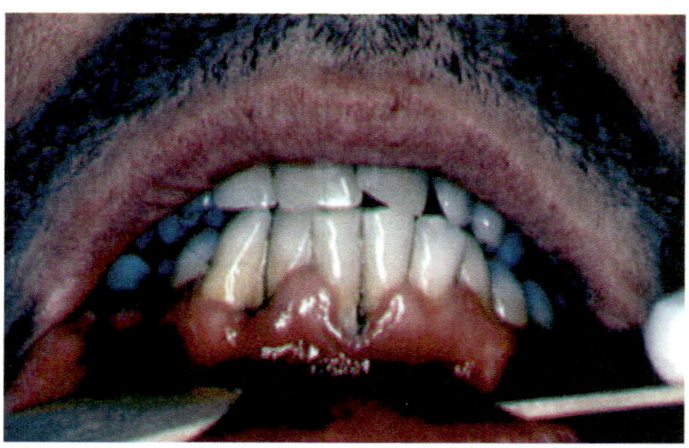

Fig. 10.2 Necrotising ulcerative periodontitis. Loss of periodontal ligament and alveolar bone in relation to lower central incisors is seen in an HIV patient. (Reproduced with permission of HIV dent.org)

Diagnosis of necrotising periodontal disease is based on history and pathognomonic clinical presentation. Further investigation for contributory underlying disease may be warranted.

Differential diagnosis of necrotising periodontal disease includes plaque-related forms of gingival and periodontal disease, primary herpetic gingivostomatitis, causes of desquamative gingivitis, gingival lesions related to haemopoietic dysfunction, and ascorbic acid deficiency.

10.3 Management and Prognosis

Recommended treatment protocol is as follows:

- Debridement of necrotic tissue, frequent mouth rinses (0.12% chlorhexidine or 3% hydrogen peroxide)
- Administration of antibiotics (metronidazole 400 mg orally, 12-hourly for 3–5 days)
- Pain management with NSAIDs or paracetamol
- Oral health care instructions, smoking cessation and follow-up, and re-evaluation

Identification and appropriate management of systemic predisposing factors in consultation with patients' physician is essential. Prognosis is good, and complete resolution is likely for periodontal disease.

Recommended Reading

Malek R, Gharibi A, Khlil N, Kissa J. Necrotizing Ulcerative Gingivitis. Contemp Clin Dent. 2017;8(3):496–500.

Oral and Dental Expert Group. Therapeutic Guidelines: Oral and Dental. Necrotising periodontal disease. Version 3. Melbourne: Therapeutic Guidelines Limited; 2019. p. 73–75.

Oral Manifestations of Actinomycosis

11

Jacinta Vu, Alissa Jacobs, and Lalima Tiwari

11.1 Introduction

Actinomycosis is an uncommon bacterial infection characterised by chronic suppurative and granulomatous inflammation typically aggressive and locally destructive. It often presents as draining sinuses, abscess formation, and fibrosis. Actinomycosis occurs in the cervicofacial region in 50–60% of cases.

Actinomycosis is caused by gram-positive anaerobic or microaerophilic *Actinomyces* species. *Actinomyces israelii* is the most frequent species detected, although actinomycotic infections are typically polymicrobial. Actinomyces species are a normal inhabitant of the oral cavity and usually have low pathogenicity. Infection usually occurs in patients with poor oral hygiene, damage to the oral mucosa, dental extractions, surgical procedures, presence of foreign bodies or devitalised tissues, or as a complication of maxillofacial trauma.

J. Vu (✉)
Perth Children's Hospital, Nedlands, WA, Australia
e-mail: jacinta.vu@health.wa.gov.au

A. Jacobs
Fiona Stanley Hospital, Murdoch, WA, Australia

L. Tiwari
UWA Dental School, The University of Western Australia, Nedlands, WA, Australia

R. Balasubramaniam et al. (eds.), *Oral Medicine - A Clinical Guide*,
https://doi.org/10.1007/978-3-031-36797-7_11

Actinomycosis typically affects males aged 20–60 years, with predisposing factors including diabetes, alcohol abuse, malnutrition, radiation therapy, chronic tonsillitis, otitis, mastoiditis, tumours, surgery, local tissue injuries, and immunosuppression. Poorer outcomes have been noted in patients with human immunodeficiency virus (HIV). There are higher rates of infections in uncontrolled diabetic patients compared to non-diabetic patients, as well as structural changes in the tissues and impaired wound healing.

11.2 Clinical Presentation and Investigations

Cervicofacial actinomycosis most frequently presents in the parotid, submandibular triangle or peri-mandibular region. There may be a chronic, less commonly painful, soft-tissue swelling, as well as draining sinus tracts with sulphur granules. There may be difficulties in chewing and often a chronic/relapsing course of infection. Isolated intra-oral lesions are rare and may present as an indurated mass, single or multiple abscesses, ulcerative lesions, or sinuses.

Diagnosis involves histopathologic examination of a biopsy specimen, microbial culture, and the presence of sulphur granules in the exudate. Culture can be problematic as Actinomyces species are difficult to grow and there is often contamination with other oral flora. Sulphur granules may not always be macroscopically evident in the exudate.

Imaging such as computed tomography (CT) scans may assist in differentiating between actinomycosis and malignancy, with a specific feature being a highly infiltrative, strongly enhancing mass with central suppurative necrosis and few, if any, reactive lymph nodes on CT scan.

Differential diagnoses include a traumatic aetiology, benign neoplasms, malignant neoplasms such as squamous cell carcinoma, granulomatous diseases, infective processes with sinus tract formation such as tuberculosis, mucormycosis, coccidioides, and nocardiosis.

11.3 Management and Prognosis

Treatment often involves surgical debridement of the lesion and a protracted course of antibiotic therapy. Penicillin is considered the drug of choice, with serious infection and bulky disease treated with intravenous penicillin for 2–6 weeks, followed by oral penicillin for 12 months. Oral disease that is of low bulk may be treated with oral antibiotics for a shorter period of between 1 and 4 weeks, or until the lesion has completely healed. For patients who are allergic to penicillin; tetracycline, erythromycin, or clindamycin may be prescribed. It is common for relapse to occur, and hence the patient requires long-term follow-up. Maintenance of excellent oral hygiene is recommended as prophylaxis against actinomycosis.

Recommended Reading

Alamillos-Granados FJ, Dean-Ferrer A, García-López A, López-Rubio F. Actinomycotic ulcer of the oral mucosa: an unusual presentation of oral actinomycosis. Br J Oral Maxillofac Surg. 2000;38(2):121–3.

Boyanova L, Kolarov R, Mateva L, Markovska R, Mitov I. Actinomycosis: a frequently forgotten disease. Future Microbiol. 2015;10(4):613–28.

Crossman T, Herold J. Actinomycosis of the maxilla - a case report of a rare oral infection presenting in general dental practice. Br Dent J. 2009;206(4):201–2.

D'Amore F, Franchini R, Moneghini L, Lombardi N, Lodi G, Sardella A, et al. Actinomycosis of the Tongue: A Case Report and Review of Literature. Antibiotics (Basel). 2020;9(3):124.

Frydrych AM, Farah CS. Non-odontogenic bacterial infections. In: Farah CS, Balasubramaniam R, McCullough MJ, editors. Contemporary oral medicine: a comprehensive approach to clinical practice. Cham: Springer International Publishing; 2019. p. 871–933.

Yadegarynia D, Merza MA, Sali S, Firuzkuhi AG. A rare case presentation of oral actinomycosis. Int J Mycobacteriol. 2013;2(3):187–9.

Oral Manifestations of Tuberculosis 12

Jacinta Vu, Alissa Jacobs, and Lalima Tiwari

12.1 Introduction

Tuberculosis (TB) is a chronic granulomatous infection caused by various strains of Mycobacteria and is still among the most life-threatening infectious diseases, with high mortality rates in adults. Tuberculosis predominantly affects the pulmonary system, but may involve extra-pulmonary sites including the head and neck region. Extra-pulmonary TB is increasing in incidence due to the emergence of drug resistant TB infections, human immunodeficiency virus (HIV), and acquired immune-deficiency syndrome (AIDS), particularly in South-East Asia and Africa.

In humans, *Mycobacterium tuberculosis* is usually the causative organism, although *Mycobacterium bovis* and atypical mycobacteria have been found. *Mycobacterium tuberculosis* is an acid-fast bacillus primarily transmitted through the respiratory tract. Primary oral TB is rare and is typically seen in young patients with associated regional lymphadenopathy. Secondary oral TB results from inoculation of the oral mucosa with infected respiratory secretions and is usually found in middle-aged and older patients, together with pulmonary involvement. Oral TB lesions may also occur due to haematogenous spread of pulmonary TB, with deposition of the mycobacteria in the submucosa, leading to proliferation of the micro-organism, and subsequent ulceration of the overlying mucosa.

J. Vu (✉)
Perth Children's Hospital, Nedlands, WA, Australia
e-mail: jacinta.vu@health.wa.gov.au

A. Jacobs
Fiona Stanley Hospital, Murdoch, WA, Australia

L. Tiwari
UWA Dental School, The University of Western Australia, Nedlands, WA, Australia

R. Balasubramaniam et al. (eds.), *Oral Medicine - A Clinical Guide*,
https://doi.org/10.1007/978-3-031-36797-7_12

In 2018, it was estimated that 10 million persons had incident TB, and there were 1.5 million TB-related deaths. South-East Asia and Africa have the highest numbers of people with incident and prevalent TB. Extra-pulmonary TB is seen in up to 25% of patients, with head and neck manifestations in 10% of cases including the lymph nodes, larynx, middle ear, oral cavity, and pharynx. Oral TB is rare, accounting for 0.05–5% of all cases.

12.2 Clinical Presentation and Investigations

Oral TB may present at nearly any mucosal site, as outlined in Table 12.1. Oral lesions often have a non-specific clinical presentation and may be apparent before pulmonary symptoms present. Table 12.2 lists the clinical presentations and

Table 12.1 Common sites for oral tuberculosis lesions

Dorsal tongue
Lateral border of the tongue
Buccal mucosa
Vestibular area near the corner of the mouth
Lower lip
Salivary glands
Uvula
Tonsils

Table 12.2 Clinical presentation and symptoms of tuberculosis in the head and neck region

Oral cavity	Ulceration Swelling Discharge Nodules (tubercles) Granulomatous enlargement Granulomatous plaques Non-healing extraction socket Diffuse inflammation
Extra-oral	Parotitis Intra-osseous lesions Preauricular swelling and trismus Tracheitis Laryngitis Lymphadenitis
Associated symptoms	Odynophagia Dysphonia Halitosis Excessive salivation
Systemic symptoms	Cough Fever Night sweats Weight loss Anorexia

symptoms of TB in the head and neck region. An ulcer is the most frequently reported lesion; often but not always painful, deep or superficial, with irregular margins, undermined borders, a granulating base, usually slowly enlarging and on occasion ragged and indurated.

For oral lesions, deep biopsy for histopathologic examination is essential, and in some cases several biopsy sites are required. Investigations for suspected oral and systemic TB are summarised in Table 12.3.

There are several clinical differential diagnoses for oral tuberculosis, outlined in Table 12.4.

Table 12.3 Investigations for tuberculosis

Oral TB	Biopsy	Histopathologic examination with haematoxylin and eosin staining demonstrates epithelioid granulomas with Langhans giant cells, with or without caseation, are suggestive of tuberculosis.
		A Ziehl–Neelsen stain for acid-fast bacilli is confirmative for tuberculosis, however only a small percentage of oral biopsies stain positively due to the paucity of tubercle microorganisms in the tissue specimen.
		Immunofluorescence using auramine-rhodamine can also demonstrate mycobacteria.
	Culture	Mycobacterial culture and detection of mycobacterial deoxyribonucleic acid (DNA) using polymerase chain reaction (PCR) also provide a definitive diagnosis.
	Fine Needle Aspiration Cytology (FNAC)	FNAC is highly specific and highly sensitive in the identification of parotitis and / or TB of the major salivary glands.
Systemic TB	Sputum analysis	The gold standard for TB screening is sputum analysis with smear, culture, and nucleic acid amplification testing,
	Radiologic examination	Posterior-anterior chest radiography is considered a safe, effective modality for TB screening and is recommended by the World Health Organization because of its high sensitivity, wide availability, and relatively low cost.
		CT is more sensitive in determining if TB infection is active or not.
	Tuberculin Skin Test (TST) Or Interferon-Gamma Release Assay	Patients who are suspected of having latent tuberculosis may undergo targeted testing with a tuberculin skin test or interferon-gamma release assay.

Table 12.4 Clinical differential diagnosis of oral tuberculosis

Infectious	Bacterial	Syphilis Leprosy Actinomycosis Cat-Scratch Disease
	Fungal	Coccidioidomycosis Blastomycosis
	Parasitic	Leishmaniasis
Malignant neoplasms	Squamous cell carcinoma Lymphoma Metastasis	
Benign neoplasms	Angiofibroma Castleman disease	
Immune-mediated	Periarteritis Nodosa Granulomatosis with polyangiitis (Wegener's granulomatosis) Sarcoidosis Melkersson-Rosenthal syndrome Behcet's disease Crohn's disease	
Other	Amyloidosis Foreign-body reaction Aphthous ulcer	
Trauma	Traumatic ulcer	

12.3 Management and Prognosis

Patients must be referred to a specialist TB physician such as a respiratory or infectious diseases physician. Treatment of oral TB is the same as systemic TB, and lesions generally respond favourably. Internationally accepted standardised drug regimens include a combination of isoniazid, rifampicin, ethambutol, and pyrazinamide, daily for 2 months, followed by isoniazid and rifampicin for a further 4–7 months. Multidrug-resistant and extremely drug resistant TB are noted, requiring alternative drug regimens as recommended by the World Health Organization.

Recommended Reading

Jain P, Jain I. Oral Manifestations of Tuberculosis: Step towards Early Diagnosis. J Clin Diagn Res. 2014;8(12):ZE18–21.

Kakisi OK, Kechagia AS, Kakisis IK, Rafailidis PI, Falagas ME. Tuberculosis of the oral cavity: a systematic review. Eur J Oral Sci. 2010;118(2):103–9.

Khateeb D, Kang M, Capitle E, Feurdean M. Oral Tuberculosis: A Rare Manifestation of Disseminated Disease in a Patient with Dermatomyositis on Chronic Corticosteroids. Case Rep Med. 2016;2016:8193178.

Kumar SN, Prasad TS, Narayan PA, Muruganandhan J. Granuloma with langhans giant cells: An overview. J Oral Maxillofac Pathol. 2013;17(3):420–3.

MacNeil A, Glaziou P, Sismanidis C, Date A, Maloney S, Floyd K. Global Epidemiology of Tuberculosis and Progress Toward Meeting Global Targets - Worldwide, 2018. MMWR Morb Mortal Wkly Rep. 2020;69(11):281–5.

Pang P, Duan W, Liu S, Bai S, Ma Y, Li R, et al. Clinical study of tuberculosis in the head and neck region-11 years' experience and a review of the literature. Emerg Microbes Infect. 2018;7(1):4.

Tandon S, Bhandari V, Kaur Lamba A, Faraz F, Makker K, Aggarwal K. Literature review of oral tuberculosis and report of a case with unique histological presentation. Indian J Tuberc. 2020;67(2):238–44.

Von Arx DP, Husain A. Oral tuberculosis. Br Dent J. 2001;190(8):420–2.

Wang WC, Chen JY, Chen YK, Lin LM. Tuberculosis of the head and neck: a review of 20 cases. Oral Surg Oral Med Oral Pathol Oral Radiol Endod. 2009;107(3):381–6.

World Health Organization. WHO consolidated guidelines on drug-resistant tuberculosis treatment. vol WHO/CDS/TB/2019.7. World Health Organization; 2019.

Oral Manifestations of Syphilis

13

Antonio Celentano

13.1 Introduction

Syphilis is a human systemic sexually transmitted infection (STI) caused by the anaerobic spirochete bacterium *Treponema pallidum*. Differently from other STIs, syphilis is also easily transmissible by close contact with an infectious lesion.

Treponema pallidum survives only a few days without a host, and humans are the only known natural reservoir. The bacterium is able to invade any organ of the body, evade the immune system, and survive in the host for decades. Syphilis can also be passed on vertically or horizontally with the main mode of transmission remaining sexual contact. More than half of new syphilis cases occur in men who have sex with men.

There are about six million new cases per year of syphilis globally in persons aged 15–49 years, and over 300,000 foetal and neonatal deaths are attributed to syphilis. The most recent global prevalence estimate from WHO for syphilis was 0.51% among men in 2016.

13.2 Clinical Presentation and Investigations

Acquired syphilis is classified into four disease stages (primary, secondary, latent, and tertiary). Firstly, the patient forms a lesion (chancre) at the site of inoculation and is highly contagious. Oral chancres are observed in 4–12% of cases, typically affecting tongue, gingiva, palate, and lips. This painless ulcer develops after

A. Celentano (✉)
Melbourne Dental School, The University of Melbourne, Carlton, VIC, Australia
e-mail: antonio.celentano@unimelb.edu.au

R. Balasubramaniam et al. (eds.), *Oral Medicine - A Clinical Guide*,
https://doi.org/10.1007/978-3-031-36797-7_13

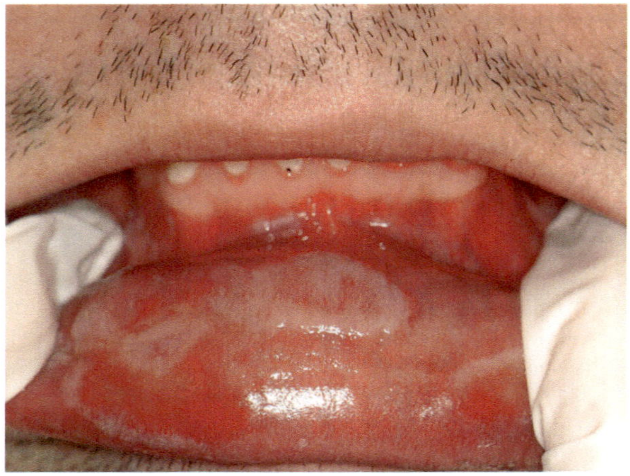

Fig. 13.1 Polymorphous syphilitic lesion (secondary syphilis): whitish and erythematous mucous patches of the lower labial mucosa

incubation (range 10–90 days), resolving 3–8 weeks later. If untreated, 25% of patients develop signs of early secondary syphilis affecting multiple organs such as a maculopapular cutaneous rash on the palms and soles, hepatitis, or uveitis. Painful oral lesions are present in at least 30% of individuals (Fig. 13.1). Secondary syphilis is the most infectious stage and resolves spontaneously in 3–12 weeks and is followed by an asymptomatic latent stage. In tertiary disease, a nodular, ulcerative lesion (gumma) may form. In the oral cavity, the gummas can affect the tongue, tonsils, palate, or lips, and involve bone.

Differential diagnoses of oral syphilis include a vast list of conditions including other infectious diseases, ulcers, autoimmune or immune-related lesion, traumatic or hyperplastic/dysplastic plaques. Diagnosis of syphilis typically relies on clinical and serological findings. Serological tests consist of two types, nontreponemal tests, e.g. Venereal Disease Research laboratory (VDRL), or rapid plasma reagin (RPR) test, and treponemal tests, e.g. those based on fluorescent treponemal antibody absorption (FTA-ABS) and agglutination (TP-PA). Tissue visualisation does require Warthin–Starry stain (WS) silver staining or immunohistochemistry specific for treponema which need to be requested in addition to standard processing.

13.3 Management and Prognosis

Syphilis is treatable with antibiotics, however dosage and length of treatment required depend on the stage and the severity of the case. Usually, long-acting penicillin G benzathine (2.4 million units for adults, 50,000 units/kg for children) in a single dose is the first-choice treatment for primary, secondary, and early latent syphilis. In absence of treatment, syphilis can result in neurological, cardiovascular, and bone diseases later in life and has a mortality rate of 8–58%, with a greater death rate among males.

Recommended Reading

Ficarra G, Carlos R. Syphilis: the renaissance of an old disease with oral implications. Head Neck Pathol. 2009;3(3):195–206.

Kelner N, Rabelo GD, da Cruz Perez DE, Assunção JN, Witzel AL, Migliari DA, et al. Analysis of nonspecific oral mucosal and dermal lesions suggestive of syphilis: a report of 6 cases. Oral Surg Oral Med Oral Pathol Oral Radiol. 2014;117(1):1–7.

Kent ME, Romanelli F. Reexamining syphilis: an update on epidemiology, clinical manifestations, and management. Ann Pharmacother. 2008;42(2):226–36.

Kojima N, Klausner JD. An Update on the Global Epidemiology of Syphilis. Curr Epidemiol Rep. 2018;5(1):24–38.

Tsuboi M, Evans J, Davies EP, Rowley J, Korenromp EL, Clayton T, et al. Prevalence of syphilis among men who have sex with men: a global systematic review and meta-analysis from 2000-20. Lancet Glob Health. 2021;9(8):e1110–e1118.

Oral Manifestations of Bacterial Infections

14

Jacinta Vu, Alissa Jacobs, and Lalima Tiwari

14.1 Introduction

Bacterial infections that cause oral mucosal lesions include scarlet fever, diphtheria, and leprosy. The features of each are outlined in Tables 14.1, 14.2, and 14.3, respectively. Impetigo and erysipelas cause cutaneous infection, and cat-scratch disease also begins in the skin but spreads to the adjacent lymph nodes and as such will not be discussed. Salivary gland bacterial infections and sexually transmitted infections will be discussed in chapters 28 and 83, respectively.

J. Vu (✉)
Perth Children's Hospital, Nedlands, WA, Australia
e-mail: jacinta.vu@health.wa.gov.au

A. Jacobs
Fiona Stanley Hospital, Murdoch, WA, Australia

L. Tiwari
UWA Dental School, The University of Western Australia, Nedlands, WA, Australia

R. Balasubramaniam et al. (eds.), *Oral Medicine - A Clinical Guide*,
https://doi.org/10.1007/978-3-031-36797-7_14

Table 14.1 Scarlet fever

Aetiology and pathogenesis	• Caused by group A ß-haemolytic streptococci • Transmitted by direct spread of large droplets
Epidemiology	• Most common between the ages of 5 and 15 years • Increased incidence during winter and spring
Systemic presentation	• Characteristic rash of confluent erythema and small papules, with a sandpaper-like texture • Flu-like symptoms
Oral manifestations	• Non-specific clinical signs • 'Strawberry tongue'—initially, a yellowish-white coating, which sheds to reveal a bright red raw appearance and enlarged papillae • Oral mucosa may be erythematous • Palatal erythema and petechiae • Enlarged uvula and tonsils • Circumoral pallor
Clinical investigations	• Rapid antigen detection tests • Throat culture • Antistreptolysin O titre
Differential diagnosis	Cutaneous rash: • Exanthem infectiosum • Measles • Rubella • Infectious mononucleosis • Viral hepatitis • Roseola infantum • Echovirus 14 • Staphylococcus infection • Secondary syphilis • Toxic shock syndrome • Rat-bite fever • *Spirillum minus* Oral presentation: • Kawasaki disease (may present with strawberry tongue)
Treatment	Antibiotics: • Oral penicillin • In penicillin allergic patients: Cephalosporin or macrolide
Prognosis	• Excellent • Complete course of antibiotics is required to prevent late complications including rheumatic fever

Table 14.2 Diphtheria

Aetiology and pathogenesis	• Acute, communicable disease caused by *Corynebacterium diphtheriae* • Spread by direct contact or by sneezing or coughing
Epidemiology	• Unimmunised or incompletely immunised children and adults
Systemic presentation	Typical mild symptoms include: • Sudden onset low-grade fever • Sore throat • Malaise In some cases: • Tender cervical lymphadenopathy • Swelling of the neck • Myocarditis • Neuritis

Table 14.2 (continued)

Oral manifestations	• Characteristic thick grey-white pseudomembrane covering the palate, uvula, and tonsils, formed of an adherent exudate of bacteria, dead cells, and fibrin • Lesions in the buccal mucosa, lips, hard and soft palate, and tongue may be seen in adult patients • Spread into the nasopharynx or larynx can cause difficulty with breathing and swallowing. • Removal of the pseudomembrane causes bleeding
Clinical investigations	• Culture of secretions from the nasopharynx and throat • Exotoxin may be detected by Elek immunodiffusion assay or polymerase chain reaction (PCR) analysis
Differential diagnosis	• Candidosis • Necrotising ulcerative gingivitis • Bacterial or viral pharyngitis • Tonsillitis • Infectious mononucleosis • Acute epiglottitis
Treatment	• Prevention through vaccination is recommended for children under 5 years • In infection: Penicillin or erythromycin may be prescribed
Prognosis	• In unimmunised, risk of respiratory failure and death

Table 14.3 Leprosy

Aetiology and pathogenesis	Granulomatous disease caused by *Mycobacterium leprae and Mycobacterium lepromatosis (*intracellular pathogens) • Long incubation period • Tropism for the skin and peripheral nerves Route of transmission: The upper airways after prolonged close contact with infected individual Classified (based on clinical manifestations and bacterial load): Lepromatous leprosy or tuberculous leprosy
Epidemiology	Affects all ages: • Most common between 10 and 14 years • Second age peak between 35 and 44 years • Rarely infants and younger children Over 80% of new cases are in developing countries such as India, Brazil, and Indonesia Prevalence has decreased dramatically over time
Systemic presentation	Lepromatous leprosy (multibacillary form): • Symmetrically distributed • Widespread cutaneous macules, plaques, papules, and nodules • Coarsening of the facial skin and earlobes (leonine facies) Tuberculoid leprosy (paucibacillary form) has fewer skin lesions than the multibacillary form Numbness due to peripheral nerve involvement present in all forms

(continued)

Table 14.3 (continued)

Oral manifestations	Oral mucosal involvement due to haematogenous or lymphatic dissemination of the bacteria, or secondary to nasal lesions Oral manifestations often asymptomatic and non-specific: • Firm reddish-yellow sessile papules or nodules. • Progression of lesions to ulceration or necrosis (usually hard and soft palate). • Destructive lesions of the palate, maxillary alveolar process or nasal bones. – Facial deformity (e.g., saddle nose) – Tooth loss • Hyperpigmentation may involve the palate, uvula, tonsils, gingivae, and posterior third of the tongue • Hypopigmentation of the oral mucosa may occur • Fissured tongue, depapillation of the tongue, and retraction of the uvula may occur • Facial nerve involvement may impair mastication and speech
Clinical investigations	Clinical: • Presence of characteristic skin lesions • Sensory loss Biopsy: • Characteristic features present in certain types of leprosy • Special stains are needed to demonstrate the bacilli • Acid-fast bacilli may or may not be present (not diagnostic) Laboratory: • Immunohistochemistry or PCR for isolation of lepromatous bacilli in the oral cavity
Differential diagnosis	• Secondary or tertiary syphilis • Cutaneous leishmaniasis • Systemic lupus erythematosus • Sarcoidosis • Deep fungal infections • Trauma • Malignancy
Treatment	• Multidrug therapy: Rifampicin, dapsone, and clofazimine for 6–12 months, dependent on the number of skin lesions present • If drug resistant: Treat with a combination ofloxacin, minocycline, and clarithromycin
Prognosis	• Curable disease

Recommended Reading

Dara SA, Babu Gadde R. Epidemiology, Prognosis, and Prevention of Leprosy Worldwide. Curr Trop Med Rep. 2016;3(4):144–8.

Hadfield TL, McEvoy P, Polotsky Y, Tzinserling VA, Yakovlev AA. The pathology of diphtheria. J Infect Dis. 2000;181 Suppl 1:S116–20.

Kumar SN, Prasad TS, Narayan PA, Muruganandhan J. Granuloma with langhans giant cells: An overview. J Oral Maxillofac Pathol. 2013;17(3):420–3.

Rodrigues GA, Qualio NP, de Macedo LD, Innocentini L, Ribeiro-Silva A, Foss NT, et al. The oral cavity in leprosy: what clinicians need to know. Oral Dis. 2017;23(6):749–56.

Shelley EW, Torgerson RR. Oral Signs of Bacterial Disease. In: Fazel N, editor. Oral Signs of Systemic Disease. Cham: Springer International Publishing; 2019. p. 169–91.
Vohra P, Rahman MSU, Subhada B, Tiwari RVC, Nabeel Althaf MS, Gahlawat M. Oral manifestation in leprosy: A cross-sectional study of 100 cases with literature review. J Family Med Prim Care. 2019;8(11):3689–94.
World Health Organization. Guidelines for the diagnosis, treatment and prevention of leprosy. New Delhi: World Health Organization, Regional Office for South-East Asia; 2017
World Health Organization. Leprosy (Hansen's disease) [Internet]. Geneva: WHO; 2023 [cited 2023 November 1]. Available from: https://www.who.int/news-room/fact-sheets/detail/leprosy.

15 Non-Candida Fungal Infections

Nirav Bhatia

15.1 Introduction

Candida is the most common fungal infection of the oral cavity, and 3–48% of healthy individuals harbour culturable levels of *Candida* in their mouth. However other fungal infections may also present in the oral cavity particularly in immunocompromised individuals. In immunocompromised individuals at a high risk of invasive fungal infections such as those undergoing a haematopoietic stem cell transplant, antifungal prophylaxis with posaconazole or fluconazole should be considered. Examples of fungal infections which affect the oral cavity are described in Table 15.1.

For further information on oral Candidal infections, see Chap. 16 on Candidosis and *Candida* related conditions.

N. Bhatia (✉)
Royal Dental Hospital of Melbourne, Carlton, VIC, Australia

Glen Iris Oral Medicine, Glen Iris, VIC, Australia
e-mail: Nirav.Bhatia@dhsv.org.au

R. Balasubramaniam et al. (eds.), *Oral Medicine - A Clinical Guide*,
https://doi.org/10.1007/978-3-031-36797-7_15

Table 15.1 Non-candida fungal infections of the orofacial region

Condition and pathogen	Epidemiology	Clinical presentation	Management and prognosis
Aspergillosis, caused by *Aspergillus* species, most commonly *Aspergillus fumigatus*	• Affects immunocompromised individuals	• Subacute to chronic infections with sinus tracts, fistula, or abscess formation due to invasion of surrounding tissue • Rarely oral infection may present as osteomyelitis • Fungal rhinosinusitis is more commonly seen as an acute or chronic presentation and may result in a fungal ball	• Primary treatment with surgical debridement of fungal mass. For sinus disease, endoscopic sinus surgery is preferred • Antimycotic therapy with voriconazole • Reduction of immunosuppression should be considered • Mortality rate of 30%
Mucormycosis, caused by multiple fungi, most commonly from Mucoraceae family	• Affects immunocompromised individuals such as uncontrolled diabetes with ketoacidosis, solid organ transplantation or severe burns • Spores found in soil, air, spoilt food, and decaying organic material	• Breach of mucocutaneous barrier is a common cause of infection • Presents as gemination of infection in paranasal sinuses that spreads to palate, orbit, and brain • Oral involvement can present as a palatal ulceration • Invasion of arteries, thrombosis and subsequent necrosis are hallmarks of infection	• Treat with surgical debridement and antimycotics • Amphotericin B is first-line treatment • Prognosis is poor and mortality rate varies from 30 to 70%
Histoplasmosis, caused by *Histoplasma capsulatum*	• Endemic in parts of USA, India, Latin America and Australia • Affects immunocompromised individuals such as those who are HIV positive • Found in bird and bat faeces	• Oral lesions are generally associated with systemic infection. Present as fungating or ulcerative lesion of oral mucosa with a nodular texture. May cause tissue destruction and bone invasion.	• In healthy individuals, may occasionally resolve spontaneously • Treat with amphotericin B and itraconazole for pulmonary involvement and disseminated disease • Prognosis is good for mild infections but for disseminated infections without treatment mortality is 80%

Condition and pathogen	Epidemiology	Clinical presentation	Management and prognosis
Cryptococcosis, caused by *Cryptococcus neoformans*	• Most common in HIV positive patients • Rare in developing countries due to antiretroviral therapies	• Primary site of infection is the lungs and then have haematogenous spread • Oral lesions such as ulcers, nodules or granules may represent a secondary site of infection	• Mild to moderate infections treated with fluconazole • Severe infections treated with amphotericin B • HIV or transplant patients use amphotericin B and flucytosine • Prognosis is good for early disease however cryptococcus meningitis is associated with 40% mortality rate
Blastomycosis caused by *Blastomyces dermatitidis*	• Found in soil in parts of USA, India, Latin America and Australia	• Primary site of infection is the lungs and oral lesions are rare and present as ulcerations or soft tissue masses	• Mild to moderate disease treated with itraconazole • Severe or disseminated disease treated with amphotericin B followed by itraconazole • Pooled mortality rate of 6.6%, 37% in immunocompromised individuals
Paracoccidioidomycosis caused by *Paracoccidioides brasiliensis*	• Occurs most commonly in Central and South America	• Primary site of infection is the lungs from where it may spread to other locations • In the oral cavity, it presents as ulceration, erythematous, and granulomatous lesions that tend to bleed. It may affect multiple sites within the oral cavity simultaneously	• Mild to moderate cases treated with itraconazole or co-trimoxazole • Severe disease treated with amphotericin B • Good prognosis in healthy individuals but mortality rate can be 40–50% in immunocompromised individuals

Recommended Reading

Fuqua TH, Sittitavornwong S, Knoll M, Said-Al-Naief N. Primary invasive oral aspergillosis: an updated literature review. J Oral Maxillofac Surg. 2010;68(10):2557–63.

Telles DR, Karki N, Marshall MW. Oral Fungal Infections: Diagnosis and Management. Dent Clin North Am. 2017;61(2):319–49.

Manfredi M, Polonelli L, Giovati L, Alnuaimi A, McCullough MJ. Oral and Maxillofacial Fungal Infections. In: Farah CS, Balasubramaniam R, McCullough MJ, editors. Contemporary Oral Medicine: A Comprehensive Approach to Clinical Practice. Cham: Springer International Publishing; 2019. p. 935–81.

Candidosis and *Candida*-Associated Conditions

16

Aileen Tyler and Michael McCullough

16.1 Introduction

Candida is an oral fungal commensal present in 70–85% of people. *Candida albicans,* the most common species, is the main cause of oral candidosis. Other, less common, species include *C. dubliniensis*, *C. tropicalis*, *C. glabrata*, and *C. krusei*.

An imbalance between candidal virulence factors and host immune response results in transition of *Candida* from commensalism to pathogenicity. Virulence factors include adherence, morphological and phenotypic switching, and proteolytic enzyme production. Predisposing host factors can be local or systemic. Local factors include denture-wearing, changes in saliva quality/quantity, use of corticosteroids, caries, and smoking. Systemic factors comprise extremes of age, immunosuppression, and medications.

16.2 Clinical Presentation and Investigations

There are three clinical presentations of oral candidosis. **Pseudomembranous candidosis** (oral thrush) appears as soft, white curd-like plaques that can be wiped off revealing erythematous mucosa. They are usually asymptomatic and typically occurring on the palate, tongue dorsum, and buccal mucosa. Rarely are they associated with burning and dysphagia. **Erythematous candidosis** (Fig. 16.1) appears as erythema on the tongue dorsum often with erythema of the opposing hard palate in patients with long-term antibiotic or corticosteroid use, or if immunocompromised. Often asymptomatic, it may be associated with burning or a metallic taste. **Chronic**

A. Tyler · M. McCullough (✉)
Melbourne Dental School, The University of Melbourne, Carlton, VIC, Australia
e-mail: m.mccullough@unimelb.edu.au

R. Balasubramaniam et al. (eds.), *Oral Medicine - A Clinical Guide*,
https://doi.org/10.1007/978-3-031-36797-7_16

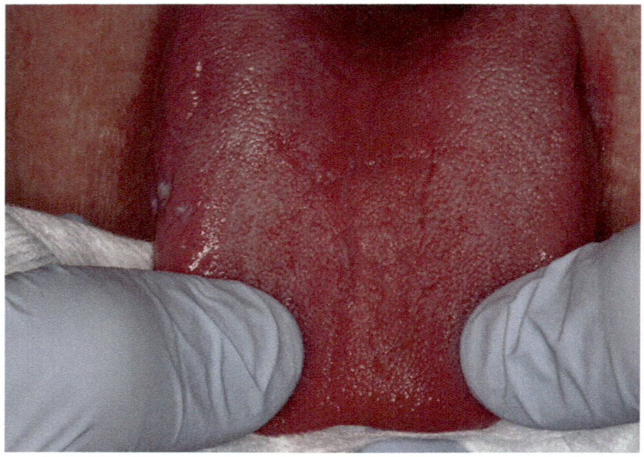

Fig. 16.1 Erythematous candidosis of the dorsal surface of the tongue

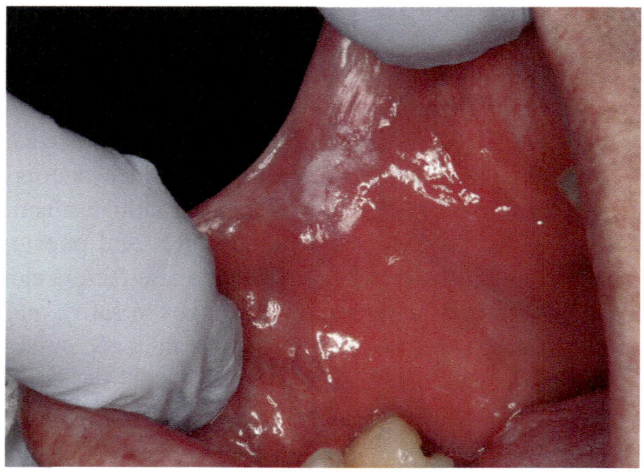

Fig. 16.2 Hyperplastic candidosis of the post-modiolus region of the buccal mucosa

hyperplastic candidosis (Fig. 16.2) appears as adherent white plaques often in the post-commissure area of the buccal mucosa, but also the lateral tongue, and palate. Usually asymptomatic, this is also often associated with smoking and has been shown to result in dysplastic epithelium.

There are several clinical entities for which *Candida* has been associated, but a definitive direct causation cannot be attributed. **Angular cheilitis**, presenting as erythematous fissuring and crusting of the labial commissures that is often painful, is more likely related to a reduced vertical dimension, diminished elasticity of the skin, and possibly associated with iron, vitamin B12, and folate deficiency. **Median rhomboid glossitis**, presenting as erythematous, atrophic, rhomboid-shaped lesion on the mid-dorsum of the tongue, is often asymptomatic, does not resolve with antifungal agents, and is often associated with smoking or inhaled corticosteroids.

Finally, erythema of the palate associated with the fitting surface of a denture has not been shown to be caused by candida. This **denture-associated erythematous stomatitis** (DAES) is usually asymptomatic and caused by poor denture hygiene.

Oral smears, swabs, rinses, salivary cultures, and histopathology have been used. However, isolation of *Candida* is not diagnostic of infection, with diagnosis based primarily on clinical signs, symptoms, and medical history. To exclude dysplasia, tissue biopsy may be indicated for hyperplastic candidosis not responding to antifungal therapy.

16.3 Management and Prognosis

Identification and elimination of predisposing local and/or systemic factors are key. Pharmacological treatment when indicated is by topical agents, including polyenes (nystatin, amphotericin B) and azoles (miconazole, clotrimazole, fluconazole) that directly or indirectly target ergosterol. Importantly, azoles interact with certain drugs, most notably warfarin, which can cause a significant increase in a patients INR, even when used topically.

Candida albicans is the most frequent species of *Candida* associated with the opportunistic infection, oral candidosis. Effective management requires rectification of predisposing factors, if possible, and administration of appropriate antifungal agents.

Recommended Reading

Alnuaimi AD, Wiesenfeld D, O'Brien-Simpson NM, Reynolds EC, McCullough MJ. Oral Candida colonization in oral cancer patients and its relationship with traditional risk factors of oral cancer: a matched case-control study. Oral Oncol. 2015;51(2):139–45.

Manfredi M, Polonelli L, Aguirre-Urizar JM, Carrozzo M, McCullough MJ. Urban legends series: oral candidosis. Oral Dis. 2013;19(3):245–61.

Mun M, Yap T, Alnuaimi AD, Adams GG, McCullough MJ. Oral candidal carriage in asymptomatic patients. Aust Dent J. 2016;61(2):190–5.

Oral Lichen Planus

17

Michelle Kang, Kenelm Kwong, and Sue-Ching Yeoh

17.1 Introduction

Oral lichen planus (OLP) is a chronic, immune-mediated condition affecting the oral mucosae (Fig. 17.1). Lichen planus may also affect the skin, nails, scalp, as well as other extra-oral mucosal surfaces including the genital mucosae and oesophagus.

M. Kang (✉)
Sydney Dental Hospital, Surry Hills, NSW, Australia

Sydney Oral Medicine, Wahroonga, NSW, Australia

Hunter New England Local Health District, Newcastle, NSW, Australia

School of Dentistry and Medical Sciences, Charles Sturt University, Orange, NSW, Australia
e-mail: michelle.kang@health.nsw.gov.au

K. Kwong
Sydney Dental Hospital, Surry Hills, NSW, Australia

S.-C. Yeoh
Sydney Oral Medicine, Wahroonga, NSW, Australia

Royal Prince Alfred Hospital, Camperdown, NSW, Australia

Chris O'Brien Lifehouse, Camperdown, NSW, Australia

Sydney Dental School, The University of Sydney, Camperdown, NSW, Australia

R. Balasubramaniam et al. (eds.), *Oral Medicine - A Clinical Guide*,
https://doi.org/10.1007/978-3-031-36797-7_17

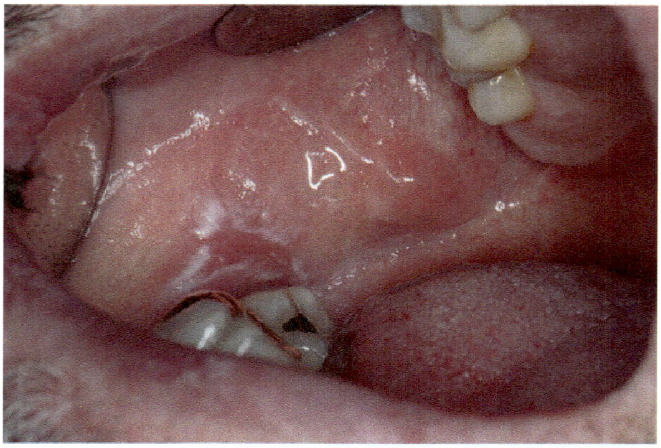

Fig. 17.1 Clinical photograph of reticular white striae on the right buccal mucosa, biopsy-proven oral lichen planus

The prevalence of OLP varies between 0.22 and 5%. Incidence is reported up to 2.2%. Female adults aged between 30 and 80 years are most commonly affected. OLP in children is rare. Patients may also present with lichen planus concurrently affecting other mucocutaneous surfaces, with 15% and 20% of oral lichen planus patients reported to also have cutaneous and genitalia lichen planus, respectively.

The aetiopathogenesis of oral lichen planus is unknown. It is widely viewed as a T lymphocyte-mediated chronic inflammatory condition, which results in cytotoxicity against epithelial basal cells. Hypotheses suggests a range of mechanisms including antigen-specific (e.g. heat shock proteins) and non-antigen specific (e.g. matrix metalloproteinases) processes.

17.2 Clinical Presentation and Investigations

Lesions usually present bilaterally or multifocally with disease severity waxing and waning over time. Clinical subtypes include reticular, plaque, papular, atrophic/erythematous, erosive, and bullous forms. Common oral mucosal locations include buccal mucosa, gingivae (as desquamative gingivitis), tongue, and labial mucosa. Koebner phenomenon (location of lesions at sites of trauma) may be observed, such as corresponding to frictional trauma from teeth cusps. There may be associated symptoms, such as pain, burning, and irritation. Mucosal bleeding may be reported with minor trauma, for example, with tooth brushing.

Diagnosis of OLP is based on clinical and histopathological features. There is variable diagnostic criteria for OLP and ongoing controversy regarding its potential malignant transformation. Histopathological features include a band-like, predominantly lymphocytic infiltrate in the lamina propria, interfacing the epithelium as well as liquefactive (hydropic) degeneration of the basal keratinocytes including the presence of civatte bodies. Other epithelial features may include hyperkeratosis,

Table 17.1 Differential diagnoses for oral lichen planus

Lichenoid reactions	Oral lichenoid drug reaction Oral lichenoid contact reaction
Dermatoses	Chronic oral graft versus host disease Lupus erythematous Mucous membrane pemphigoid Pemphigus vulgaris Chronic ulcerative stomatitis
Idiopathic	Oral lichenoid lesion Leukoplakia Erythroleukoplakia Erythroplakia Proliferative verrucous leukoplakia

acanthosis, atrophy, and ulceration. The wide range of clinical and histopathological presentations may resemble other conditions (Table 17.1).

17.3 Management and Prognosis

Patients should be educated regarding the commonality of chronicity of the diagnosis and the low but existent risk of malignant transformation. Management of OLP is target as symptomatic control and reduction of impact on quality of life. Modalities includes non-pharmacological measures (e.g. optimising oral hygiene, eliminate local sources of mechanical irritation) as well as pharmacological measures (e.g. topical and/or systemic immunomodulatory agents). Ongoing follow-up and surveillance is required to ensure symptomatic disease control and monitor for changes in mucosal appearance which may be asymptomatic.

Recommended Reading

Cheng YS, Gould A, Kurago Z, Fantasia J, Muller S. Diagnosis of oral lichen planus: a position paper of the American Academy of Oral and Maxillofacial Pathology. Oral Surg Oral Med Oral Pathol Oral Radiol. 2016;122(3):332–54.

Sugerman PB, Savage NW, Walsh LJ, Zhao ZZ, Zhou XJ, Khan A, et al. The pathogenesis of oral lichen planus. Crit Rev Oral Biol Med. 2002;13(4):350–65.

Lupus Erythematosus 18

Kenelm Kwong, Michelle Kang, and Sue-Ching Yeoh

18.1 Introduction

There are systemic and limited forms of lupus erythematosus (LE). Systemic lupus erythematosus (SLE) is a chronic auto-immune disease with a heterogenous clinical presentation, characterised by multi-system involvement and progressive organ damage. SLE is associated with increased morbidity and mortality. Limited forms of lupus erythematosus (LE) with cutaneous involvement may be acute, subacute, or chronic and associate with varying risks of progression to SLE.

K. Kwong (✉)
Sydney Dental Hospital, Surry Hills, NSW, Australia
e-mail: Kenelm.kwong@health.nsw.gov.au

M. Kang
Sydney Dental Hospital, Surry Hills, NSW, Australia

Sydney Oral Medicine, Wahroonga, NSW, Australia

Hunter New England Local Health District, Newcastle, NSW, Australia

School of Dentistry and Medical Sciences, Charles Sturt University, Orange, NSW, Australia

S.-C. Yeoh
Sydney Oral Medicine, Wahroonga, NSW, Australia

Royal Prince Alfred Hospital, Camperdown, NSW, Australia

Chris O'Brien Lifehouse, Camperdown, NSW, Australia

Sydney Dental School, The University of Sydney, Camperdown, NSW, Australia

R. Balasubramaniam et al. (eds.), *Oral Medicine - A Clinical Guide*,
https://doi.org/10.1007/978-3-031-36797-7_18

LE is a multifactorial disease characterised by the formation and deposition of anti-nuclear auto-antibody immune-complexes, which activate the complement cascade resulting in an inappropriate inflammatory response. LE is associated with over 150 genetic and epigenetic alterations, which result in decreased auto-antigen clearance or lowered immune activation threshold.

Environmental factors may serve as a trigger, either by promoting apoptosis (UV light) or by increasing pro-inflammatory immune signalling (smoking, Epstein Barr virus, drugs). The net effect is presentation of endogenous nuclear material to receptive adaptive immune cells, resulting in B-cell activation and production of auto-antibodies. Hormonal factors such as upregulation of T-cell activation by oestrogen may provide explanation for female predisposition.

Tissue damage occurs throughout multiple organ systems, mediated by auto-antibodies via pro-inflammatory immune complex deposition (vascular damage), inappropriate activation of receptors, or binding of phospholipids.

The prevalence of SLE in adults ranges from 30 to 150 per 100,000, with a higher prevalence amongst non-Caucasian (African, Asian, Indigenous Australian) populations. Adult women in the third to seventh decade of life are predominantly affected, at a ratio of 9:1 compared to men.

18.2 Clinical Presentation and Investigations

Multiple organ systems may be affected in SLE (Table 18.1). SLE may cause constitutional symptoms of fatigue, weight loss, and fever. Oral lesions are found in 9–45% of SLE patients and 3–20% in chronic forms of cutaneous LE, particularly the discoid subcategory. Commonly affected sites are the buccal mucosa, lip vermillion, palate, and gingiva. Lesions can appear as painful, often annular (discoid) erosions or ulcerations with peripheral or adjacent white radiating striae or plaques. LE lesions can look similar to oral lichen planus. Oral LE lesions harbour potential to progress to oral squamous cell carcinoma (OSCC). SLE can also be associated with salivary hypofunction.

Diagnosis of LE is difficult given the great range of clinical manifestations. Multiple research-based classification criteria have been used to guide diagnosis, including the 2019 European League Against Rheumatism/American College of Rheumatology Criteria (EULAR/ACR 2019).

Investigations such as serology (ANA, anti-Smith, anti-dsDNA, complement C3 and C4) can assist with diagnosis.

Orally, a biopsy can assist in excluding lesions with oral epithelial dysplasia or overt OSCC. Histopathology suggestive of SLE includes hyperkeratosis, basal keratinocyte degeneration, deep diffuse perivascular inflammatory infiltrate, subepithelial oedema, and vesiculation. Oral LE lesions may share similar histopathological features with oral lichen planus. Direct immunofluorescence testing may show immunoreactants (C3, IgM, IgG) deposited in a granular pattern along the basement membrane zone.

Table 18.1 Manifestations of systemic lupus erythematosus

Organ system	Presentation
Cutaneous	• Alopecia • Photosensitive rash • Malar rash[a] • Discoid rash[b]
Rheumatological	• Arthritis (including temporomandibular joint) • Arthralgia • Myalgia
Cardiac	• Pericarditis • Pericardial effusion
Respiratory	• Pleuritis
Renal	• Lupus nephritis • Renal failure
CNS	• Cognitive dysfunction • Depression • Psychosis • Epilepsy
Haematological	• Haemolytic anaemia • Leukopenia • Thrombocytopenia • Thrombosis
Reproductive	• Pre-term birth • Pre-eclampsia

[a] Butterfly rash over the malar eminences that spares the nasolabial folds
[b] Well defined, erythematous raised patches

18.3 Management and Prognosis

Prognosis and management of LE vary greatly depending on disease severity and organ involvement. Patients should be counselled on modification of risk factors, such as sun protection and smoking cessation. Cutaneous and/or oral lesions can be managed symptomatically, often with topical corticosteroids or systemic hydroxychloroquine. SLE is managed by a rheumatologist and specific organ specialists and are often treated with systemic immunosuppressive agents such as glucocorticoids, mycophenolate, cyclophosphamide, or biological agents such as rituximab.

Recommended Reading

Fanouriakis A, Kostopoulou M, Alunno A, Aringer M, Bajema I, Boletis JN, et al. 2019 update of the EULAR recommendations for the management of systemic lupus erythematosus. Ann Rheum Dis. 2019;78(6):736–45.

Lisnevskaia L, Murphy G, Isenberg D. Systemic lupus erythematosus. Lancet. 2014 Nov 22;384(9957):1878–88.

Louis PJ, Fernandes R. Review of systemic lupus erythematosus. Oral Surg Oral Med Oral Pathol Oral Radiol Endod. 2001;91(5):512–6.

Rees F, Doherty M, Grainge MJ, Lanyon P, Zhang W. The worldwide incidence and prevalence of systemic lupus erythematosus: a systematic review of epidemiological studies. Rheumatology (Oxford). 2017;56(11):1945–61.

Tsokos GC. Systemic lupus erythematosus. N Engl J Med. 2011;365(22):2110–21.

Wahren-Herlenius M, Dörner T. Immunopathogenic mechanisms of systemic autoimmune disease. Lancet. 2013;382(9894):819–31.

Pemphigus Vulgaris

19

Sue-Ching Yeoh, Michelle Kang, and Kenelm Kwong

19.1 Introduction

Pemphigus is a group of chronic autoimmune blistering diseases affecting mucosal and/or cutaneous sites. Lesions result from autoantibodies binding to specific desmosomal proteins on the surface of the keratinocytes, resulting in loss of cell-cell adhesion (acantholysis). Pemphigus vulgaris (PV) is the most common and clinically aggressive variant.

S.-C. Yeoh (✉)
Sydney Oral Medicine, Wahroonga, NSW, Australia

Royal Prince Alfred Hospital, Camperdown, NSW, Australia

Chris O'Brien Lifehouse, Sydney, NSW, Australia

Sydney Dental School, The University of Sydney, Camperdown, NSW, Australia
e-mail: Sue-Ching.Yeoh@sydney.edu.au

M. Kang
Sydney Oral Medicine, Wahroonga, NSW, Australia

Sydney Dental Hospital, Surry Hills, NSW, Australia

Hunter New England Local Health District, Newcastle, NSW, Australia

School of Dentistry and Medical Sciences, Charles Sturt University, Orange, NSW, Australia

K. Kwong
Sydney Dental Hospital, Surry Hills, NSW, Australia

R. Balasubramaniam et al. (eds.), *Oral Medicine - A Clinical Guide*,
https://doi.org/10.1007/978-3-031-36797-7_19

Patients with PV generate autoantibodies against the desmosomal cadherins, typically desmoglein-3 (dsg-3) and also to dsg-1 (in 50% of cases), both of which are adhesion molecules found in stratified squamous epithelium. Dsg-3 is found primarily in the suprabasal layer of the epithelium. This antigen–antibody interaction activates complement, releasing inflammatory mediators and recruiting activated T cells. Damage to the intercellular area induces apoptosis, acantholysis, and eventually intraepithelial blistering.

The incidence of PV varies among populations from 0.1 to 0.5 patients per 100,000 population per year. However this increases in the Ashkenazi Jew, Mediterranean, and South Asian populations, where it is associated with HLA subtypes HLA-DR4 (DRB1*0402), DR13 (DRB1*1401), and DQB1*0503, respectively. The mean age of onset is between 50 and 60 years.

19.2 Clinical Presentation and Investigations

Oral lesions develop in up to 70% of cases, appearing as fluid filled vesicles and bullae that rupture easily to form tender, irregular mucosal erosions, and ulcers. These may affect any mucosal site; however, most commonly involve the buccal and palatal mucosa, and gingiva (desquamative gingivitis) (Fig. 19.1). Oral lesions may be the initial and only sign of disease. Other mucosal sites, such as the oropharynx, larynx, conjunctiva, oesophagus, and genitalia, may also become involved. Cutaneous involvement may develop several months after oral lesions are observed. A positive "Nikolsky sign" is often noted in association with both mucosal and cutaneous lesions, wherein blisters may be induced or expanded when lateral pressure is applied to the tissue.

Biopsy of perilesional tissue demonstrates intraepithelial vesicle formation, with intercellular oedema and loss of intercellular attachment in the basal layer. Common features include "clefting" as the suprabasal keratinocytes separate from the basal cells, with a "tombstone" appearance at the floor of the blister as the basal cells separate from each other but remain attached to the basement membrane. Acantholytic (Tzanck) cells may be seen within the vesicle. Direct immunofluorescence performed on fresh tissue shows a "basket weave" staining pattern for IgG, reflecting its deposition on the surface of the keratinocytes. IgM and complement components such as C3 are also often present. Intercellular cement substance antibody (ICSA) may be detected by indirect immunofluorescence in 80–90% of patients with PV. Enzyme-linked immunosorbent assay using human recombinant dsg-1 and dsg-3 protein detects circulating autoantibodies.

Differential diagnosis of oral PV includes mucous membrane pemphigoid and epidermolysis bullosa.

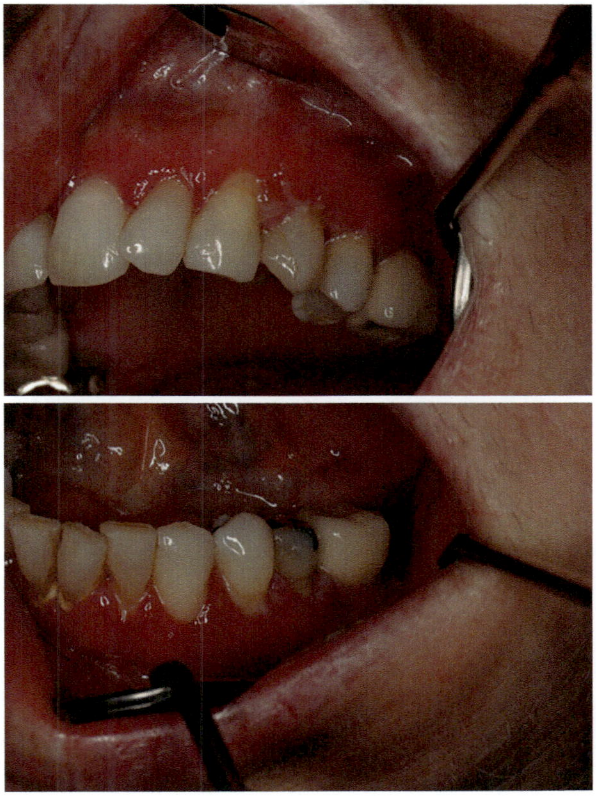

Fig. 19.1 Gingival lesions of pemphigus vulgaris showing erythema and ulceration with some blistering along the gingival margins

19.3 Management and Prognosis

Management of PV relies primarily on corticosteroids and steroid sparing agents such as azathioprine or mycophenolate. Emerging therapies include intravenous immunoglobulin, plasmapheresis, extracorporeal photochemotherapy, and cholinergic agonists. The use of rituximab, a monoclonal anti-CD20 antibody as well as TNF-antagonists, has shown great success in suppressing PV.

The natural course and disease severity of PV are variable, with the mortality rate ranging from 5 to 15%. Complications secondary to the use of immunosuppressive therapy may also contribute to the reported mortality rate.

Levels of dsg autoantibodies are reported to fluctuate with disease activity; therefore, titres of ICSA and levels of anti-dsg IgG tend to fall with successful therapy and rise with relapse. The incidence and duration of true remission in PV are uncertain.

Recommended Reading

Bhol K, Mohimen A, Ahmed AR. Correlation of subclasses of IgG with disease activity in pemphigus vulgaris. Dermatology. 1994;189 Suppl 1:85–9.
Black M, Mignogna MD, Scully C. Number II. Pemphigus vulgaris. Oral Dis. 2005;11(3):119–30.
Schmidt E, Hunzelmann N, Zillikens D, Bröcker EB, Goebeler M. Rituximab in refractory autoimmune bullous diseases. Clin Exp Dermatol. 2006;31(4):503–8.
Schmidt E, Dähnrich C, Rosemann A, Probst C, Komorowski L, Saschenbrecker S, et al. Novel ELISA systems for antibodies to desmoglein 1 and 3: correlation of disease activity with serum autoantibody levels in individual pemphigus patients. Exp Dermatol. 2010;19(5):458–63.

Mucous Membrane Pemphigoid 20

Michelle Kang, Kenelm Kwong, and Sue-Ching Yeoh

20.1 Introduction

Mucous membrane pemphigoid (MMP) is an immunobullous, subepithelial blistering disease with autoantibodies against components of the epithelial-connective tissue junction. MMP predominantly affects the mucous membranes. Previous terms used include cicatricial pemphigoid and benign mucous membrane pemphigoid.

The incidence is approximately 1.3 to 2 per one million. MMP has a predilection for adult females, with an age of onset around the sixth and seventh decade.

M. Kang (✉)
Sydney Dental Hospital, Surry Hills, NSW, Australia

Sydney Oral Medicine, Wahroonga, NSW, Australia

Hunter New England Local Health District, Newcastle, NSW, Australia

School of Dentistry and Medical Sciences, Charles Sturt University, Orange, NSW, Australia
e-mail: Michelle.Kang@health.nsw.gov.au

K. Kwong
Sydney Dental Hospital, Surry Hills, NSW, Australia

S.-C. Yeoh
Sydney Oral Medicine, Wahroonga, NSW, Australia

Royal Prince Alfred Hospital, Camperdown, NSW, Australia

Chris O'Brien Lifehouse, Camperdown, NSW, Australia

Sydney Dental School, The University of Sydney, Camperdown, NSW, Australia

R. Balasubramaniam et al. (eds.), *Oral Medicine - A Clinical Guide*,
https://doi.org/10.1007/978-3-031-36797-7_20

The aetiopathogenesis of MMP is unclear; however, it is likely due to a combination of genetic, immunological, and environmental factors. Genetic HLA associations have been reported (including DQB1*0301, DRB04, DRB11); however, the role and significance remain to be elucidated. Environmental factors can include exposure to certain drugs, such as penicillamine, which can rarely cause drug-induced MMP.

MMP is an immune-mediated disease, with autoantibodies directed against epithelial structural proteins, leading to detachment of basal cells from the basement membrane zone, as well as recruitment of inflammatory cells such as neutrophils and eosinophils. Identified target antigens in MMP include:

- Bullous pemphigoid antigen 1 (BP230).
- Bullous pemphigoid antigen 2 (BP180)/type XVII collagen.
- Laminin 332/laminin 5.
- Laminin 6.
- Type VII collagen.
- Integrin β4 subunit.

20.2 Clinical Presentation and Investigations

MMP has a variable clinical presentation and disease severity; however, most patients have a gradual onset. Clinically MMP presents as blisters, which rupture quickly to form ulcers (Fig. 20.1). It predominantly involves the mucous membrane. Oral mucosal lesions are seen in most patients and often is the initial site of involvement. Other mucosal sites which may be affected include ocular, nasal, nasopharyngeal, anogenital, laryngeal, and oesophageal surfaces.

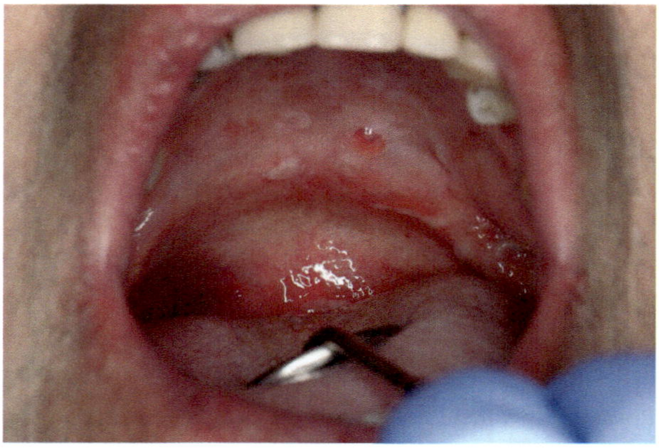

Fig. 20.1 Clinical photograph demonstrating oral blister and ulcerations associated with mucous membrane pemphigoid

Skin involvement may also be seen. Blisters may involve single or multiple (most patients) mucosal or cutaneous sites.

Healing can occur with scarring. Importantly, conjunctival fibrosis (symblepharon) may lead to blindness in severe cases. Scarring is not clinically evident with oral mucosal lesions.

Oral MMP most commonly presents as desquamative gingivitis. Other oral mucosal sites which may be involved include buccal mucosa, hard and soft palate, alveolar ridge, tongue and labial mucosa. Symptoms of oral MMP include pain, odynophagia, dysphagia, and oral bleeding with minor trauma.

Diagnosis is based on clinical, histopathological, and serological features. A combination of clinical assessment and direct immunofluorescence is widely considered as the gold standard. Biopsy of fixed and fresh perilesional tissue allows assessment via haematoxylin and eosin staining and direct immunofluorescence, respectively. Histopathological features include subepithelial junctional separation at the basement membrane zone, possible mild chronic inflammation (including eosinophils, lymphocytes, neutrophils) in the lamina propria. Direct immunofluorescence demonstrates linear deposition of immunoreactants (most commonly IgG, C3) along the basement membrane zone. Serology allows semi-quantitative (such as indirect immunofluorescence) and/or quantitative (such as enzyme-linked immunosorbent assay) assessment of autoantibodies. Antibody titres may play a role in initial diagnosis and ongoing monitoring, including response to therapy, correlation with disease activity, and possible prediction of remission/relapse.

20.3 Management and Prognosis

Management for MMP includes the following:

- Referral to ophthalmologist and other specialists as appropriate for baseline examination and management of extra-oral lesions.
- Patient education and symptom control (topical and/or systemic analgesia).
- Topical and/or systemic corticosteroids, immunosuppressants (such as mycophenolate, azathioprine, rituximab).

Depending on the location and extent of the disease, there may be severe consequences such as change in vision and blindness, airway constriction secondary to laryngeal webs, odynophagia, dysphagia, and secondary infection which may lead to death.

Recommended Reading

Cai F, Dai P, Jiang J, Lee AY, Emerson J, Taylor MS, et al. Treatment and monitoring of an immunobullous disease cohort in a single centre in Sydney Australia. Australas J Dermatol. 2021;62(3):e437–e440.

Chan LS, Ahmed AR, Anhalt GJ, Bernauer W, Cooper KD, Elder MJ, et al. The first international consensus on mucous membrane pemphigoid: definition, diagnostic criteria, pathogenic factors, medical treatment, and prognostic indicators. Arch Dermatol. 2002;138(3):370–9.

Knudson RM, Kalaaji AN, Bruce AJ. The management of mucous membrane pemphigoid and pemphigus. Dermatol Ther. 2010;23(3):268–80.

Schmidt E, Zillikens D. Pemphigoid diseases. Lancet. 2013;381(9863):320-32.

Epidermolysis Bullosa

21

Kenelm Kwong, Michelle Kang, and Sue-Ching Yeoh

21.1 Introduction

Hereditary epidermolysis bullosa (EB) is a heterogenous group of geno-dermatoses, characterised by skin and mucosal fragility of variable severity and extent. This leads to formation of blisters and erosions. Epidermolysis bullosa acquisita differs markedly from hereditary EB, as it is an immune mediated vesiculobullous disease; therefore, it is not discussed here.

In hereditary EB, genetic mutations lead to decreased, absent or malfunctioning proteins that normally comprise the molecular supra-structures functionally linking keratinocytes to one another, the keratinocytes to the underlying basement membrane, and the basement membrane to the underlying connective tissue. This results

K. Kwong (✉)
Sydney Dental Hospital, Sydney, NSW, Australia

Sydney Dental Hospital, Surry Hills, NSW, Australia
e-mail: Kenelm.Kwong@health.nsw.gov.au

M. Kang
Sydney Dental Hospital, Sydney, NSW, Australia

Sydney Oral Medicine, Wahroonga, NSW, Australia

Hunter New England Local Health District, Newcastle, NSW, Australia

School of Dentistry and Medical Sciences, Charles Sturt University, Orange, NSW, Australia

S.-C. Yeoh
Sydney Oral Medicine, Wahroonga, NSW, Australia

Sydney Dental School, The University of Sydney, Camperdown, NSW, Australia

Chris O'Brien Lifehouse, Camperdown, NSW, Australia

Royal Prince Alfred Hospital, Camperdown, NSW, Australia

R. Balasubramaniam et al. (eds.), *Oral Medicine - A Clinical Guide*,
https://doi.org/10.1007/978-3-031-36797-7_21

in mucocutaneous fragility. These proteins also have a role in cellular signalling, with defects variably leading to impaired wound healing, chronic inflammation, scar formation, as well as a permissive tumour micro-environment.

The prevalence of EB is 11.1 cases/million in the USA, with similar prevalence rates worldwide. Markedly higher prevalence rates have been reported in Scotland and Norway.

Classification of hereditary EB is based on the ultra-structural level of skin cleavage, with further subtyping by clinical severity, the molecular defect including the relative levels of protein expression and disease-causing sequence variants, as well as inheritance pattern. The main types are EB simplex (EBS), junctional EB (JEB), kindler EB (KEB), and dystrophic EB that is further subtyped into dominant dystrophic EB (DDEB) and recessive dystrophic EB (RDEB).

21.2 Clinical Presentation and Investigations

Mucocutaneous presentation of EB includes epithelial fragility with bullae formation, erosion, and ulceration of variable severity and extent. Locations may include epithelial linings of the eyes, mouth, oesophagus, respiratory tract, and genitourinary systems. Healing can be delayed, particularly for JEB with pathognomonic exuberant granulation tissue. Scarring often occurs, which can lead to significant functional deficit, such as ocular symblepharon, microstomia, oesophageal strictures, urethral or vaginal stenosis, and pseudosyndactyly (particularly for RDEB). Alopecia, nail dystrophy, and nail loss are also common. Squamous cell carcinoma (SCC) occurs often in RDEB, with 67.8% having developed at least 1 SCC by age 35, and 90.1% by age 55%. These generally develop within chronic cutaneous wounds or scars. Cutaneous and mucosal SCC has also been associated with JEB and KEB.

Oral manifestations include mucosal bullae formation, erosion, ulceration, and microstomia. JEB is additionally associated with enamel hypoplasia and gingival hypertrophy. RDEB is associated with progressive microstomia, vestibular depth loss, ankyloglossia, as well as oral SCC, most commonly of the tongue. KEB is associated with rapidly progressive periodontal disease, childhood dentoalveolar bone loss and tooth exfoliation, and oral SCC. Extra-cutaneous manifestations can include muscular dystrophy, cardiomyopathy, and renal failure.

Following a compatible history, family pedigree mapping, and clinical examination, diagnosis is confirmed through immunofluorescence antigen mapping or transmission electron microscopy of a skin biopsy in conjunction with genetic sequencing and analysis.

21.3 Management and Prognosis

Management is largely supportive including avoidance of mucocutaneous trauma, wound management, prevention and management of mucocutaneous infection, and pain management. Emerging therapies include off-label prescribing (i.e. aminoglycosides for enhancing gene transcription), gene therapies, and allogenic bone marrow transplantation. Increased mortality is seen in severe forms of JEB and RDEB. Squamous cell carcinoma is a leading cause of death in patients with RDEB.

Recommended Reading

Bardhan A, Bruckner-Tuderman L, Chapple ILC, Fine JD, Harper N, Has C, et al. Epidermolysis bullosa. Nat Rev Dis Primers. 2020;6(1):78.

Fine JD, Bauer EA, Briggaman RA, Carter DM, Eady RA, Esterly NB, et al. Revised clinical and laboratory criteria for subtypes of inherited epidermolysis bullosa. A consensus report by the Subcommittee on Diagnosis and Classification of the National Epidermolysis Bullosa Registry. J Am Acad Dermatol. 1991;24(1):119–35.

Fine JD, Johnson LB, Weiner M, Li KP, Suchindran C. Epidermolysis bullosa and the risk of life-threatening cancers: the National EB Registry experience, 1986-2006. J Am Acad Dermatol. 2009;60(2):203–11.

Has C, Liu L, Bolling MC, Charlesworth AV, El Hachem M, Escámez MJ, et al. Clinical practice guidelines for laboratory diagnosis of epidermolysis bullosa. Br J Dermatol. 2020;182(3):574–92.

Krämer S, Lucas J, Gamboa F, Peñarrocha Diago M, Peñarrocha Oltra D, Guzmán-Letelier M, et al. Clinical practice guidelines: Oral health care for children and adults living with epidermolysis bullosa. Spec Care Dentist. 2020;40(Suppl 1):3–81.

Titeux M, Bonnet des Claustres M, Izmiryan A, Ragot H, Hovnanian A. Emerging drugs for the treatment of epidermolysis bullosa. Expert Opin Emerg Drugs. 2020;25(4):467–89.

Erythema Multiforme, Stevens-Johnson Syndrome and Toxic Epidermal Necrolysis

22

Michelle Kang, Kenelm Kwong, and Sue-Ching Yeoh

22.1 Introduction

Erythema multiforme (EM), Stevens-Johnson syndrome (SJS) and toxic epidermal necrolysis (TEN) are rare, mucocutaneous hypersensitivity reactions characterised by extensive detachment of full-thickness epithelium. Although oral presentations share many common features across the conditions, cutaneous findings are diverse and EM is not considered a continuous spectrum with SJS and TEN. EM is most commonly associated with infections (such as HSV-1), whereas SJS and TEN are more commonly induced by drugs. SJS/TEN is a type of severe cutaneous adverse reaction (SCAR) and is associated with high morbidity and mortality. Both reactions are part of a disease continuum, distinguished based on severity and percentage of body surface area involvement (Table 22.1).

M. Kang (✉)
Sydney Dental Hospital, Surry Hills, NSW, Australia

Sydney Oral Medicine, Wahroonga, NSW, Australia

Hunter New England Local Health District, Newcastle, NSW, Australia

School of Dentistry and Medical Sciences, Charles Sturt University, Orange, NSW, Australia
e-mail: Michelle.Kang@health.nsw.gov.au

K. Kwong
Sydney Dental Hospital, Surry Hills, NSW, Australia

S.-C. Yeoh
Sydney Oral Medicine, Wahroonga, NSW, Australia

Royal Prince Alfred Hospital, Camperdown, NSW, Australia

Chris O'Brien Lifehouse, Camperdown, NSW, Australia

Sydney Dental School, The University of Sydney, Camperdown, Australia

R. Balasubramaniam et al. (eds.), *Oral Medicine - A Clinical Guide*,
https://doi.org/10.1007/978-3-031-36797-7_22

Table 22.1 Disease classification based on body surface area involvement

SJS	Skin detachment <10% of body surface area
SJS/TEN overlap	Skin detachment between 10 and 30%
TEN	Skin detachment >30% of body surface area

Table 22.2 Possible aetiologies of Stevens-Johnson syndrome and toxic epidermal necrolysis

Drugs	Most common trigger for SJS/TEN. Associated medications include NSAIDs, anticonvulsants (including as carbamazepine, lamotrigine, phenytoin), allopurinol, amoxicillin, sulphonamides, nevirapine
Genetics	HLA associations which significantly increased the risk of EM/SJS/TEN, such as HLA-B*15:02 with carbamazepine (SJS/TEN, in South-East Asian populations), HLA-DQ3 and HLA-DQB1*0402 with EM
Infections	Mycoplasma pneumoniae (mainly in children), viruses especially herpes simplex virus (most common cause of EM)

The annual incidence of EM is estimated at around 1%. The incidence is approximately 1–2 cases per million/year for SJS and 0.4–1.2 cases per million/year for TEN. EM typically affects young adults, whereas the reported median age of SJS/TEN is 49 years but the conditions may be seen in any age group. EM/SJS/TEN have been reported in a wide range of ethnicities.

The pathogenesis of EM and SJS/TEN is incompletely understood but both are widely viewed as a type IV delayed hypersensitivity reaction. Apoptosis of keratinocytes is mediated by cytotoxic CD8+ T cells and natural killer cells. Possible aetiology of SJS/TEN is summarised in Table 22.2.

22.2 Clinical Presentation and Investigations

With EM, there are 2 types based on the degree of mucosal involvement, nature, and distribution of cutaneous lesions (Table 22.3). SJS/TEN occurs 4–28 days after drug exposure. Symptoms include fevers, general physical deterioration, influenza-like illness, and skin pain. These symptoms can frequently present prior to cutaneous clinical signs. Initial erythematous, irregularly-shaped macules on the skin are distributed commonly on the upper trunk, face, and proximal extremities (see Fig. 22.1). Further progression leads to painful, generalised vesiculobullous eruption and desquamation with full-thickness epidermal necrosis. In 80% of cases, 2 or more mucous membranes are involved (often prior to cutaneous presentation) which may include the oral cavity, oropharynx, nasopharynx, eyes, and/or genitalia.

Oral presentation of EM and /SJS/TEN can include haemorrhagic crusting of the lips, as well as widespread inflammation, blistering, and ulcer formation of the

Table 22.3 Types of erythema multiforme (EM) and their clinical presentation

Type	Clinical presentation
EM major	• Involvement of 2 or more mucous membrane surfaces such as oral, ocular, genital, laryngeal, and oesophageal • Variable cutaneous presentation, from typical target lesions (regular round shape, well defined border, concentric rings) to raised atypical target lesions • May have internal organ involvement
EM minor	• More common form • Milder presentation • Involvement of no more than 1 mucous membrane surface • May have symmetrical target lesions on extremities (arms, legs)

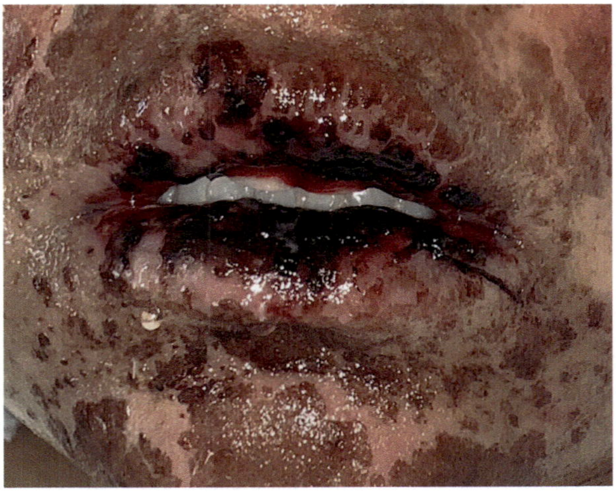

Fig. 22.1 Extra-oral presentation of toxic epidermal necrolysis

intra-oral mucosa. Associated odynophagia and dysphagia may lead to poor oral food and/or fluid intake.

Other possible systemic involvement includes arthralgia, myalgia, pneumonia, myocarditis, nephritis, hepatic cytolysis, and cholestasis.

Diagnosis is mainly a combination of clinical and histopathological features. Differential diagnoses include:

- Other SCAR.
  - Drug reaction with eosinophilia and systemic symptoms (DRESS) syndrome.
  - Acute generalised exanthematous pustulosis (AGEP).
- Linear IgA bullous dermatosis.
- Generalised fixed drug eruption.
- Acute graft versus host disease.
- Superficial burns.

Histopathological features may include full-thickness epithelial necrosis and keratinocyte apoptosis, sub and intra-epithelial vesiculation, and perivascular lymphocytic infiltrate. Direct immunofluorescence demonstrates no specific features.

22.3 Management and Prognosis

Early diagnosis and management are critical. Main considerations for management are supportive care and adjuvant medications/therapies. Critical supportive care includes discontinuing potential causative medications, wound care, monitoring and management of fluids, temperature, nutrition, pain, infection, airway, and systemic co-morbidities. Adjuvant medications and therapies considered include corticosteroids, IVIg, cyclosporin, TNF-alpha inhibitors, and antivirals.

Prognosis of SJS/TEN is assessed using SCORTEN, an illness severity score used at the time of admission to predict mortality. Reported mortality rates for SJS are 10% (after 1 year, around 24%) and at least 30% for TEN (around 49% after 1 year).

Trigger identification for prevention of recurrence is important. For EM, recurrence is reported in up to 25% cases. HSV-1 suppression may be appropriate for some patients.

Recommended Reading

Charlton OA, Harris V, Phan K, Mewton E, Jackson C, Cooper A. Toxic epidermal necrolysis and Steven-Johnson syndrome: a comprehensive review. Adv Wound Care (New Rochelle). 2020;9(7):426–39.

Coias JL, Abbas LF, Cardones AR. Management of Stevens-Johnson syndrome/toxic epidermal necrolysis: a review and update. Curr Dermatol Rep. 2019;8(4):219–233.

Duong TA, Valeyrie-Allanore L, Wolkenstein P, Chosidow O. Severe cutaneous adverse reactions to drugs. Lancet. 2017;390(10106):1996–2011.

Forman R, Koren G, Shear NH. Erythema multiforme, Stevens-Johnson syndrome and toxic epidermal necrolysis in children: a review of 10 years' experience. Drug Saf. 2002;25(13):965–72.

Oral Graft-Versus-Host Disease

23

Katrusha Hull

23.1 Introduction

Graft-versus-host disease (GvHD) is one of the most common transplant-associated complication after allogeneic haematopoietic stem cell transplantation (HSCT), second only to infection. Beyond survival, GvHD can decrease the success of transplantation by increasing morbidity. GvHD is, as its name suggests, a battle between graft (donor stem cells) and the host themselves (the patient).

The oral cavity is frequency impacted by chronic GvHD. Although most chronic GvHD arises in the first year after HSCT it may present and will often persist for many years after transplant. This chapter will focus on oral mucosal GvHD. The remaining oral effects of transplant are discussed in Chap. 67.

The incidence of oral GvHD varies widely in the available literature; 45–80% of patients who develop chronic GvHD at other sites show features of oral involvement.

23.2 Clinical Presentation and Investigations

Clinical presentations diagnostic for oral mucosal GvHD are lichen planus-like features, specifically hyperkeratotic non-removable white lines and lacy/mesh-like patterns (Fig. 23.1). Isolated plaque-like changes are no longer considered diagnostic for GvHD due to the inability to differentiate between those associated with oral mucosal GvHD and leukoplakia and therefore, malignant potential.

K. Hull (✉)
Royal Dental Hospital of Melbourne, Carlton, VIC, Australia

Monash Medical Centre, Clayton, VIC, Australia
e-mail: katrusha.hull@dhsv.org.au

R. Balasubramaniam et al. (eds.), *Oral Medicine - A Clinical Guide*,
https://doi.org/10.1007/978-3-031-36797-7_23

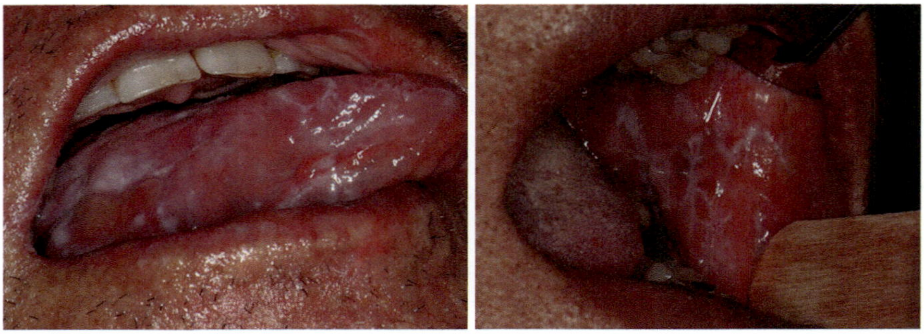

Fig. 23.1 Ulceration in oral GvHD and lichen planus-like lesions of oral GvHD

Table 23.1 Clinical features of oral mucosal chronic GvHD

		Clinical feature	Symptoms
Diagnostic features[a]	Lichen planus-like changes	Hyperkeratotic non-removable white lines and lacy-appearing lesions over the oral mucosa. Any intra-oral site can be affected, buccal mucosa and tongue most common	When erythema and ulceration are absent, symptoms may be minor. Patients may report an altered texture or roughness over affected sites
Distinctive features[b]	Xerostomia	Oral mucosa may appear glazed with ropy white saliva. Overlying features of candidosis is possible	Subjective report of oral dryness. Inability to chew and swallow dry foods with the aid of liquids/sauces
	Superficial mucoceles	Usually painless, smooth surfaced clear fluid-filled swellings which may be solitary or multiple	Occasionally tender and sensitive when burst. May be accompanied by xerostomia
	Mucosal atrophy, ulceration and pseudomembranes	Areas of erythema, ulceration which can affect any intra-oral site	Usually pain and sensitive oral mucosa to spiced and acidic foods and beverages. May report intolerance of SLS-containing toothpastes and alcohol-containing mouthwashes

SLS sodium lauryl sulphate

[a] Diagnostic features—As defined by the national institute of health consensus conference 2005

[b] Distinctive features—Features not ordinarily found in acute GVHD but not considered sufficient in isolation to establish a diagnosis of chronic GVHD

Oral mucosal lesions can be a significant source of pain and restrict oral aperture. When extensive, they may limit nutritional intake, affect the success of daily oral hygiene and hinder the provision of routine dental care (Table 23.1). Significant erythema and ulceration (Fig. 23.1) will usually result in significant discomfort and an inability to tolerate flavoured foods or dentifrices.

Considerations at the time of diagnosis should include exclusion of infection, particularly of fungal and viral origin; and mucosal trauma especially in those presenting with salivary hypofunction. Biopsy may be required to exclude malignancy.

23.3 Management and Prognosis

Topical preparations are often the sole therapy for symptomatic oral chronic GvHD with topical corticosteroid preparations the mainstay of local therapy. Transient burning and the development of secondary oral candidosis are the most common adverse effects. For this reason, topical antifungal agents are often prescribed in a prophylactic capacity to reduce the probability of overlying candidosis. Topical analgesics, such as lignocaine viscous (2% solution), may be helpful when daily activities and nutritional intake are impeded.

Non-pharmacological management strategies include the avoidance of known irritants, such as SLS-containing toothpaste and alcohol-containing mouthwashes. A bland diet is better tolerated with avoidance of spiced or acidic foods and beverages during symptomatic phases.

Where oral GvHD is severe and recalcitrant to topical preparations systemic therapy may be required and is typically prescribed in concert with the overseeing transplant physicians.

Survivors of HSCT face a significant risk of secondary malignancies at a rate of 2–6% in 10 years. Squamous cell carcinoma of the skin and oral cavity account for one-third of all secondary solid tumours, 50% of these arising within the oral cavity, history of oral chronic GvHD being a major risk factor (Fig. 23.2). Long-term specialist oral mucosal screening and addressing modifiable risk factors are an essential part of the long-term care for patients with oral chronic GvHD.

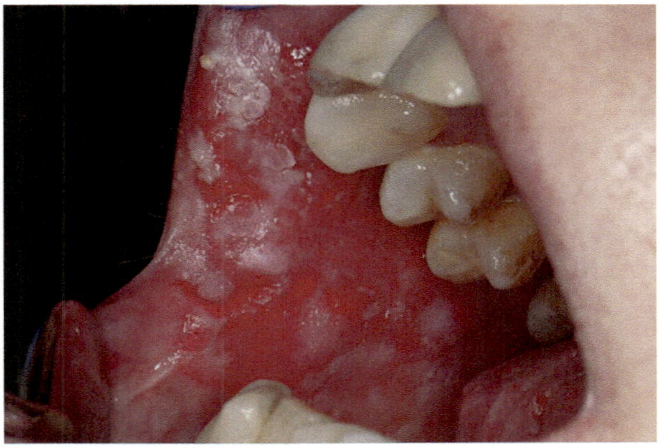

Fig. 23.2 Oral squamous cell carcinoma developing within oral GvHD

Recommended Reading

Jagasia MH, Greinix HT, Arora M, Williams KM, Wolff D, Cowen EW, et al. National Institutes of Health Consensus Development Project on Criteria for Clinical Trials in Chronic Graft-versus-Host Disease: I. The 2014 Diagnosis and Staging Working Group report. Biol Blood Marrow Transplant. 2015;21(3):389–401.e1.

Majhail NS, Rizzo JD, Lee SJ, Aljurf M, Atsuta Y, Bonfim C, et al. Recommended screening and preventive practices for long-term survivors after hematopoietic cell transplantation. Biol Blood Marrow Transplant. 2012;18(3):348–71.

Mawardi H, Elad S, Correa ME, Stevenson K, Woo SB, Almazrooa S, et al. Oral epithelial dysplasia and squamous cell carcinoma following allogeneic hematopoietic stem cell transplantation: clinical presentation and treatment outcomes. Bone Marrow Transplant. 2011;46(6):884–91.

Schubert MM, Correa ME. Oral graft-versus-host disease. Dent Clin North Am. 2008;52(1):79–109, viii-ix.

Allergy and the Mouth

24

Guangzhao Guan and Ajith Polonowita

24.1 Introduction

Allergic/hypersensitivity reactions are inappropriate responses of the immune system against an antigen or allergen. Allergic reactions of the oral mucosa may comprise wide spectrum of clinical presentations, induced by type I to type IV hypersensitivity. For example, allergic angioedema is a severe type I hypersensitivity reaction (mediated by IgE) that can cause acute angioedema and urticaria. A number of autoimmune diseases, such as pemphigus vulgaris and mucous membrane pemphigoid, occur in which antibodies against tissue antigens cause immunopathological damage by activation of type II hypersensitivity. Oral lichenoid lesion and allergic contact stomatitis/cheilitis are considered to be type IV hypersensitive reactions. These oral mucosa changes are the first sign of a contact allergy and are discussed in other chapters. Reactions to the dental materials, food, food additive, drugs and oral hygiene products are common aetiological factors. The patient's history is necessary to establish the diagnosis. Sometimes histopathology might be useful. Prick, patch and intradermal tests are commonly used in allergy testing. Particularly, patch test is a simple and non-invasive diagnostic method in many oral diseases including hypersensitivity to dental materials.

A typical allergy of the oral mucosa is plasma cell gingivitis (PCG) which will be the focus of this chapter. PCG is a rare benign inflammatory condition of gingiva characterised by diffuse and dense infiltration of plasma cells into the connective tissue. It was first reported in the early 1970s and is also known by other names,

G. Guan (✉)
Faculty of Dentistry, University of Otago, Dunedin, New Zealand
e-mail: simon.guan@otago.ac.nz

A. Polonowita
Faculty of Dentistry, University of Otago, Dunedin, New Zealand

Christchurch Hospital, Canterbury, New Zealand

R. Balasubramaniam et al. (eds.), *Oral Medicine - A Clinical Guide*,
https://doi.org/10.1007/978-3-031-36797-7_24

such as plasmacytosis of the gingiva, idiopathic gingivostomatitis, plasma cell gingivostomatitis, atypical gingivostomatitis and allergic gingivostomatitis.

The aetiology of PCG remains unknown, but due to the obvious presence of plasma cells, it is postulated to be a hypersensitivity reaction to allergens such as the ingredients of dentifrices (flavours or preservatives), chewing gum, cosmetics, Colocasia (arbi) leaves, khat leaves (Catha edulis) and other substances which may contact the oral cavity.

PCG is uncommon. The average age of the reported cases is 25.8 ± 13.4 years, and there is a female predominance of 1.8:1.

24.2 Clinical Presentation and Investigations

Clinically, PCG appears as a diffuse, friable, erythematous and oedematous enlargement of the gingiva and may extend from the free marginal gingivae onto the attached gingiva and often painful (Fig. 24.1). Gingival bleeding may occur with the slightest provocation. This contrasts with plaque-induced gingivitis that usually involves the marginal gingiva alone and not the entire width of the attached gingiva with a predilection for the maxillary anterior gingival. Involvement of sites such as lip, tongue and palate, other than the attached gingiva, has been reported. PCG is typically painless although may sometimes sting or burn, particularly with spicy food.

Differential diagnosis may include some common solitary gingival epulides (pyogenic granuloma, peripheral giant cell granuloma), desquamative gingivitis associated with dermatoses, granulomatous gingivitis, drug-related gingival hyperplasia, haematological malignancy (acute leukaemia, myeloma, plasmacytoma) and granulomatosis with polyangiitis (Wegener's granulomatosis). Thorough patient

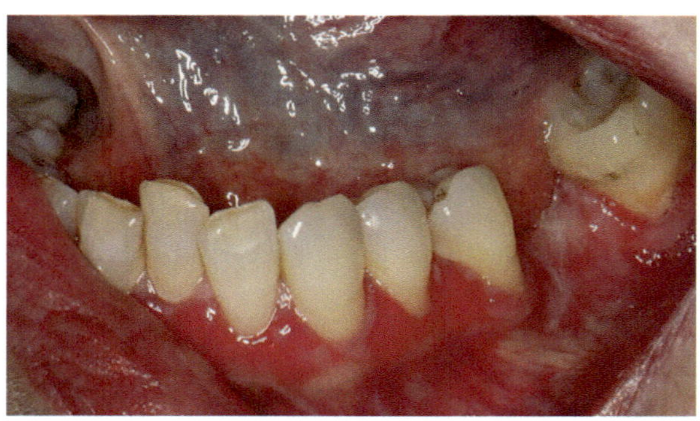

Fig. 24.1 Buccal gingival lesion showing diffuse inflammatory gingival enlargement with erythematous overlying epithelium

history should be taken to exclude any potential causative factors predisposing to PCG. Haematological screening is required to exclude haematologic malignancies, connective tissue diseases and systemic vasculitis. The diagnosis of PCG is made on the basis of histopathological examination. Histopathology demonstrates surface epithelium with psoriasiform hyperplasia, spongiosis and exocytosis. Multiple neutrophilic micro-abscesses can be identified in the parakeratin layer. The connective tissues exhibited dense chronic inflammatory infiltrate consisting predominantly of plasma cells.

24.3 Management and Prognosis

Once the diagnosis of PCG is made, the screening of various antigenic substances must be done. This includes a complete dietary history with careful investigation of any consistent exposure to new medications (including non-prescribed supplements) and dietary contents. Patch testing may be indicated. PCG involves both medical and surgical approaches. Standard professional oral hygiene and non-surgical periodontal protocol can serve as first-line therapy, especially in paediatric cases. Several treatments have been trialled, including topical (betamethasone rinses, fluocinonide gel, triamcinolone paste), systemic corticosteroids and topical fusidic acid with variable results. Treatment may also include excision by laser, scalpel or electrosurgery. No treatment clearly stands out as consistently effective with tailored management required. After the withdrawal of the offending agent, the resolution of PCG may take days to weeks and it depends on the extent and severity of the disease.

Recommended Reading

Arduino PG, D'Aiuto F, Cavallito C, Carcieri P, Carbone M, Conrotto D, et al. Professional oral hygiene as a therapeutic option for pediatric patients with plasma cell gingivitis: preliminary results of a prospective case series. J Periodontol. 2011;82(12):1670–5.

Gargiulo AV, Ladone JA, Ladone PA, Toto PD. Case report: plasma cell gingivitis A. CDS Rev. 1995;88(3):22–3.

Kerr DA, McClatchey KD, Regezi JA. Idiopathic gingivostomatitis. Cheilitis, glossitis, gingivitis syndrome; atypical gingivostomatitis, plasma-cell gingivitis, plasmacytosis of gingiva. Oral Surg Oral Med Oral Pathol. 1971;32(3):402–23.

Marker P, Krogdahl A. Plasma cell gingivitis apparently related to the use of khat: report of a case. Br Dent J. 2002;192(6):311–3.

Negi BS, Kumar NR, Haris PS, Yogesh JA, Vijayalakshmi C, James J. Plasma-Cell Gingivitis a Challenge to the Oral Physician. Contemp Clin Dent. 2019;10(3):565–70.

Thomson A, Polina H, Pannu M, Bhatt V. Diagnosis of plasma cell gingivitis from a dental referral of 'strawberry gingivitis': the diagnostic challenges associated with the diagnosis and treatment of plasma cell gingivitis. Int J Oral Maxillofac Surg. 2019;48(Suppl 1):104–5.

25 Orofacial Granulomatosis

Ahmed Sultan and Maryam Jessri

25.1 Introduction

The term orofacial granulomatosis (OFG) was first introduced in 1985 to encompass non-infectious, non-necrotising granulomatous enlargement of the lips, face, and oral cavity in the absence of other systemic disease (such as Crohn's disease or sarcoidosis). The term OFG includes cheilitis granulomatosa and Melkersson–Rosenthal syndrome.

Although OFG is largely considered to be idiopathic, cell-medicated hypersensitivity to a variety of inciting agents such as foods, food additives, or certain flavouring agents have been implicated in the aetiopathogenesis of the disease. Additionally, a number of systemic granulomatous diseases such as Crohn's disease, sarcoidosis and tuberculosis and other triggering local causes such as chronic infections, foreign materials and allergy to cosmetics may present similarly to OFG.

OFG has no gender predilection and can present at any age; however, the peak prevalence is in the second and third decades. OFG is rare and reliable epidemiologic data does not yet exist. Most cases series are from single-centres and include mixed cohorts that do not differentiate the oral presentations of various granulomatous diseases.

A. Sultan
School of Dentistry, The University of Maryland, Baltimore, MD, USA

Greenbaum Comprehensive Cancer Centre, The University of Maryland, Baltimore, MD, USA

M. Jessri (✉)
School of Dentistry, The University Of Queensland, Herston, QLD, Australia

Metro North Hospital and Health Services, Queensland Health, Herston, QLD, Australia
e-mail: m.jessri@uq.edu.au

R. Balasubramaniam et al. (eds.), *Oral Medicine - A Clinical Guide*,
https://doi.org/10.1007/978-3-031-36797-7_25

25.2 Clinical Presentation and Investigations

OFG has a highly variable course and tends to be chronic. The lips are the most common site of involvement; and the condition presents as nontender, persistent swelling which may involve one or both lips and can be unilateral or bilateral (Fig. 25.1). When OFG is limited to the lips only, it may be referred to as granulomatous cheilitis or Miescher cheilitis. Melkersson–Rosenthal syndrome is a rare variety of OFG which involves granulomatous cheilitis in combination with facial nerve palsy and a fissured tongue. Since neither Melkersson–Rosenthal syndrome and cheilitis granulomatosa represent specific entities, it is recommended to include both these conditions under the unifying term of OFG.

In addition to the lips, intraoral sites including the gingivae, buccal mucosae, tongue and palatal mucosae may also be involved. The clinical presentations are variable ranging from oedema and swelling to focal areas of submucous enlargement and hyperplastic tissues, ulceration and fissured tongue.

Upon identifying non-necrotising granulomatosis inflammation in the absence of microorganisms and foreign body material, a diagnosis of OFG should be considered. A range of systemic granulomatous disorders, including Crohn's disease, sarcoidosis, leprosy, tuberculosis, chronic granulomatous disease and possibly deep fungal infections have been reported to cause oral manifestations. Hence, once a working diagnosis of OFG is proposed, these conditions should be fully investigated and therefore OFG is strictly a diagnosis of exclusion.

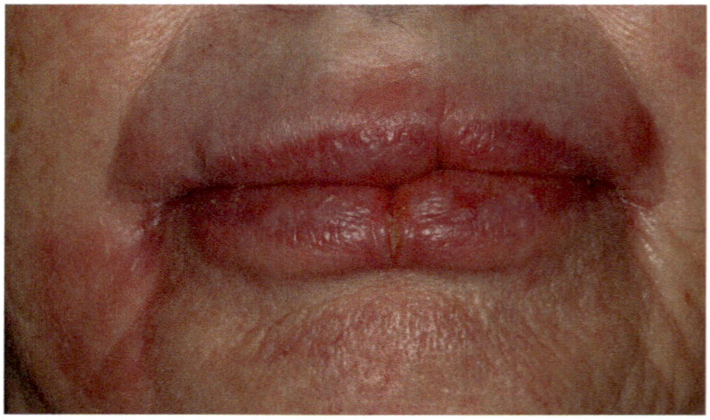

Fig. 25.1 A 63-year-old female with orofacial granulomatosis

25.3 Management and Prognosis

Although not always feasible, the first goal of therapy should be identification and elimination of the inciting agent. Allergy testing and a diet diary may be beneficial in attempting to identify and eliminate the offending agent/allergen. Common allergens include cinnamon and benzoate compounds; less commonly reported allergens include but are not limited to dental restorative materials, dentifrice, chocolate, carvone, aspartate, monosodium glutamate and tartrazine. Topical application or intralesional injection of corticosteroids has been shown to be effective. In recurrent or recalcitrant cases, systemic treatment with corticosteroids, dapsone, clofazimine, hydroxychloroquine, thalidomide, sulfasalazine, azathioprine, and tumour necrosis factor (TNF)-α inhibitors may be beneficial. Surgical recontouring is seldom necessary.

Recommended Reading

Allen CM, Camisa C, Hamzeh S, Stephens L. Cheilitis granulomatosa: report of six cases and review of the literature. J Am Acad Dermatol. 1990;23(3 Pt 1):444–50.

Campbell HE, Escudier MP, Patel P, Challacombe SJ, Sanderson JD, Lomer MC. Review article: cinnamon- and benzoate-free diet as a primary treatment for orofacial granulomatosis. Aliment Pharmacol Ther. 2011;34(7):687–701.

Busingye D, Pollack A, Chidwick K. Prevalence of inflammatory bowel disease in the Australian general practice population: A cross-sectional study. PLoS One. 2021;16(5):e0252458.

Crohn's Disease

26

Ahmed Sultan and Maryam Jessri

26.1 Introduction

Crohn's disease otherwise referred to as regional ileitis or regional enteritis is a chronic granulomatous disorder of the gastrointestinal tract of unknown cause. Crohn's disease and ulcerative colitis are the two major categories of inflammatory bowel disease. A close assessment of clinical features, biochemical studies, endoscopic findings, histology, and luminal imaging help differentiate between Crohn's disease and ulcerative colitis. Inflammatory bowel diseases are discussed further in Chap. 78.

The exact cause of inflammatory bowel disease is not known. Current literature suggests a combination of genetic susceptibility, and dietary and environmental factors in the aetiopathogenesis of both Crohn's disease and ulcerative colitis.

A. Sultan (✉)
School of Dentistry, The University of Maryland, Baltimore, MD, USA

Greenbaum Comprehensive Cancer Centre, The University of Maryland, Baltimore, MD, USA
e-mail: asultan@umaryland.edu

M. Jessri
School of Dentistry, The University of Queensland, Herston, QLD, Australia

Metro North Hospital and Health Services, Queensland Health, Herston, QLD, Australia

R. Balasubramaniam et al. (eds.), *Oral Medicine - A Clinical Guide*,
https://doi.org/10.1007/978-3-031-36797-7_26

The peak prevalence of Crohn's disease diagnosis is in the second and third decades; a second diagnostic peak has been noted in patients over 50–60 years of age. Women are slightly more likely to develop Crohn's disease, and the crude prevalence of Crohn's disease is estimated to be 306 per 100,000 in Australia.

26.2 Clinical Presentation and Investigations

Up to 15% of patients with Crohn's disease exhibit oral presentations, which are varied and in parts, non-specific. The most common intra-oral presentations include diffuse or nodular swelling of the oral and perioral tissues (identical to orofacial granulomatosis), a cobblestone appearance of the mucosa, and deep, granulomatous-appearing linear fissures, hyperplastic folds, and/or ulcers of the buccal vestibule. Mucogingivitis which is defined as erythematous patches and plaques of the attached gingivae has also been reported in Crohn's disease.

Pyostomatitis vegetans which is characterised by yellowish, slightly raised, irregular shaped pustules of the oral cavity is a relatively rare condition associated with inflammatory bowel disease that is more common in ulcerative colitis than Crohn's disease.

Obtaining a biopsy (oral and colonic) for microscopic examination, in addition to a comprehensive clinical history and possible diagnostic serology including haematological evaluation for gastrointestinal malabsorption and anti-saccharomyces cerevisiae antibodies should be considered in the diagnostic work-up. Histopathological examination may reveal noncaseating granulomatous inflammation with or without the presence of multinucleated giant cells. Special stains for microorganisms should be negative, and no foreign-body material should be present.

26.3 Management and Prognosis

As soon as a diagnosis of oral Crohn's has been made, the patient must be referred to a gastroenterologist to investigate the extent of systemic involvement. Oral lesions typically respond to the systemic therapy provided for the bowel disease (such as, e.g. sulfasalazine, TNF-α inhibitors, systemic corticosteroids, or other biologics); however, in some cases localised adjunctive treatment with topical or intralesional corticosteroids may be beneficial.

Recommended Reading

Adams SM, Bornemann PH. Ulcerative colitis. Am Fam Physician. 2013;87(10):699–705.

Veauthier B, Hornecker JR. Crohn's Disease: Diagnosis and Management. Am Fam Physician. 2018;98(11):661–9.

Other Granulomatous Disorders 27

Ahmed Sultan and Maryam Jessri

27.1 Introduction

This chapter discusses granulomatous diseases (granulomatosis with polyangiitis, sarcoidosis, and tuberculosis) apart from Crohn's disease and orofacial granulomatosis that may exhibit oral presentations. Table 27.1 summarises these granulomatous disorders.

A. Sultan
School of Dentistry, The University of Maryland, Baltimore, MD, USA

Greenbaum Comprehensive Cancer Centre, The University of Maryland, Baltimore, MD, USA

M. Jessri (✉)
School of Dentistry, The University of Queensland, Herston, QLD, Australia

Metro North Hospital and Health Services, Queensland Health, Herston, QLD, Australia
e-mail: m.jessri@uq.edu.au

R. Balasubramaniam et al. (eds.), *Oral Medicine - A Clinical Guide*,
https://doi.org/10.1007/978-3-031-36797-7_27

Table 27.1 Clinicopathological features of granulomatous disorders and their treatment strategy

	Granulomatosis with polyangiitis	Sarcoidosis	Tuberculosis
Aetiology and pathogenesis	Environmental exposure in genetically susceptible individuals	Environmental exposure in genetically susceptible individuals	*Mycobacterium tuberculosis*
Epidemiology	Seventh decade	Fifth and sixth decades	All age groups
Oral presentations	Strawberry gingivitis and painful necrotic oral ulcerations	Nodules in salivary glands or enlarged lymph nodes and oral ulcers (less common)	Non-specific presentations such as ulcers and swelling
Treatment strategy	Immunosuppression	Immunosuppression	Multiagent therapy

27.2 Granulomatosis with Polyangiitis

Granulomatosis with polyangiitis (GPA), formerly known as Wegener's granulomatosis, is an uncommon necrotising granulomatous disorder that is often fatal. GPA is a midline destructive disease and involves every organ system including the head and neck and oral cavity. Based on its extent, GPA has been classified into limited and generalised.

GPA is an idiopathic condition with possible autoimmune components triggered by environmental factors and genetic susceptibility.

GPA can involve children and older adults with a mean age at presentation in the seventh decade with no gender predilection.

27.2.1 Clinical Presentation and Investigations

The most common oral presentation of GPA is diffuse gingival hyperplasia characterised by pebbly, granular, red, friable gingivae with pinpoint haemorrhage (strawberry gingivitis). Other oral presentations may include painful necrotic oral ulcerations.

Histopathologic examination, a thorough history and laboratory findings (anti-neutrophil cytoplasm antibodies, ANCAs) aid in the diagnosis of GPA.

27.2.2 Management and Prognosis

Long-term, high-dose corticosteroids and cyclophosphamide are the first-line medications for GPA. Other medications include methotrexate, azathioprine, and mycophenolate mofetil.

27.3 Sarcoidosis

Sarcoidosis is a multiorgan granulomatous disease with a variable course and mostly affects the lungs, lymph nodes, and skin with lymphoid tissues being most commonly involved.

Similar to GPA, the exact cause of sarcoidosis is not completely understood. Environmental activation of autoimmune pathways in genetically susceptible individuals has been suggested to play a causative role.

Sarcoidosis generally affects those under the age of 50 years with a high affinity for females and African Americans.

27.3.1 Clinical Presentation and Investigations

Depending on the involved organs and severity, the clinical presentations vary. Most oral presentations of sarcoidosis are limited to the lymphoid tissues and salivary glands. In rare non-lymphoid, non-salivary gland sarcoidosis cases, a submucosal oral mass can signify the disease.

Histopathologic examination and identifying the characteristics of sarcoidosis (e.g. granulomatous inflammation, Langerhans giant cells, schaumann bodies, and asteroid bodies) aid in the diagnosis of sarcoidosis. In addition, serum angiotensin-converting enzyme levels and possible chest X-ray are necessary to rule out other possible granulomatous conditions.

27.3.2 Management and Prognosis

Sarcoidosis is managed through immunosuppression, although not all patients need to be treated actively. Treatment is generally reserved for those with organ dysfunction, involvement of vital organs, or progressive disease.

27.4 Tuberculosis

Tuberculosis is a chronic infectious disease which results in formation of granulomata with central necrosis.

Tuberculosis is an infectious disease caused by *Mycobacterium tuberculosis* which is a pathogenic bacterium in the family Mycobacteriaceae.

It is estimated that one-third of the world's population is infected by *Mycobacterium tuberculosis* and 2–3 million deaths a year are attributed to tuberculosis. Although antibacterial medications are improving outcomes, the success of this decline is hindered by the emergence of drug-resistant strains.

27.4.1 Clinical Presentation and Investigations

Enlarged cervical lymph nodes are the most common extrapulmonary presentation of tuberculosis in the head and neck. TB-associated lymphadenopathy is usually characterised by multiple, matted, hard to fluctuant and in later stages with draining sinuses. Oral lesions are uncommon in tuberculosis and may include non-specific presentations such as chronic swelling and ulceration.

Special mycobacterial stains and culture of infected tissue confirm a diagnosis of tuberculosis.

27.4.2 Management and Prognosis

Multi-agent therapy is the treatment of choice for patients with diagnosed active tuberculosis.

Recommended Reading

Greco A, Marinelli C, Fusconi M, Macri GF, Gallo A, De Virgilio A, et al. Clinic manifestations in granulomatosis with polyangiitis. Int J Immunopathol Pharmacol. 2016;29(2):151–9.

Llanos O, Hamzeh N. Sarcoidosis. Med Clin North Am. 2019;103(3):527–34.

Thakkar K, Ghaisas SM, Singh M. Lymphadenopathy: Differentiation between Tuberculosis and Other Non-Tuberculosis Causes like Follicular Lymphoma. Front Public Health. 2016;4:31.

Benign Salivary Gland Disorders 28

Anastasia Georgiou

28.1 Introduction

Common benign salivary gland disorders include salivary gland infections (sialadenitis), salivary calculi or stones (sialolithiasis) and mucoceles. Less common is necrotising sialometaplasia. Salivary gland infections and sialolithiasis typically involve the major salivary glands, mucoceles involve both major and minor salivary glands and necrotising sialometaplasia typically associated with minor salivary glands.

Common causes of salivary gland infections are shown in Table 28.1. Sialolithiasis is characterised by the development of calcified stones from calcium deposition around a nidus of bacteria or mucus plug often leading to obstruction of the salivary duct.

Mucoceles are result of accumulation of mucous either in the surrounding connective tissues or within the salivary duct (Table 28.2), often resulting from trauma. Necrotising sialometaplasia is a benign inflammatory condition resulting from ischaemic necrosis of minor salivary glands from local trauma or surgical manipulation of the area.

The exact incidence of acute and chronic bacterial sialadenitis is not clear but it is often seen in older adults with no gender predilection. The prevalence of epidemic parotitis (mumps) is declining in countries with established immunisation programs for mumps. Viral sialadenitis commonly affects children and young

---

A. Georgiou (✉)
Macquarie Oral and Maxillofacial Specialists, Sydney, NSW, Australia

Sydney Skin, Newtown, NSW, Australia
e-mail: admin@sydneyskin.com

R. Balasubramaniam et al. (eds.), *Oral Medicine - A Clinical Guide*,
https://doi.org/10.1007/978-3-031-36797-7_28

adults. Sialolithiasis is common in the fourth and fifth decades of life. Mucoceles are the most common benign salivary gland disorder, common in young individuals with a history of trauma. Necrotising sialometaplasia is rare and observed in adults who are usually smokers.

Table 28.1 Common causes of salivary gland infections

	Aetiology	Causative agent	Clinical presentation
Acute suppurative sialadenitis	Bacterial	*Staphylococcus aureus* *Streptococcus viridians* *Streptococcus pneumoniae* *Haemophilus influenzae* *Escherichia coli* *Bacteroides*	• Sudden onset, pain and swelling of affected gland • Usually unilateral, often parotid gland • Purulence from duct • Redness of overlying skin
Chronic/ recurrent sialadenitis	Obstruction	Sialolith or stricture of the salivary gland duct Repeated episodes of acute sialadenitis	• Swollen or firm gland • Often not tender to palpation • No systemic signs of infection • Pain and increased swelling often with mealtimes
Viral sialadenitis	Viral	Epstein–Barr virus Parainfluenza HIV	• Often bilateral swelling • May be tender
Epidemic parotitis	Viral	Paramyxovirus (mumps virus)	• Acute onset, bilateral enlarged parotid glands • Overlying skin normal • Fever and malaise
Recurrent parotitis of childhood	Possible congenital duct anomalies	Non-obstructive sialectasis (dilatation of the parotid ducts)	• Intermittent, unilateral parotid swelling • 2 to 3 weeks duration

Table 28.2 Classification of mucoceles

Type	Pathogenesis	Aetiology	Location
Mucous extravasation phenomenon	Mucus extravasating into surrounding connective tissue	Trauma to, or transection of minor salivary gland duct	Often lower lip
Mucous retention cyst	Accumulation of saliva within duct	Duct obstruction often of sublingual or submandibular gland	Floor of mouth
Ranula Plunging/cervical	Accumulation of saliva within duct	Herniation of mucous retention cyst through mylohyoid muscle and along fascial planes of neck	Floor of mouth and neck
Superficial mucocele	Extravasation of mucus beneath epithelium	Trauma to minor salivary gland duct	Soft palate

28.2 Clinical Presentation and Investigations

The clinical presentation of sialadenitis is summarised in Table 28.1.

Sialolithiasis often affect the submandibular gland, within Wharton's duct. Majority of sialoliths are asymptomatic until obstruction of the gland results in episodic pain and swelling around mealtimes. The sialolith might be visible as a pale yellow or white hard mass within the duct (Fig. 28.1).

Mucoceles present as a bluish, fluid-filled sessile lesion most commonly involving the lower lip but can develop at any oral site. Mucous retention cysts of submandibular or sublingual gland will cause swelling in the floor of the mouth (Fig. 28.2). Small superficial mucoceles and often seen involving the soft palate (Fig. 28.3).

Necrotising sialometaplasia presents as an asymptomatic swelling of sudden onset followed by painful solitary ulceration often of the hard palate.

Diagnosis of acute bacterial and viral sialadenitis is often made clinically. Pus should be sent for culture and sensitivity testing. The underlying cause should be identified to prevent further episodes.

Some sialoliths can be identified on a radiograph. Other imaging modalities might demonstrate dilatation of duct.

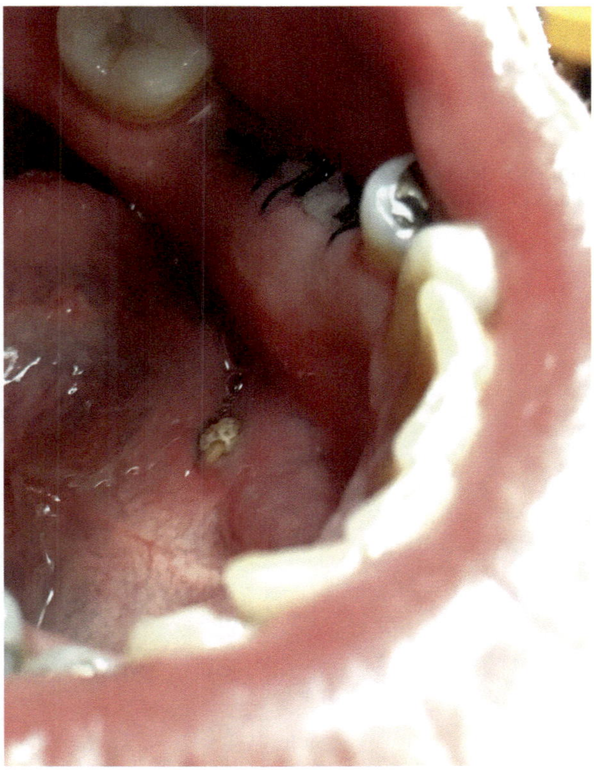

Fig. 28.1 Sialolith within submandibular gland duct

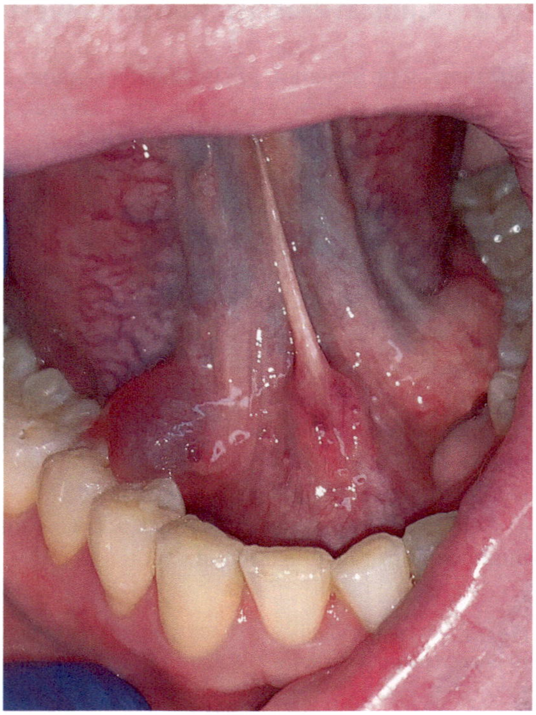

Fig. 28.2 Retention cyst

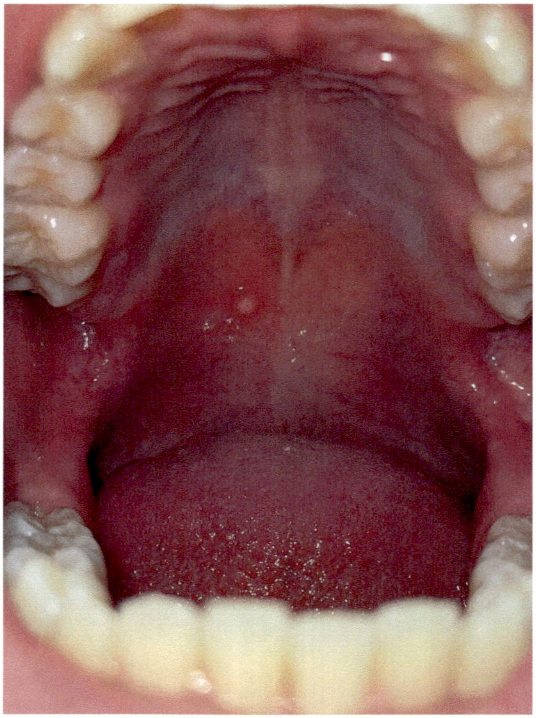

Fig. 28.3 Superficial mucocele of the soft palate

Diagnosis of a mucocele is usually made on history and examination. Simple and plunging ranula require further imaging such as CT scan.

Biopsy is a required investigation for necrotising sialometaplasia with histopathology demonstrating pseudoepitheliomatous hyperplasia, often mistaken for squamous cell carcinoma.

28.3 Management and Prognosis

Management of bacterial sialadenitis consists of empirical antibiotic therapy with beta-lactamase activity such as flucloxacillin; in addition to other supportive measures saliva flow should be encouraged by chewing gum or drinking lemon juice. Abscess formation may require surgical drainage. Salivary hypofunction may predispose patients to recurrence. Bacterial sialadenitis may also become chronic. Management of viral sialadenitis is predominantly symptomatic care with rest, fluid and analgesia.

Small sialoliths or those located within the duct can be manipulated to facilitate removal (Fig. 28.1). Sialoliths located within the body of the gland causing repeated obstruction and sialadenitis may require complete surgical removal of the gland.

Mucous extravasation phenomenon is often treated by surgical excision to include the lesion and associated minor salivary glands. Mucous retention cysts are treated by complete excision of the lesion and the affected gland.

Establishing diagnosis of necrotising sialometaplasia from histology is important in the management of this self-limiting condition. Regular antibacterial mouthwash can be used for pain relief. The ulcer tends to heal within 6–10 weeks with recurrence unusual.

Recommended Reading

Jansen SB, Vissink A, Firth N. Salivary gland disorders and diseases. In Farah CS, Balasubramaniam R, McCullough MJ. Contemporary Oral Medicine, Volume 2 Springer 2019. P. 1437–1521.

Van der Waal, I. (1997a). Cysts. In: Diseases of the Salivary Glands Including Dry Mouth and Sjögren's Syndrome. Springer, Berlin, Heidelberg. P 51–59.

Van der Waal, I. (1997b). Inflammatory Diseases. In Diseases of the Salivary Glands including Dry Mouth and Sjögren's Syndrome. Springer, Berlin, Heidelberg. P. 24–36.

Salivary Hypofunction and Xerostomia

29

Anastasia Georgiou

29.1 Introduction

Dry mouth is the most common salivary-related presenting complaint. Interestingly, dry mouth can occur without a significant reduction in salivary quantity and may be secondary to qualitative changes in salivary composition. Salivary gland hypofunction refers to objective decrease in either whole and/or individual salivary gland flow rates, whereas xerostomia is the subjective sensation of oral dryness.

The aetiology of salivary gland dysfunction is variable, often multifactorial and can be both salivary and non-salivary related (Table 29.1). A detailed medical and symptom history, along with a clinical examination will identify whether the sensation of dry mouth is accompanied by reduced salivary flow. These findings will often guide clinicians to an underlying associated cause. Some causes are easily identified whereas others, such as contribution of polypharmacy and anxiety, are more difficult to discern. Several hundred medications list 'dry mouth' as an adverse drug effect (Table 29.2). The most common xerogenic medications are anticholinergic in action.

The prevalence of xerostomia and salivary hypofunction in the general population has great variation from 5.5% to 46% due to differences in study design, populations, and diagnostic nomenclature. Overall, the prevalence of salivary dysfunction increases with age and affects >30% of the population aged 65 years and older. The prevalence of salivary dysfunction can be close to 100% in patients who have received radiotherapy for head and neck tumours over 25 Gy and individuals with Sjögren's syndrome.

A. Georgiou (✉)
Macquarie Oral and Maxillofacial Specialists, Sydney, NSW, Australia

Sydney Skin, Newtown, NSW, Australia
e-mail: admin@sydneyskin.com

R. Balasubramaniam et al. (eds.), *Oral Medicine - A Clinical Guide*,
https://doi.org/10.1007/978-3-031-36797-7_29

Table 29.1 Causes of salivary gland dysfunction

Salivary	Autoimmune/chronic inflammatory	Sjögren's syndrome Systemic lupus erythematosus Rheumatoid arthritis Scleroderma and CREST Mixed connective tissue disease Primary biliary cirrhosis Sarcoidosis Graft versus host disease
	Infectious	Hepatitis C Human immunodeficiency virus Herpes virus family • Cytomegalovirus • Epstein–Barr virus
	Endocrine/metabolic	Diabetes mellitus Diabetes insipidus Hypothyroidism End-stage renal disease Eating disorders • Anorexia nervosa • Bulimia nervosa Alcohol abuse
	Neoplastic	Salivary gland tumour Lymphoma
Non-salivary	Iatrogenic	Graft versus host disease External beam radiation therapy Radioactive iodine therapy (^{131}I) Medications (see Table 29.2)
	Physiological	Dehydration Mouth breathing
	Neurological/psychiatric	Cognitive alteration Neurological dysfunction Oral sensory dysfunction Psychological/anxiety

Table 29.2 Medication classes associated with dry mouth

Mechanism of action	Pharmacologic group of medication	Examples
Anticholinergic	Tricyclic antidepressants	Amitriptyline, nortriptyline, imipramine
	Antipsychotics	Phenothiazines, haloperidol, clozapine, olanzapine, quetiapine, lithium, risperidone
	Anticonvulsants	Gabapentin, pregabalin, sodium valproate
	Anti-Parkinson's medications	Levodopa
	Antihistamines	Doxylamine, cetirizine, loratadine, promethazine, diphenhydramine
	Antiemetics	Scopolamine
	Antimuscarinic	Atropine, oxybutynin
	Bronchodilators	Tiotropium bromide, ipratropium bromide
	Diuretics	Furosemide
Sympathomimetic	Antidepressants (selective serotonin reuptake inhibitors, serotonin, and noradrenaline reuptake inhibitors)	Bupropion, venlafaxine, duloxetine
	Antihypertensives	Methyldopa, clonidine
	Appetite suppressants	Phentermine, orlistat
	CNS stimulants	Methylphenidate, dexmethylphenidate, lisdexamfetamine
	Decongestants	Doxylamine, cetirizine
	Skeletal muscle relaxants	Baclofen
	Antimigraine agents	Clonidine, rizatriptan
Synergistic action (anticholinergic and sympathomimetic)	Opioids	Morphine, codeine, tramadol, tapentadol
	Non-benzodiazepine	Zolpidem, zopiclone
	Benzodiazepine	Diazepam, temazepam, lorazepam, alprazolam
	Drugs of abuse	Methamphetamine, heroin, cocaine
Others	Antibiotics	Amoxycillin, tetracycline, metronidazole
	Antivirals	Saquinavir, lamivudine
	Anti-neoplastic/cytotoxic medications	Bevacizumab, interferon alpha,
	Proton pump inhibitors	Omeprazole
	Retinoids	Isotretinoin

29.2 Clinical Presentation and Investigations

The primary presentation of a patient with xerostomia is the complaint of a dry mouth. The clinical presentation can vary with amount of saliva present. Some patients complain of a dry mouth but on clinical examination, normal range of saliva is present. A myriad of clinical complications resulting from salivary hypofunction may present in affected individuals (Table 29.3). Symptoms that cause considerable distress are mucosal discomfort, burning sensation, altered taste, halitosis, difficulty swallowing, and the tongue stuck to the roof of the mouth on nocturnal or morning waking.

A thorough medical history and evaluation of symptoms are crucial for diagnosis of salivary dysfunction. Specific questionnaires have been developed for use in this setting. The combination of history and clinical findings can determine if patient has either xerostomia, salivary hypofunction, or both. Unstimulated and stimulated saliva can be collected chairside, with an unstimulated whole saliva flow rate <0.1 mL/min suggestive of salivary gland hypofunction. Once salivary dysfunction has been established, the differential diagnosis relates to the underlying cause. Laboratory blood tests may suggest related systemic disease, and minor salivary gland biopsy may support a diagnosis of Sjögren's syndrome.

Table 29.3 Clinical presentation of salivary hypofunction

	Clinical presentation
Extra-oral findings	• Dry and cracked lips • Angular cheilitis (Fig. 29.1) • Enlarged major salivary glands
Intra-oral findings	• Low resting salivary flow rate • Frothy saliva • No saliva pooling in floor of mouth (Fig. 29.2) • Desiccated mucosal surfaces • Fissured, red or depapillated tongue (Fig. 29.3) • Oral candidosis—pseudomembranous and erythematous types • Caries—root surface, new carious lesions on surfaces not normally affected (incisal edges of anterior teeth), recurrent lesions on restoration margins • Reduced denture retention • Denture-related trauma • Food retention and debris

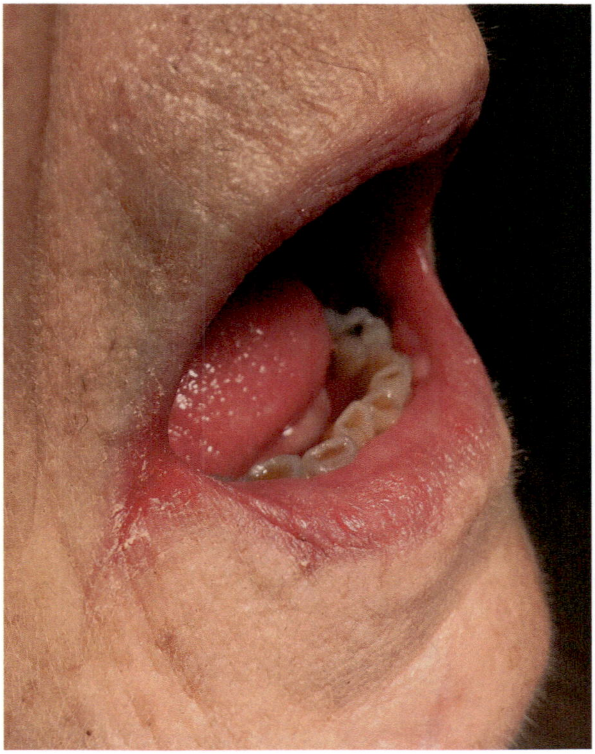

Fig. 29.1 Angular cheilitis

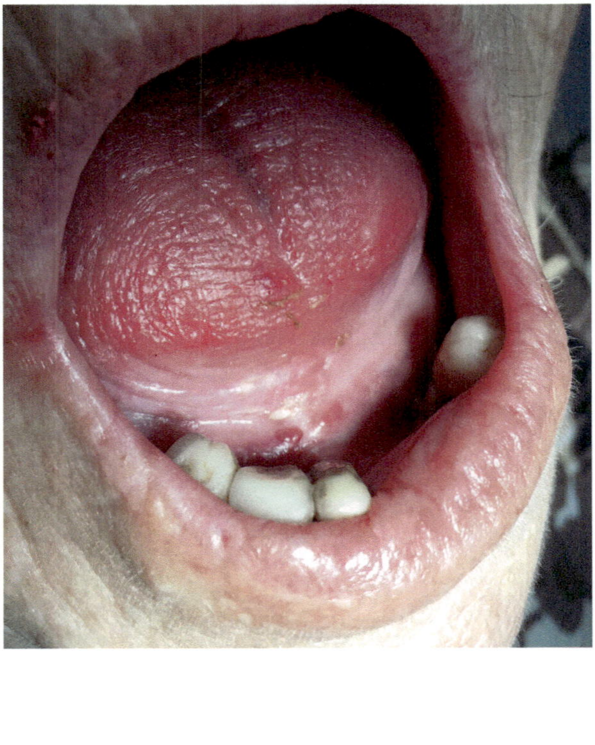

Fig. 29.2 Absence of saliva pooling floor of mouth

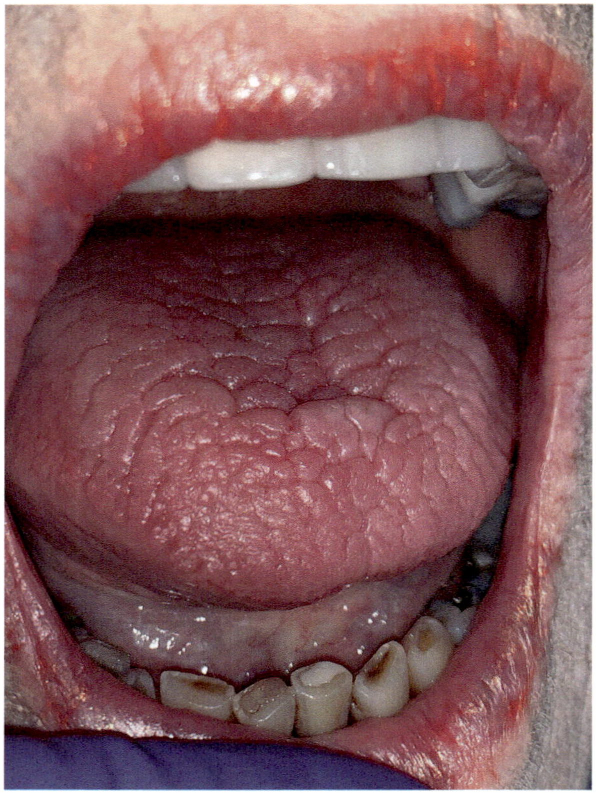

Fig. 29.3 Depapillated fissured tongue

29.3 Management and Prognosis

There is no single treatment to relieve the symptoms of salivary dysfunction. Often, various methods are employed to optimise relief of symptoms and manage any resultant oral complications (Table 29.4). Patient education and understanding are critical in the successful long-term management of salivary dysfunction. Interventions may provide temporary but crucial symptomatic relief. Strategies to reduce oral complications with a preventive emphasis require critical collaborative commitment from both the patient and oral health care team.

Table 29.4 Management of salivary gland hypofunction and xerostomia

Key Area	
Patient education	• Understanding xerostomia vs salivary hypofunction • Potential causes of salivary dysfunction • Role of saliva and sequelae of salivary dysfunction
Systemic disease and medication use	• Consultation with patient's medical team to ensure underlying disease identified and managed • Discussing substitution of xerogenic medication
Salivary replacement	• Use of saliva substitutes/artificial saliva • Appropriate hydration
Salivary stimulation	• Mechanical stimulation with chewing sugar-free gum • Chemical stimulation with muscarinic medications in individuals with residual functional glandular tissue such as pilocarpine titrated up to 5 mg 3–4 times a day • Electrical stimulation with use of hand-held device • Acupuncture
Protection of dentition	• Regular dental visits with establishment of preventive regime • Regular fluoride therapy including at home regime • Use of remineralising products • Ensure dentures are well fitting and comfortable • Removing dentures at night
Treatment of sequalae	• Oral candidosis which can occur, particularly after topical corticosteroids or antibiotics use • Caries
Modifiable avoidance	• Medications that can make dry mouth worse • Excessive caffeine • Alcohol • Smoking • High sugar and acidic drinks • Eating dry foods without liquids

Recommended Reading

Edgar M, Dawes C, O'Mullane D, editors. Saliva and Oral Health. 4th ed. Stephen Hancocks Limited; 2012.

Plemons JM, Al-Hashimi I, Marek CL. Managing xerostomia and salivary gland hypofunction: executive summary of a report from the American Dental Association Council on Scientific Affairs. J Am Dent Assoc. 2014;145(8):867–73.

Sreebny LM. Xerostomia: diagnosis, management and clinical complications. In: Edgar WM, O'Mullane DM, editors. Saliva and Oral Health. 2nd ed. London: British Dental Association; 1996. p. 43–66.

Sreebny LM, Vissink, editors. Dry Mouth – the malevolent symptoms: a clinical guide. Iowa: Wiley-Blackwell; 2010.

Thomson WM, Chalmers JM, Spencer AJ, Williams SM. The Xerostomia Inventory: a multi-item approach to measuring dry mouth. Community Dent Health. 1999;16(1):12–7.

Wolff A, Joshi RK, Ekström J, Aframian D, Pedersen AM, Proctor G, et al. A Guide to Medications Inducing Salivary Gland Dysfunction, Xerostomia, and Subjective Sialorrhea: A Systematic Review Sponsored by the World Workshop on Oral Medicine VI. Drugs R D. 2017;17(1):1–28.

Sjögren's Syndrome

30

Anastasia Georgiou

30.1 Introduction

Sjögren's syndrome was named after Swedish ophthalmologist, Henrik Sjögren (1899–1986) who identified a group of women presenting with characteristic triad of dry eyes (keratoconjunctivitis sicca), subjective dry mouth (xerostomia) and polyarthritis in 1933. It is now recognised as a chronic autoimmune condition with progressive focal lymphocytic cell infiltration and destruction of exocrine glands. Sjögren's syndrome affects females more than males, and can present alone (referred to as *primary*) or associated with other autoimmune conditions (referred to as *secondary*).

Sjögren's syndrome is an autoimmune disorder caused by lymphocytic infiltration of exocrine glands, salivary and lacrimal, resulting in glandular dysfunction. Genetic predisposition, an environmental or hormonal trigger may be contributing factors that lead to glandular infiltration of T lymphocytes (predominantly CD4+). The T lymphocytes are cytotoxic, destroying the epithelial cells of the lacrimal glands.

Females are affected more than males with a mean ratio of 9:1 and although any age can be affected, the median age of onset is 50–60 years. The variability of presentation has led to development of several classification criteria for Sjögren's syndrome; the prevalence rates vary considerably depending on classification criteria used. Sjögren's syndrome is one of the more prevalent autoimmune disorders.

A. Georgiou (✉)
Macquarie Oral and Maxillofacial Specialists, Sydney, NSW, Australia

Sydney Skin, Newtown, NSW, Australia
e-mail: admin@sydneyskin.com

R. Balasubramaniam et al. (eds.), *Oral Medicine - A Clinical Guide*,
https://doi.org/10.1007/978-3-031-36797-7_30

30.2 Clinical Presentation and Investigations

The most common clinical symptoms are dry mouth (Fig. 30.1) and dry eyes, often with associated fatigue (Table 30.1). Extra-glandular manifestations include lymphadenopathy, joint and muscle pain, dryness of the nose, throat and vagina in women. If Sjögren's syndrome is associated with other autoimmune conditions, features of those organ-specific conditions will also be noted.

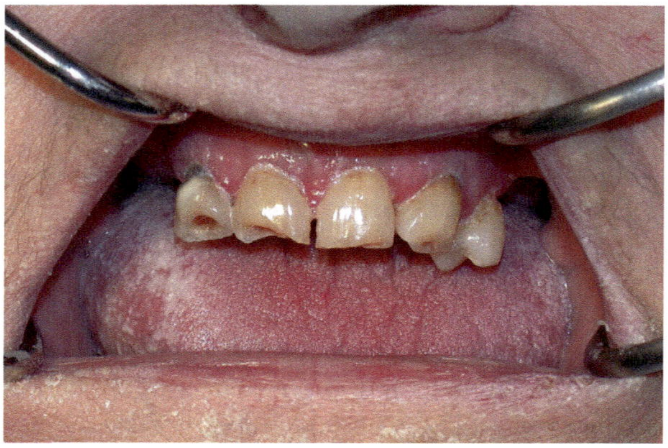

Fig. 30.1 Dry mouth and associated dental decay in Sjögren's syndrome

Table 30.1 Glandular manifestations of Sjögren's syndrome

Exocrine gland	Symptom	Complaints	Presentation
Salivary glands	Dry mouth	• Sore mouth • Difficult eating dry foods (biscuits/crackers) • Drinking fluids to aid in swallowing food • Waking through night to drink water • Unpleasant or loss of taste • Mouth sticks together • Sensation of swollen glands • Speech difficulties • Unable to wear dentures	• Reduced saliva with appearance dry, glazed mucosa • Minimal saliva expressed from saliva glands with external palpation • Dorsal surface of tongue red, lobulated with partial or complete depapillation (Fig. 30.2) • Lack of saliva pooling in the floor of the mouth • Frothy or viscous saliva • Food debris may be present • Mucosa can stick to dental mirror • Resultant debris on dental mirror • Smooth surface caries • Angular cheilosis
Lacrimal glands	Dry eyes	• Sensation of 'gritty', burning or itchy eyes • Sore and irritated eyes • Blurred vision • Light sensitivity • Discomfort wearing contact lenses	• Red eyes • Dry eyes • Infection of conjunctivae • Soft crusts forming at angles of eyes • Swelling of lacrimal glands

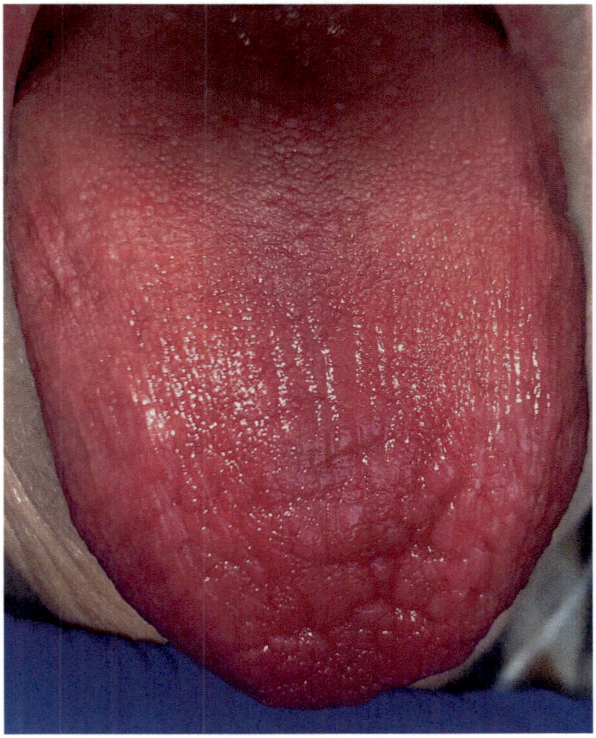

Fig. 30.2 Erythema, depapillation and lobulation of atrophic tongue in Sjögren's syndrome

Diagnosis of Sjögren's syndrome includes a comprehensive history and clinical examination in addition to utilisation of diagnostic tests. Diagnosis can be difficult because of the varied presentation and not all patients are positive for the diagnostic tests. Published classification criteria are often utilised for patients entering clinical trials rather than a diagnosis for patient care in clinical practice.

Ophthalmic tests include Schirmer test, ocular surface staining and tear break-up time with the last two requiring slit-lamp examination by an ophthalmologist. Tests for quantifying salivary hypofunction include salivary gland scintigraphy and whole sialometry. Individuals with Sjögren's syndrome often have antibodies to anti-Ro/SSA or anti-LA/SSB, some have both. Labial biopsy of minor salivary glands for histopathology may demonstrate focal lymphocytic sialadenitis. Salivary gland ultrasonography may show characteristic structural changes of the glands, and sialography may show sialectasis.

Differential diagnosis includes investigations to determine if Sjögren's syndrome is associated with other autoimmune conditions.

Table 30.2 Management of dry mouth symptoms in Sjögren's syndrome

Key Area	
Replacement	• Use of saliva substitutes/artificial saliva • Frequent sips of water
Stimulation	• Mechanical stimulation with chewing sugar-free gum • Chemical stimulation with muscarinic medications in individuals with residual functional glandular tissue such as pilocarpine titrated up to 5 mg 3–4 times day • Electrical stimulation with use of hand-held device
Protection	• Regular dental visits with establishment of preventive regime • Regular fluoride therapy including at home regime • Use of remineralising products • Removing dentures at night
Treat	• Oral candidosis which can occur, particularly after topical corticosteroids or antibiotics use • Caries
Avoid	• Medications that can make dry mouth worse • Excessive caffeine • Alcohol • Smoking • High sugar and acidic drinks • Eating dry foods without liquids

30.3 Management and Prognosis

There is no cure for Sjögren's syndrome. The aims of treatment are to manage symptoms and treat complications resulting from glandular manifestations (Table 30.2). Patients should be reviewed by an ophthalmologist for appropriate management of dry eyes. Management of extra-glandular manifestations will involve physicians with special interest in autoimmune condition. Given the chronic nature of this condition, long-term follow-up is important. Patients with Sjögren's syndrome are at an increased risk of lymphoma, usually MALT.

Recommended Reading

Jensen SB, Vissink A, Firth N. Salivary gland disorders and diseases. In Farah CS, Balasubramaniam R, McCullough MJ. Contemporary Oral Medicine, Volume 2. Springer 2019. P. 1437–1521.

Shiboski CH, Shiboski SC, Seror R, Criswell LA, Labetoulle M, Lietman TM, et al. 2016 American College of Rheumatology/European League Against Rheumatism classification criteria for primary Sjögren's syndrome: A consensus and data-driven methodology involving three international patient cohorts. Ann Rheum Dis. 2017;76(1):9–16.

Vitali C, Bombardieri S, Jonsson R, Moutsopoulos HM, Alexander EL, Carsons SE, et al. Classification criteria for Sjögren's syndrome: a revised version of the European criteria proposed by the American-European Consensus Group. Ann Rheum Dis. 2002;61(6):554–8.

Benign Mucosal Lesions

31

Sarah Chaw and Anura Ariyawardana

31.1 Introduction

There are a number of benign mucosal lesions that present frequently in the oral cavity. These commonly appear as asymptomatic soft tissue masses of variable size, with normal overlying mucosa. The aetiology of these lesions is varied. Clinical features, investigations and management of these lesions are outlined in Table 31.1.

S. Chaw (✉)
School of Dentistry, The University of Queensland, Brisbane, QLD, Australia
e-mail: s.chaw@uq.edu.au

A. Ariyawardana
College of Medicine and Dentistry, James Cook University, Cairns, QLD, Australia

School of Medicine and Dentistry, Griffith University, Southport, QLD, Australia

Metro South Oral Health, Queensland Health, Brisbane, QLD, Australia

R. Balasubramaniam et al. (eds.), *Oral Medicine - A Clinical Guide*,
https://doi.org/10.1007/978-3-031-36797-7_31

Table 31.1 Summary of aetiopathogenesis, clinical features, diagnostic features and management of select benign lesions affecting the head and neck

Lesion	Aetiopathogenesis and epidemiology	Clinical features	Diagnosis	Treatment and prognosis
Fibroepithelial polyp (fibroma, traumatic fibroma, fibrous hyperplasia, leaf fibroma, epulis fissuratum)	• Reactive lesion which leads to collagen deposition by an inflammatory process • May display significant chronic inflammation, ulceration, and epithelial hyperplasia	• Sessile or pedunculated, soft or firm nodules • Normal, ulcerated or hyperkeratinised overlying mucosa (Fig. 31.1) • Most less than 1 cm • Most common sites: buccal mucosa, tongue	• Relies on history, clinical findings and histopathology	• Conservative surgical excision • Address causative factors • Low recurrence rate
Lipoma	• Benign neoplasm of adipose tissue • Most common benign mesenchymal tumour • 20% presenting in the head and neck region; of those, 1–4% occur intra-orally • Bimodal peak of presentation in children and during the fifth or sixth decades of life	• Asymptomatic soft, mobile submucosal mass, covered by normal overlying mucosa/skin • Yellow hue (Fig. 31.2) • Average size: 2 cm • Most commonly site: buccal mucosa (45.7%) • Usually solitary; if multiple, may be syndromic (e.g. Cowden syndrome)	• Relies on history, clinical findings and histopathology • Imaging studies such as CT or MRI may assist diagnosis, particularly for large or deep lesions	• Surgical excision • Recurrence is unlikely
Neurilemmoma (Schwannoma)	• Benign peripheral nerve sheath neoplasm derived from Schwann cells • 25–45% located in the head, intraoral rare • Commonly present in middle age, no gender predilection	• Solitary, slow growing, firm encapsulated mass • Common on the tongue, usually painless • Bilateral schwannomas are usually related to neurofibromatosis type 2 syndrome (NF2)	• Relies on history, clinical findings and histopathology • Imaging studies such as MRI may assist diagnosis, particularly for large of deep lesions	• Complete resection • Unlikely to recur • If arising from cranial nerve, paraesthesia, or palsy may occur

Table 31.1 (continued)

Lesion	Aetiopathogenesis and epidemiology	Clinical features	Diagnosis	Treatment and prognosis
Granular cell tumour (GCT)	• Aetiology unclear; most likely Schwann cell origin • May affect all age groups with peak incidence between 40 and 60 years of age • F:M ratio = 2:1	• Solitary and slow growing, small (<3 cm), firm subcutaneous or submucosal lump • Overlying mucosa may be normal, yellow, pale or white • Most common on tongue • If multiple may be syndromic (e.g. LEOPARD syndrome, NF1, Noonan syndrome)	• Relies on history, clinical findings and histopathology	• Wide local excision • Recurrence is low if excised completely • Rarely, malignant GCTs have been reported
Congenital Epulis (congenital granular cell lesion)	• Unknown aetiology • Hypothesised to arise from undifferentiated mesenchymal cells, odontogenic epithelial remnants, neuroendocrine progenitor cells, pericytes, histiocytes, fibroblasts, or myofibroblasts • Newborns; females > males	• Lesions typically present at birth as a solitary pink pedunculated growth with smooth or lobulated surface • Size: millimetres to >7 cm • Typically presents lateral to midline of anterior maxilla • Large lesions may cause feeding difficulties	• Generally made based on clinical features and confirmed histopathologically if the lesion is excised	• Surgical excision if necessary • Very small lesions may regress spontaneously • Excellent prognosis
Oral haemangiomas (OHs)	• Benign vascular lesions that develop from endothelial cell proliferation during infancy followed by rapid proliferation phase during the first year of life • Infancy, 4–5% are observed <1st year of life • F:M ratio = 3:1	• >80% solitary • Lips, tongue, buccal mucosa, and palate • Rapid growth initially followed by stabilisation • Asymptomatic sessile or pedunculated soft lobulated and compressible masses with red hue • Occasionally intra-bony	• Relies mainly on history, clinical features (blanching when compressed) • Intra-bony lesions may be visualised during routine dental radiography	• Regress spontaneously and no active intervention is required

(continued)

Table 31.1 (continued)

Lesion	Aetiopathogenesis and epidemiology	Clinical features	Diagnosis	Treatment and prognosis
Lymphangiomas	• Hamartomas of lymphatic vessels which enlarge to a tumour-like mass with accumulation of lymph • Commonly occur in the head and neck region (50–75%) • 50% of diagnosed at birth; 90% <2 years	• Lesions in tongue may result in macroglossia • Slow growing, painless soft tissue mass (Fig. 31.3) • Nodular surface with pebbly appearance, translucent hue when superficial; soft and diffuse masses with normal appearing mucosa when deep	• Relies mainly on history, clinical features and histopathology • Imaging studies such as CT, MRI or ultrasonography may be necessary in some cases	• Removal may be for cosmetic reasons • Surgical excision • Nonsurgical therapies: cryotherapy, injecting sclerosants
Melanotic neuroectodermal tumour of infancy (MNTI)	• Rare, locally invasive, benign neoplasm of neural crest cell origin • Aetiology unknown • <12 months old • Slight female predilection	• Majority in anterior maxilla • Bluish painless swelling • Rapid growth with destruction of bone and displacement of teeth • May cause facial asymmetry and feeding problems • 3.5–20 cm	• Relies on history, clinical features and histopathology	• Surgical excision • Better prognosis with early detection and excision

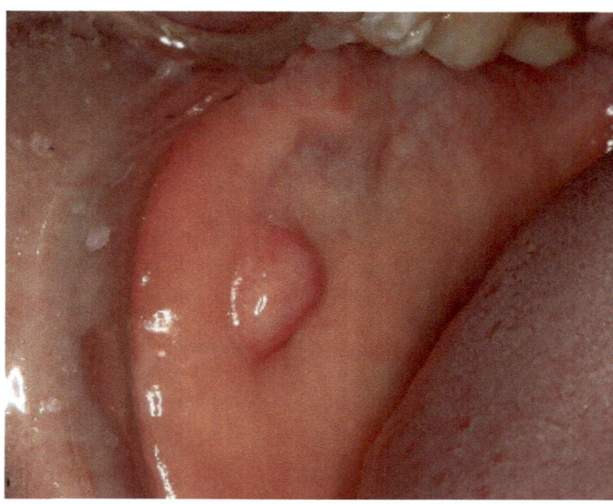

Fig. 31.1 A fibroepithelial polyp presenting as a pink sessile nodule on the right buccal mucosa

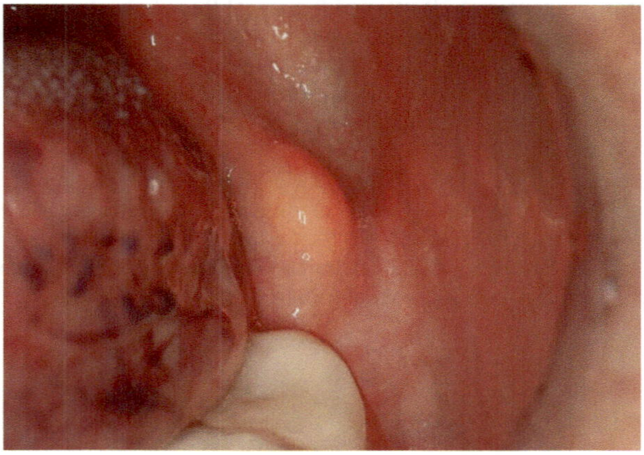

Fig. 31.2 An intraoral lipoma presenting as a soft, discrete submucosal mass on the left retromolar trigone. Note the yellow hue visible through the normal overlying mucosa

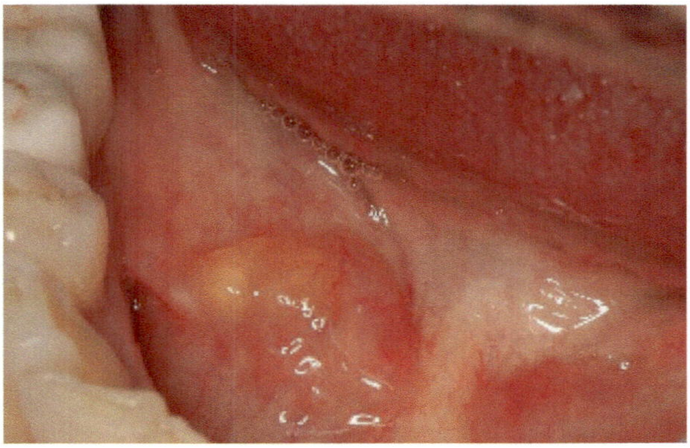

Fig. 31.3 An intraoral lymphangioma, presenting as a soft, fluctuant, yellow-coloured soft tissue swelling in the right anterior floor of mouth

31.2 Conclusion

These benign oral mucosal lesions present with common clinical features. History, examination and diagnostic investigations such as histopathology will allow for an accurate diagnosis and guide appropriate management.

Recommended Reading

Colreavy MP, Lacy PD, Hughes J, Bouchier-Hayes D, Brennan P, O'Dwyer AJ, et al. Head and neck schwannomas—a 10 year review. J Laryngol Otol. 2000;114(2):119–24.

Fregnani ER, Pires FR, Falzoni R, Lopes MA, Vargas PA. Lipomas of the oral cavity: clinical findings, histological classification and proliferative activity of 46 cases. Int J Oral Maxillofac Surg. 2003;32(1):49–53.

Jain N, Sinha P, Singh L. Large Congenital Epulis in a Newborn: Diagnosis and Management. Ear Nose Throat J. 2020;99(7):NP79–NP81.

Kolay SK, Parwani R, Wanjari S, Singhal P. Oral lymphangiomas - clinical and histopathological relations: An immunohistochemically analyzed case series of varied clinical presentations. J Oral Maxillofac Pathol. 2018;22(Suppl 1):S108–S111.

Lyssy LA, Puckett Y. Oral Hemangiomas. In: StatPearls [Internet]. Treasure Island(FL): StatPearls Publishing; 2022. Available at: https://www.ncbi.nlm.nih.gov/books/NBK560768/. Accessed 10 Dec 2022.

Rachidi S, Sood AJ, Patel KG, Nguyen SA, Hamilton H, Neville BW, et al. Melanotic Neuroectodermal Tumor of Infancy: A Systematic Review. J Oral Maxillofac Surg. 2015;73(10):1946–56.

van der Wal JE. Granular cell tumor, dental and oral. In: Encyclopedia of Pathology. Switzerland: Springer International Publishing; 2016. p. 202–4.

Oral Potentially Malignant Disorders 32

Renae Alford and Omar Kujan

32.1 Introduction

Oral potentially malignant disorders (OPMD) are "any oral mucosal abnormality that is associated with a statistically increased risk of developing oral cancer" (Table 32.1). This chapter will also describe palatal lesions in reverse smokers and dyskeratosis congenita (DC).

Table 32.1 WHO classification of oral potentially malignant disorders

Leukoplakia
Proliferative verrucous leukoplakia
Erythroplakia
Oral submucous fibrosis
Oral lichen planus
Actinic keratosis/actinic cheilitis
Palatal lesions in reverse smokers
Oral lupus erythematosus
Dyskeratosis congenita
Oral lichenoid lesions
Oral graft vs host disease

R. Alford · O. Kujan (✉)
UWA Dental School, The University of Western Australia, Nedlands, WA, Australia
e-mail: omar.kujan@uwa.edu.au

R. Balasubramaniam et al. (eds.), *Oral Medicine - A Clinical Guide*,
https://doi.org/10.1007/978-3-031-36797-7_32

OPMDs have no clear pathogenesis as the natural history and aetiology of these disorders are yet to be elucidated. Nevertheless, tobacco consumption, alcohol consumption and areca nut chewing are the predominant risk factors, while human papillomavirus (high-risk types) has a low prevalence in OPMDs.

Palatal lesions in reverse smokers are a reactive lesion on the palate due to the lit end of the cigar/cigarette being held inside the mouth. These lesions range from reactive lesions to overt carcinoma.

DC is a hereditary condition with a triad of skin hyperpigmentation, nail dystrophy and oral leukoplakia, as well as other systemic complications.

The overall prevalence of OPMDs worldwide is 4.47%, with a global malignant transformation rate (MTR) of all OPMDs groups was 7.9%. OPMD prevalence and MTR depend on type, gender, and geographic region.

Palatal lesions in reverse smokers are seen in India, Central America and some Pacific Islands, being more prevalent in females, with some studies suggesting a rate of 6.23 times higher than in males.

DC predominantly occurs in males (13:1), with an incidence of 1 in 1,000,000 births. Oral leukoplakia has been reported in 65–80% of DC patients with an MTR of up 30%.

32.2 Clinical Presentation and Investigations

OPMDs have a variable mucosal presentation of colour change; white, red or mixed with variable surface texture (Fig. 32.1). Palatal lesions in reverse smokers may appear as white thickened plaques, red or ulcerated areas on a hyperpigmented/depigmented background. These lesions occur in the posterior hard palate and will classically demonstrate excrescences marking the orifices of mucosal glands.

DC clinically presents in the oral cavity as a leukoplakic lesion most commonly on the dorsal tongue, but can also occur on the buccal mucosa, palate and any other mucous membrane.

Comprehensive oral and neck examinations and detailed patient and lesion histories are essential to make a provisional clinical diagnosis of an OPMD. Biopsy and histopathological examination is the gold standard for definitive diagnosis with presence of oral epithelial dysplasia (OED) paramount (Fig. 32.2).

Oral leukoplakia in a child calls for consideration for investigation of other DC features associated with DC, namely nail dystrophy and reticular skin pigmentation. The differential diagnosis includes other oral leukoplakic conditions such as white sponge nevus, frictional keratosis, lichen planus, lichenoid lesions, chemical injury and chronic hyperplastic candidiasis.

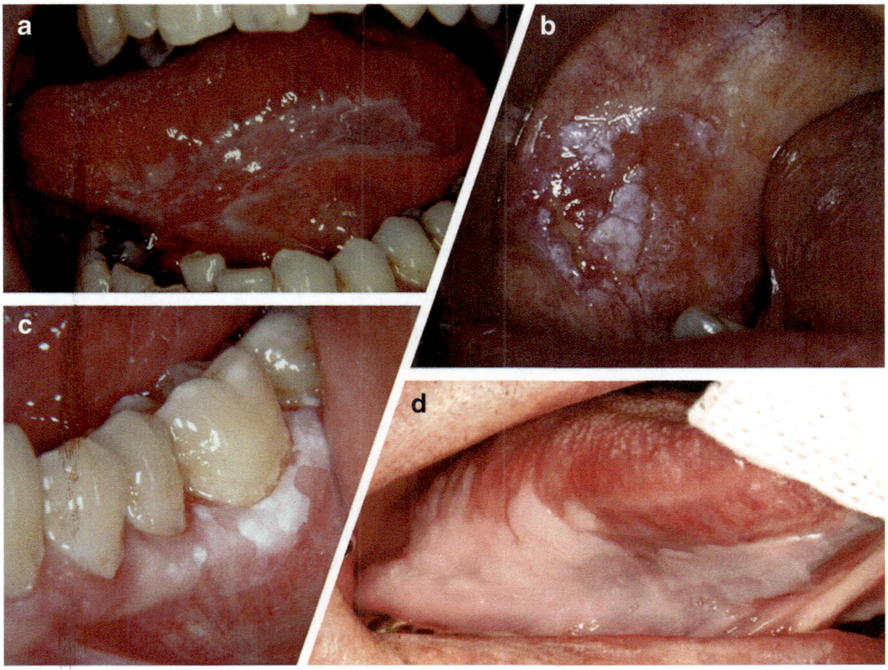

Fig. 32.1 Various clinical presentations of oral potentially malignant disorders: (**a**) Predominantly white lesion of the lateral tongue with striated areas. (**b**) Speckled leukoplakia on the buccal mucosa showing non-homogenous white plaque with erythematous areas. (**c**) Multifocal white lesion of the posterior buccal gingivae. (**d**) Homogenous leukoplakia of the lateral tongue

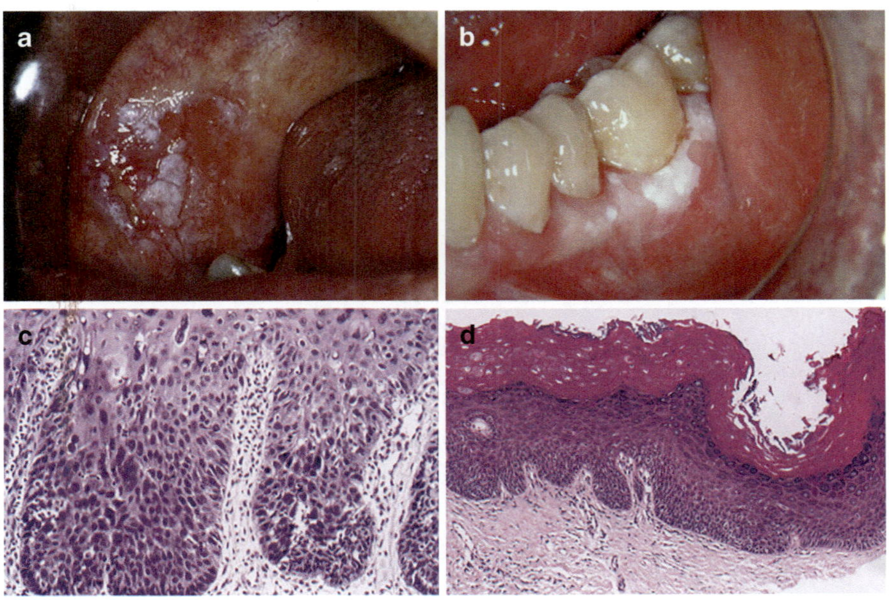

Fig. 32.2 Non-homogenous white lesions (**a**, **b**) with their corresponding microscopic features (**c**, **d**). The histopathological evaluation demonstrates severe/moderate epithelial dysplasia in both OPMD lesions

32.3 Management and Prognosis

The clinical course of OPMDs lesions is unpredictable because they may persist, regress or progress to malignancy. OPMD management depends on accurate diagnosis, risk stratification, risk factors counselling, selecting the appropriate intervention and surveillance. DC treatment consists of monitoring patients for oral malignancy. Prognosis is poor, with most patients succumbing in their third or fourth decade due to bone marrow failure, cancer or lung disease.

Recommended Reading

Kerr AR, Lodi G. Management of oral potentially malignant disorders. Oral Dis. 2021;27(8):2008–25.

Speight PM, Khurram SA, Kujan O. Oral potentially malignant disorders: risk of progression to malignancy. Oral Surg Oral Med Oral Pathol Oral Radiol. 2018;125(6):612–27.

Warnakulasuriya S. Oral potentially malignant disorders: A comprehensive review on clinical aspects and management. Oral Oncol. 2020;102:104550.

Warnakulasuriya S, Kujan O, Aguirre-Urizar JM, Bagan JV, González-Moles MÁ, Kerr AR, et al. Oral potentially malignant disorders: A consensus report from an international seminar on nomenclature and classification, convened by the WHO Collaborating Centre for Oral Cancer. Oral Dis. 2021;27(8):1862–80.

Oral Leukoplakia

33

Lalima Tiwari, Benjamin Gad, and Maryam Jessri

33.1 Introduction

The World Health Organization (WHO) Classification for Head and Neck Tumours 5th edition nests oral leukoplakia (OLK) within the broader category of oral potentially malignant disorders (OPMDs). OPMDs encapsulate a heterogenous group of conditions associated with a variable risk of progression to oral squamous carcinoma (OSCC). WHO defines OLK as "a predominantly white plaque of questionable risk having excluded (other) known diseases or disorders that carry no increased risk for cancer."

The aetiology and pathogenesis of OLK are multifactorial and similar to OSCC. Known risk factors for the development of OLK include tobacco use, excessive alcohol consumption, areca nut use, history of cancer or cancer therapy, immunosuppression and older age.

OLK is considered the most common OPMD. The worldwide prevalence of OLK ranges from 1% to 4%; however, a higher prevalence is observed in south-east Asia. OLK is more common in men (M:F approximately 2–3:1). It usually affects individuals in the sixth decade of life or older and shows increased prevalence with age.

L. Tiwari
UWA Dental School, The University of Western Australia, Nedlands, WA, Australia

B. Gad
School of Dentistry, The University of Queensland, Herston, QLD, Australia

M. Jessri (✉)
School of Dentistry, The University of Queensland, Herston, QLD, Australia

Metro North Hospital and Health Services, Queensland Health, Herston, QLD, Australia
e-mail: m.jessri@uq.edu.au

R. Balasubramaniam et al. (eds.), *Oral Medicine - A Clinical Guide*,
https://doi.org/10.1007/978-3-031-36797-7_33

33.2 Clinical Presentation and Investigations

OLK most frequently presents as a localised white plaque. Diffuse, multifocal presentations are less common. It is generally asymptomatic and most often found on the tongue, gingiva, buccal mucosa, floor of mouth and palatal mucosa. Homogenous leukoplakia presents as a well demarcated, white plaque, uniform in colour and texture. The surface is often described as smooth, fissured or wrinkled (Fig. 33.1). Non-homogenous leukoplakia can vary in colour and texture and be described as verrucous (Fig. 33.2), nodular and speckled (also referred to as erythroleukoplakia) (Fig. 33.3). All lesions should be palpated to evaluate for induration, which may indicate the presence of an invasive carcinoma.

Proliferative verrucous leukoplakia (PVL) is a distinct form of multifocal OLK and is characterised by relentlessly, although protracted, progressive involvement of the oral mucosa by verrucous and fissured white plaques. Clinically, this OPMD begins as one or more OLK and evolves to affect multiple locations, changing in clinical and histopathological features. PVL differentiates itself from localised OLK in that it is more common in women and is not strongly associated with smoking. Despite its high transformation to carcinoma, biopsies of early lesions often show little to no evidence of dysplasia.

A clinical diagnosis of OLK is arrived at after excluding other clinically recognisable white lesions (Table 33.1). A diagnostic biopsy is necessary to assess for the presence or absence of epithelial dysplasia or malignancy.

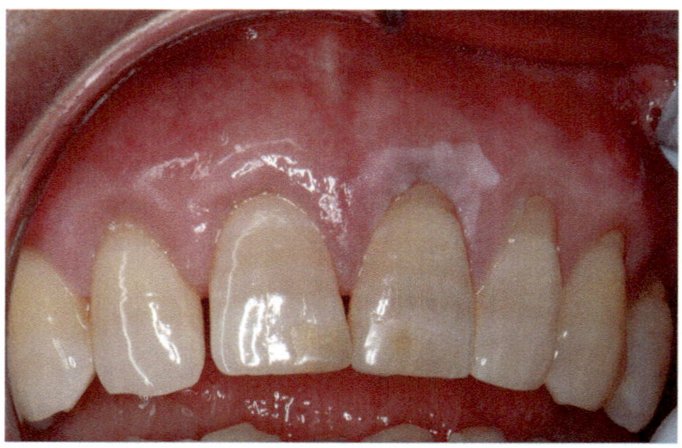

Fig. 33.1 Homogenous leukoplakia on labial gingiva of tooth 21

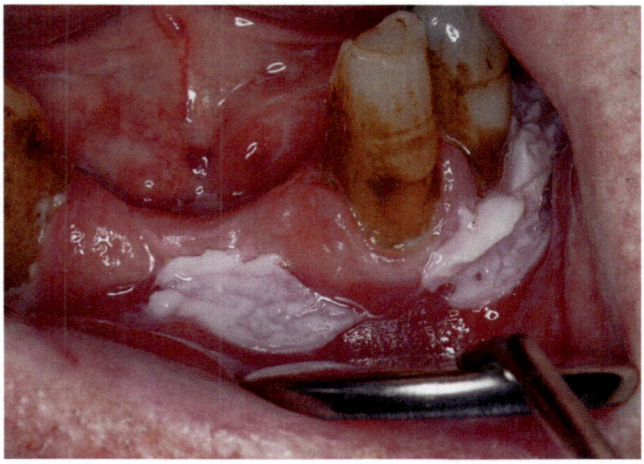

Fig. 33.2 Verrucous leukoplakia extending

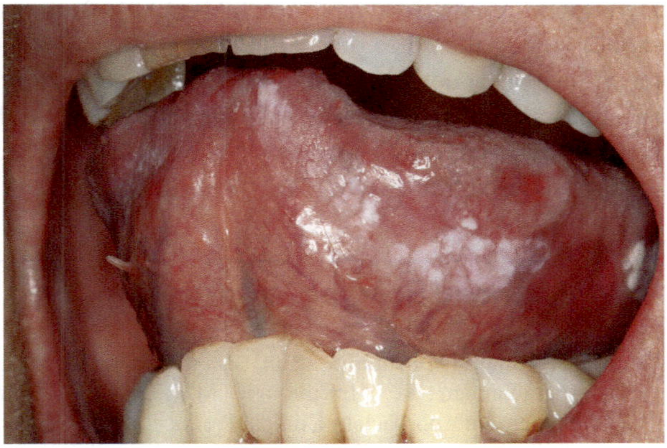

Fig. 33.3 Erythroleukoplakia on left ventral tongue

Table 33.1 Differential diagnoses for oral leukoplakia

Developmental	Fordyce's granules
	White sponge nevus
Reactive	Frictional keratosis
	Chemical injury
	Morsicatio mucosae oris
	Leukoedema
	Smokeless tobacco keratosis
	Nicotinic stomatitis
Infections	Candidosis
	Oral hairy leukoplakia
Immune-mediated and Autoimmune conditions	Geographic tongue
	Oral lichen planus and lichenoid reactions

Table 33.2 Factors associated with increased risk of malignant transformation of oral leukoplakia

Advanced age
Female gender
Lesion exceeding 200 mm^2
Non-homogenous
High-risk site (e.g. tongue, floor of mouth)
Non-smoker, non-drinkers
History of previous oral carcinoma
Presence of epithelial dysplasia

33.3 Management and Prognosis

The annual global transformation rate of OLK and PVL is 1.56% and 9.3%, respectively. Management of OLK is targeted towards prevention of cancer development and to reduce mortality and morbidity. Patients should be counselled on modifiable risk factors including smoking and areca nut cessation, and alcohol reduction. Predictive factors for cancer development are based on clinical features and histologic grades epithelial dysplasia (Table 33.2). Both lesion and patient risk factors should be considered in proceeding to excision. Excision does not guarantee non-recurrence or reduction of risk of OSCC. Close monitoring instead of excision may be appropriate for low-grade dysplastic lesions in individuals with minimal risk factors. OLK with high-grade dysplasia usually undergoes narrow excision with patients requiring indefinite follow-up.

Recommended Reading

Farah CS, Kujan O, Prime S, Binti Zain R. Oral Mucosal Malignancies. In: Farah C, Balasubramaniam R, McCullough MJ, editors. Contemporary Oral Medicine—A Comprehensive Approach to Clinical Practice. Cham: Springer; 2019. p. 1249–436.

Iocca O, Sollecito TP, Alawi F, Weinstein GS, Newman JG, De Virgilio A, et al. Potentially malignant disorders of the oral cavity and oral dysplasia: A systematic review and meta-analysis of malignant transformation rate by subtype. Head Neck. 2020;42(3):539–55.

Jessri M, Mawardi H, Farah CS, Woo S. White and Red Lesions of the Oral Mucosa. In: Farah C, Balasubramaniam R, McCullough MJ, editors. Contemporary Oral Medicine—A Comprehensive Approach to Clinical Practice. Cham: Springer; 2019. p. 1207–48.

Kerr AR, Lodi G. Management of oral potentially malignant disorders. Oral Dis. 2021;27(8):2008–25.

Odell E, Muller S, Wanninayake T. Oral Cavity and Mobile Tongue—Oral potentially malignant disorders. WHO Classification of Tumours Online. 5th ed: World Health Organization; 2022.

Warnakulasuriya S, Ariyawardana A. Malignant transformation of oral leukoplakia: a systematic review of observational studies. J Oral Pathol Med. 2016;45(3):155–66.

Warnakulasuriya S, Kujan O, Aguirre-Urizar JM, Bagan JV, González-Moles MÁ, Kerr AR, et al. Oral potentially malignant disorders: A consensus report from an international seminar on nomenclature and classification, convened by the WHO Collaborating Centre for Oral Cancer. Oral Dis. 2021;27(8):1862–80.

Erythroplakia

34

Renae Alford and Omar Kujan

34.1 Introduction

Erythroplakia is a clinical term defined as a "predominantly fiery red patch that cannot be characterised clinically or pathologically as any other definable disease." Queyrat first used the term erythroplasia to describe a red, precancerous lesion of the penis, and oral erythroplakia is clinically and histopathologically similar. It is regarded as an oral potentially malignant disorder (OPMD) as nearly 90% of erythroplakia at the time of diagnostic biopsy show either severe epithelial dysplasia or invasive oral squamous cell carcinoma (OSCC).

The aetiology for this condition is largely unknown; however, several risk factors may contribute to its development. Heavy alcohol consumption and tobacco use are the main contributing factors. It has one of the highest malignant transformation rates (MTR) of all OPMDs (33.1%), with an annual MTR of 2.7%. The pathogenesis for this condition is also unclear, with debate over whether these lesions develop *de novo* or as a progression from leukoplakia/erythroleukoplakia.

Oral erythroplakia is a rare lesion, and its prevalence has been reported between 0.02% and 0.83%. It is commonly overlooked compared to other OPMDs, particularly leukoplakias. There is no gender predilection and is commonly diagnosed between 40 and 70 years of age.

R. Alford · O. Kujan (✉)
UWA Dental School, The University of Western Australia, Nedlands, WA, Australia
e-mail: omar.kujan@uwa.edu.au

R. Balasubramaniam et al. (eds.), *Oral Medicine - A Clinical Guide*,
https://doi.org/10.1007/978-3-031-36797-7_34

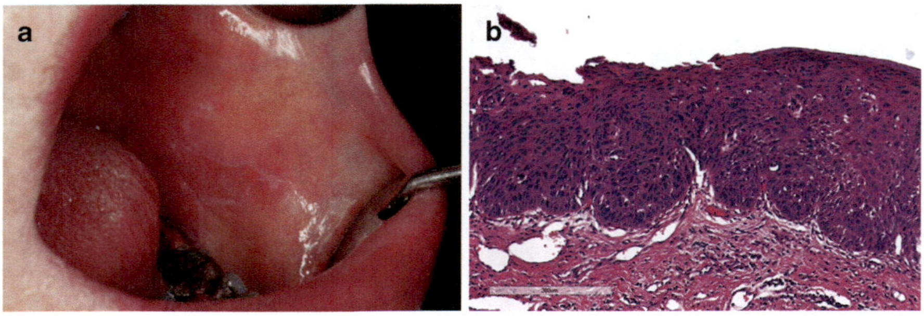

Fig. 34.1 Clinical and microscopic features of oral erythroplakia. (**a**) Clinical presentation of defined velvety plaque with granular surface (Courtesy of Dr. Sue-Ching Yeoh, The University of Sydney). (**b**) The histopathology revealed severe epithelial dysplasia (carcinoma in situ). Note the marked disordered epithelial architecture and severe cellular atypia

34.2 Clinical Presentation and Investigations

Oral erythroplakia presents most often as a solitary, well-defined flat or depressed erythematous lesion with a smooth or velvety surface (Fig. 34.1). The most common sites are the floor of the mouth, ventral tongue and soft palate. It is frequently asymptomatic, although discomfort such as tingling and sensitivity to hot/spicy foods may develop.

As erythroplakia is a diagnosis of exclusion, it is important to eliminate inflammatory conditions due to local irritation, such as denture stomatitis and lichenoid reactions. Other lesions including erosive lichen planus, erythematous candidiasis, erythema migrans, discoid lupus erythematosus, pemphigoid, pemphigus vulgaris, vascular hamartomas, vascular neoplasms and desquamative gingivitis can be considered. However, erythroplakias are typically more widespread in distribution and have more diffuse borders. Due to the high MTR, a lesion that is still present after two weeks should be biopsied.

34.3 Management and Prognosis

Surgical excision is the treatment of choice for erythroplakic lesions due to their high rate of dysplasia and progression to OSCC. However, this does not guarantee that recurrence or de novo OSCC in adjacent areas will not occur. Therefore, addressing modifiable risk factors and regular follow-up is crucial.

Recommended Reading

Lorenzo-Pouso AI, Lafuente-Ibáñez de Mendoza I, Pérez-Sayáns M, Pérez-Jardón A, Chamorro-Petronacci CM, Blanco-Carrión A, et al. Critical update, systematic review, and meta-analysis of oral erythroplakia as an oral potentially malignant disorder. J Oral Pathol Med. 2022;51(7):585–93.

Speight PM, Khurram SA, Kujan O. Oral potentially malignant disorders: risk of progression to malignancy. Oral Surg Oral Med Oral Pathol Oral Radiol. 2018;125(6):612–27.

Warnakulasuriya S, Kujan O, Aguirre-Urizar JM, Bagan JV, González-Moles MÁ, Kerr AR, et al. Oral potentially malignant disorders: A consensus report from an international seminar on nomenclature and classification, convened by the WHO Collaborating Centre for Oral Cancer. Oral Dis. 2021;27(8):1862–80.

Oral Submucous Fibrosis

35

Elizabeth Fitriana Sari

35.1 Introduction

Oral submucous fibrosis (OSMF) is a chronic inflammatory mucosal disease characterised by fibrosis of the submucosa leading to progressive inability of mouth opening in advanced stage. OSMF is an irreversible condition strongly associated with betel quid chewing and considered a potentially malignant disorder. The main components of betel quid (BQ) are areca nut (*Areca catechu*), slaked lime, and betel leaf or stem inflorescence of *Piper betle*, with some spices might be added as per preference.

OSMF is common in South, Southeast Asia, and the Asia Pacific Region. Most cases of OSMF are concentrated in the Indian subcontinent with an estimated prevalence range of 0.2–2.3% in males and 1.2–4.6% in females. It is further considered a leading explanation for Papua New Guinea experiencing the highest rate per population in oral cancer. OSMF occurs commonly in young adults with age range 20–40 years. The global South Asian and subsets of Southeast Asian diaspora particularly experience OSMF cases reported in the United Kingdom, USA, South Africa, many European countries, and increasingly in Australia. Malignant transformation of OSMF rates range from 1.9% to 9%.

E. F. Sari (✉)
Dentistry and Oral Health Discipline, Latrobe Rural Health School, Bendigo, VIC, Australia

Faculty of Dentistry, Universitas Padjadjaran, Bandung, Indonesia
e-mail: E.Sari@latrobe.edu.au

R. Balasubramaniam et al. (eds.), *Oral Medicine - A Clinical Guide*,
https://doi.org/10.1007/978-3-031-36797-7_35

The cause of OSMF is multifactorial. Studies have shown that the strongest risk factor associated with OSMF is chewing of betel quid containing areca nut. The major alkaloid in areca nut called *Arecoline* stimulates the fibroblast cells to express growth factors and cytokines that enhance the collagen deposition and inhibit collagen degradation. The chemical interaction between alkaloid, phenolics, and slaked lime (calcium hydroxide) in BQ are believed to responsible in the development of OSMF. Contributory risk factors include chewing of smokeless tobacco, high intake of chilies, toxic levels of copper in foodstuffs and masticatories, vitamin deficiencies, and malnutrition resulting in low levels of serum proteins, anaemia, and genetic and immunological predisposition.

35.2 Clinical Presentation and Investigations

The clinical presentation of OSMF depends on the stage of the disease. Symptoms in early stage include excessive salivation, burning sensation, blanching of the oral mucosa, blister formation, presence of thin palpable fibrous bands, and sparse brown/black pigmentation. In the moderate stage, gradual decrease in mouth opening occurs. Difficulty in whistling, formation of vesicles, petechiae, rigid oral mucosa, defective gustatory sensation, and blanching of oral mucosa are prominent features. Soft palate, buccal mucosa (Fig. 35.1), labial mucosa, tongue, floor of mouth, and faucial pillars appear pale and blanched with some degree of rigidity. In advanced stage, mottled or opaque or white marble-like appearance of oral mucosa, and thick palpable fibrous bands on buccal and labial mucosae are common. Restricted mouth opening, depapillation of the tongue, shrunken, inverted, hockey-stick shaped deviated uvula, and involvement of the pharyngeal and oesophageal mucosa are also prominent features in advanced stage of the disease. Differential diagnosis includes mucosal involvement in scleroderma, amyloidosis, generalised fibromatosis, and oral reticular lichen planus.

Diagnosis of OSMF is based on clinical signs and symptoms. Histopathological findings are essential to rule out dysplastic changes. Histologically, OSMF is characterised by epithelial atrophy with loss of rete ridges and hyalinization of the lamina propria and the underlying muscle. Mixed inflammation, oedema, and large fibroblasts in early stages and collagen bundles with hyalinisation and lack of vascularity in late stages are common.

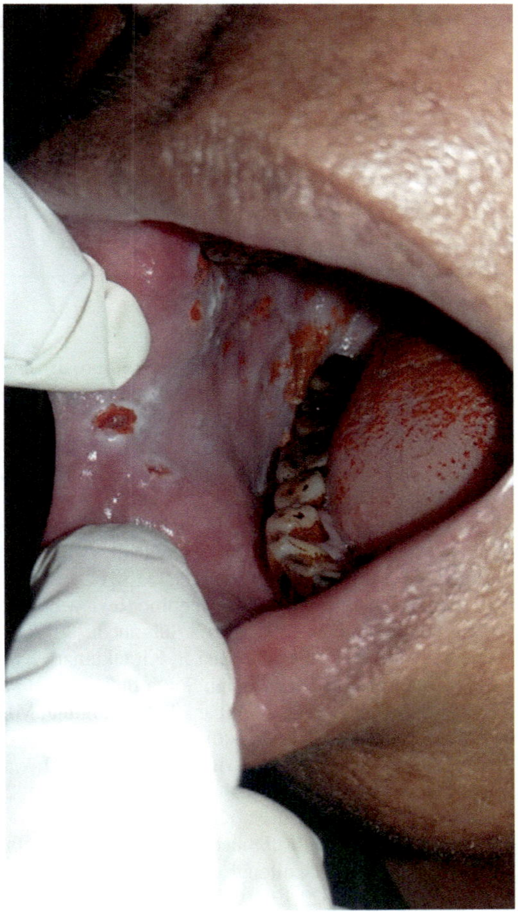

Fig. 35.1 Blanching of oral mucosa in patient consuming betel quid

35.3 Management and Prognosis

OSMF is irreversible in moderate and late stages. Quitting betel quid chewing is the best strategy to attempt prevention of OSMF and malignancy to progress. There are three treatment modalities available for OSMF. These include medical, surgical, and physical therapy. Surgical treatments may be necessary to improve mouth opening and movements. Medical therapy includes steroids (4 mg/mL/biweekly injections of betamethasone diluted in 1.0 mL of 2% xylocaine given on buccal mucosa, bilaterally for 6 months), or hyaluronidase, (intralesional injection of hyaluronidase 1500 IU mixed in 1.5 mL of dexamethasone and 0.5 mL of lignocaine hydrochloride biweekly for 4 weeks). Other agents include interferon gamma, placental extracts, immunized milk, pentoxifylline, buflomedil hydrochloride, nylidrin, isoxsuprine, $_{\beta}$-carotene, lycopene, vitamins, micronutrients, collagenase, chymotrypsin, and aloe vera. Other treatments include physical exercise regimens, oral splints, or other mouth opening devices and microwave diathermy.

Recommended Reading

Arakeri G, Brennan PA. Oral submucous fibrosis: an overview of the aetiology, pathogenesis, classification, and principles of management. Br J Oral Maxillofac Surg. 2013;51(7):587–93.

Arakeri G, Patil SG, Aljabab AS, Lin KC, Merkx MAW, Gao S, et al. Oral submucous fibrosis: An update on pathophysiology of malignant transformation. J Oral Pathol Med. 2017;46(6):413–7.

Betel-quid and areca-nut chewing and some areca-nut derived nitrosamines. IARC Monogr Eval Carcinog Risks Hum. 2004;85:1–334.

Chattopadhyay A, Ray JG. Molecular Pathology of Malignant Transformation of Oral Submucous Fibrosis. J Environ Pathol Toxicol Oncol. 2016;35(3):193–205.

Cirillo N, Duong PH, Er WT, Do CTN, De Silva MEH, Dong Y, et al. Are There Betel Quid Mixtures Less Harmful than Others? A Scoping Review of the Association between Different Betel Quid Ingredients and the Risk of Oral Submucous Fibrosis. Biomolecules. 2022;12(5):664.

Ranganathan K, Devi MU, Joshua E, Kirankumar K, Saraswathi TR. Oral submucous fibrosis: a case-control study in Chennai, South India. J Oral Pathol Med. 2004;33(5):274–7.

Sari EF, Prayogo GP, Loo YT, Zhang P, McCullough MJ, Cirillo N. Distinct phenolic, alkaloid and antioxidant profile in betel quids from four regions of Indonesia. Sci Rep. 2020;10(1):16254.

Tilakaratne WM, Klinikowski MF, Saku T, Peters TJ, Warnakulasuriya S. Oral submucous fibrosis: review on aetiology and pathogenesis. Oral Oncol. 2006;42(6):561–8.

Trivedy CR, Warnakulasuriya KA, Peters TJ, Senkus R, Hazarey VK, Johnson NW. Raised tissue copper levels in oral submucous fibrosis. J Oral Pathol Med. 2000;29(6):241–8.

Zhang P, Sari EF, McCullough MJ, Cirillo N. Metabolomic Profile of Indonesian Betel Quids. Biomolecules. 2022;12(10):1469.

Malignant Potential of Oral Lichen Planus

36

Lalima Tiwari, Janani Ravichandran, and Maryam Jessri

36.1 Introduction

Oral lichen planus (OLP) is a chronic, immune-mediated, inflammatory disorder commonly presenting as white reticular lesions or as desquamative gingivitis and is classified by the World Health Organisation (WHO) as an oral potentially malignant disorder (OPMD). The malignant potential of OLP however has been debated over several decades, with wide ranges of malignant transformation rates (MTR) being reported (0–12.5%). Some studies argue that oral lichenoid lesions (OLL), rather than OLP, carry malignant transformation potential, while others suggest both are associated with malignant transformation, with a higher MTR for OLL than OLP. These controversies exist largely due to the challenges in distinguishing OLL from OLP as they share similar clinical and histopathological features. This results in the use of varying diagnostic criteria and populations being studied, making comparison among studies difficult and creating further confusion in OPMD classification.

Currently, WHO classifies OLP and OLL as distinct OPMDs with MTR ranging from 0.44% to 1.4% for OLP and slightly higher from 1.71% to 1.88% for OLL. While further studies with quality methodology are needed to establish accuracy in MTR, there is now sound evidence that OLP is an OPMD. The features of OLP are discussed in Chap. 17.

L. Tiwari (✉)
UWA Dental School, The University of Western Australia, Nedlands, WA, Australia
e-mail: lalima.tiwari@uwa.edu.au

J. Ravichandran
NSW Health, Sydney, NSW, Australia

M. Jessri
School of Dentistry, The University of Queensland, Herston, QLD, Australia

Metro North Hospital and Health Services, Queensland Health, Herston, QLD, Australia

R. Balasubramaniam et al. (eds.), *Oral Medicine - A Clinical Guide*,
https://doi.org/10.1007/978-3-031-36797-7_36

Table 36.1 Endorsed diagnostic criteria of oral lichen planus based on oral potentially malignant disorders: a consensus report from an international seminar on nomenclature and classification, convened by the WHO Collaborating Centre for Oral Cancer

Clinical criteria	Presence of bilateral, more or less symmetrical white lesions affecting buccal mucosa, and/or tongue, and/or lip, and/or gingiva
	Presence of white papular lesions and lace-like network of slightly raised white lines (reticular, annular, or linear pattern) with or without erosions and ulcerations
	Sometimes presents as desquamative gingivitis
Histopathological criteria	Presence of a well-defined band-like predominantly lymphocytic infiltrate that is confined to the superficial part of the connective tissue
	Liquefactive degeneration of the basal cell layer
	Epithelial thinning and sometimes ulceration caused by failure of epithelial regeneration as a result of basal cell destruction seen in atrophic subtype. Mixed inflammatory infiltrate may be found

36.2 Differentiating Oral Lichen Planus from Oral Lichenoid Lesions

The current WHO diagnostic criteria for OLP requires the combined classical findings on clinical and histopathological examination (Table 36.1). Conditions which share features compatible with OLP are labelled OLL. By definition, OLL may encompass oral lichenoid contact reactions (OLCR), lichenoid drug reactions (LDR), and oral lesions associated with graft-versus-host disease (OGVHD) which can be consider distinct sub-entities of their own.

36.3 Oral Lichen Planus and Epithelial Dysplasia

The diagnostic criteria exclude a diagnosis of OLP in the presence of epithelial dysplasia on histopathology. Lesions with lichenoid features which also have epithelial dysplasia are classified as OLL or controversially, a separate entity termed "lichenoid dysplasia". The presence of epithelial dysplasia is associated with a higher MTR of 6.81%, with grading of dysplasia used for risk stratification. Molecular studies have not supported "lichenoid dysplasia" as a separate entity, and the WHO recommends the wording "dysplasia with lichenoid features" or "OLP with dysplasia". The underlying biology of how dysplastic changes develop in OLP lesions is not well understood but has been attributed to chronic mucosal inflammation. All suspicious cases suggestive of lichenoid features should be biopsied to evaluate for the presence of underlying epithelial dysplasia.

36.4 Clinical Factors Increasing Malignant Potential of Oral Lichen Planus

Higher rates of malignant transformation have been reported in OLP patients who smoke (OR = 4.62) and consume excessive amounts of alcohol (OR = 3.22). Hepatitis C-virus (HCV) has also been associated with increased malignant

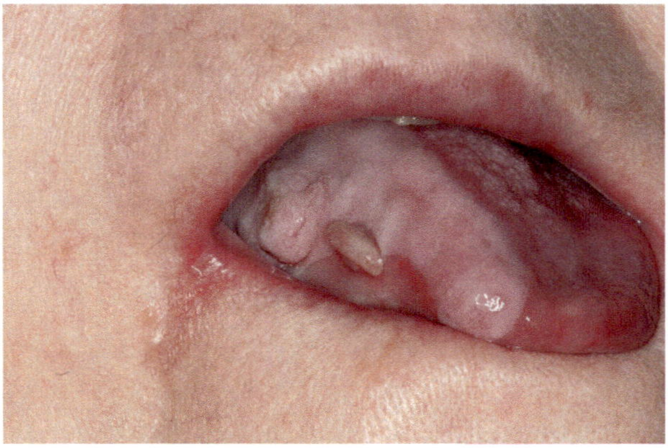

Fig. 36.1 Squamous cell carcinoma arising from right lateral tongue in a patient with oral lichen planus

potential of OLP (OR = 3.77), although the pathological association is controversial. Higher MTR has been reported in erythematous or atrophic-erosive lesions (OR = 0.37), particularly those of the tongue (Fig. 36.1). Modification of risk factors includes smoking cessation, moderation of alcohol intake, and avoidance of alcohol containing mouth rinses. Symptomatic management for OLP and OLL are similar with topical corticosteroids used as first--line management. Long-term follow-up is advised to monitor disease activity and symptoms. Further discussion of this topic is presented in Chap. 17.

Recommended Reading

González-Moles MÁ, Ruiz-Ávila I, González-Ruiz L, Ayén Á, Gil-Montoya JA, Ramos-García P. Malignant transformation risk of oral lichen planus: A systematic review and comprehensive meta-analysis. Oral Oncol. 2019;96:121–30.

Idrees M, Kujan O, Shearston K, Farah CS. Oral lichen planus has a very low malignant transformation rate: A systematic review and meta-analysis using strict diagnostic and inclusion criteria. J Oral Pathol Med. 2021;50(3):287–98.

McCullough MJ, Alrashdan MS, Cirillo N. Oral lichen planus. In: Farah CS, Balasubramaniam R, McCullough MJ, eds. Contemporary oral medicine: A comprehensive approach to clinical practice. Cham, CH: Springer International Publishing AG, 2019; p. 1043–82.

Ramos-García P, González-Moles MÁ, Warnakulasuriya S. Oral cancer development in lichen planus and related conditions-3.0 evidence level: A systematic review of systematic reviews. Oral Dis. 2021;27(8):1919–35.

Warnakulasuriya S, Kujan O, Aguirre-Urizar JM, Bagan JV, González-Moles MÁ, Kerr AR, et al. Oral potentially malignant disorders: A consensus report from an international seminar on nomenclature and classification, convened by the WHO Collaborating Centre for Oral Cancer. Oral Dis. 2021;27(8):1862–80.

Oral Lichenoid Lesions

37

Renae Alford and Omar Kujan

37.1 Introduction

Oral lichenoid lesions (OLL) are defined as "oral lesions with lichenoid features but lacking the typical clinical or histopathological appearances of oral lichen planus (OLP), that is, may show asymmetry or are reactions to dental restorations or are drug-induced". They are now regarded as an oral potentially malignant disorder (OPMD). These lesions can be classified into four broad groups: (1) lichenoid contact reactions (e.g. dental restorative materials); (2) lichenoid drug reactions (e.g. NSAIDs); (3) chronic graft vs host disease (GvHD) oral lesions; and (4) idiopathic.

In contrast to OLP, lichenoid lesions can be associated with a specific causative agent or systemic disease although are commonly idiopathic. Almost all dental restorative materials except titanium, palladium, and zirconium have been implicated, the most common being components of amalgam. It appears the degradation by products of these materials causes an allergic hypersensitivity reaction. Similarly, some foods (e.g. cinnamon) can cause a similar reaction, but the exact pathogenesis is unclear. An increasing list of drugs, including NSAIDs, antihypertensives, and antimalarials have also been shown to cause lichenoid lesions.

Due to the extensive overlap with OLP and differences in definition and criteria for OLL diagnosis, it is difficult to estimate the incidence of OLL as a separate entity. It appears, like OLP, with a slightly higher incidence in females and most often present in middle age.

R. Alford · O. Kujan (✉)
UWA Dental School, The University of Western Australia, Nedlands, WA, Australia
e-mail: omar.kujan@uwa.edu.au

R. Balasubramaniam et al. (eds.), *Oral Medicine - A Clinical Guide*,
https://doi.org/10.1007/978-3-031-36797-7_37

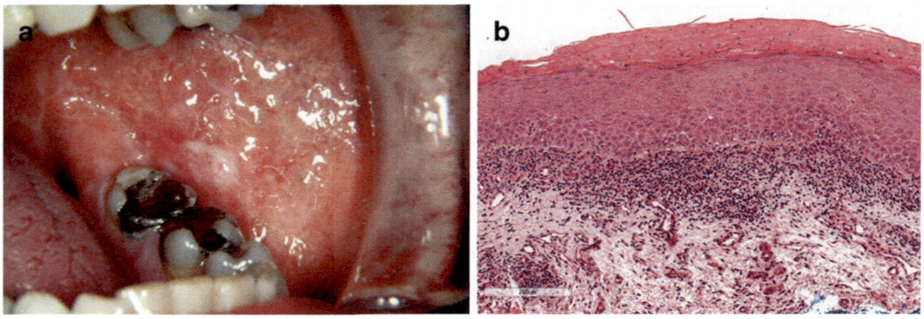

Fig. 37.1 Clinical and microscopic presentation of oral lichenoid contact reaction due to contact with amalgam restoration. (**a**) Clinical photograph showing predominantly reticular lesion on the left buccal mucosa with a white plaque adjacent to the dental amalgam restorations of 36 and 37. (**b**) A photomicrograph showing intense lymphocytic infiltrate closely associated with the hyperkeratotic squamous epithelium that shows basal cell degeneration and thickened basement membrane. Note a focus of lymphocytes aggregate in the lamina propria

37.2 Clinical Presentation and Investigations

OLL share significant attributes with OLP, and often clinical differentiation is difficult. Clinically they may appear as plaque-like or erosive lesions with striae. In contact allergy, they are often unilateral or solitary, or demonstrate an intimate relationship to dental restorations (Fig. 37.1). The most common sites are the buccal mucosa, tongue, or gingiva. With GvHD, there is a history of autologous haemopoietic cell transplant and other associated features may be present.

Diagnosis of an OLL requires clinical and histopathological correlation. A detailed and thorough clinical presentation, medical history, and biopsy are required. Histologically there is overlap with OLP, but some studies report further immune cell infiltration into the lamina propria, particularly perivascularly, with the presence of eosinophils and plasma cells in OLL associated with allergy. Although lichenoid contact reactions, lichenoid drug reactions, and GvHD lichenoid lesions may be diagnosed with clinical and careful histological evaluation, a debate exists on the distinction between OLP and idiopathic OLL lesions because of the overlap in the clinical and histopathological findings and the lack of clear aetiology. Other inflammatory mucosal conditions, such as ulcerative oral stomatitis, mucous membrane pemphigoid, and discoid lupus erythematosus, should be considered.

37.3 Management and Prognosis

Treatment depends on the type of OLL. Lichenoid contact reactions and lichenoid drug reactions fully resolve on the removal of the offending agent; however, while it may be easy to assess and remove the causative agent with contact reactions, the same cannot be said of drug reactions. There are several drugs implicated in OLL, and the list is increasing. Often drug reactions can appear weeks, months, or longer after commencing the drug and may take just as long to resolve after cessation. The medical risk versus benefit of ceasing the drug must be considered. Topical corticosteroids are used to treat symptomatic OLL. A recent meta-analysis reported an overall OLL-associated malignant transformation rate of 3.8%; thus, it has recently been regarded as an OPMD in the latest WHO Collaborating Centre for Oral Cancer classification. As such, regular follow-up is essential.

Recommended Reading

Al-Hashimi I, Schifter M, Lockhart PB, Wray D, Brennan M, Migliorati CA, et al. Oral lichen planus and oral lichenoid lesions: diagnostic and therapeutic considerations. Oral Surg Oral Med Oral Pathol Oral Radiol Endod. 2007;103 Suppl:S25.e1–12.

van der Meij EH, van der Waal I. Lack of clinicopathologic correlation in the diagnosis of oral lichen planus based on the presently available diagnostic criteria and suggestions for modifications. J Oral Pathol Med. 2003;32(9):507–12.

van der Waal I. Oral lichen planus and oral lichenoid lesions; a critical appraisal with emphasis on the diagnostic aspects. Med Oral Patol Oral Cir Bucal. 2009;14(7):E310–4.

Warnakulasuriya S, Kujan O, Aguirre-Urizar JM, Bagan JV, González-Moles MÁ, Kerr AR, et al. Oral potentially malignant disorders: A consensus report from an international seminar on nomenclature and classification, convened by the WHO Collaborating Centre for Oral Cancer. Oral Dis. 2021;27(8):1862–80.

Actinic Cheilitis

38

Neil Savage and Sarah Chaw

38.1 Introduction

Actinic Cheilitis (AC) is a direct effect of actinic radiation and a potentially malignant condition of the lower lip vermilion. The upper lip is generally spared with exceptions being an everted vermilion and reflective exposure (water/snow/ice).

Onset is insidious, unpredictable and unnoticed until advanced changes present. Careful lip examination should occur at every dental visit.

Actinic cheilitis is the result of cumulative ultraviolet (UV) radiation damage to the vermilion of the lip with the boundaries being the mucocutaneous junction and the vermilion border. These boundaries are often breached and so in each individual case the initial anatomical definition is important, as well as the loss of that anatomical definition. Broad lips are more likely to be affected, and AC has a strong predilection for males.

Individual susceptibility depends upon complexion, actinic exposure, preexisting conditions (actinic prurigo), tobacco or alcohol use and medications causing photosensitivity (e.g. tetracyclines, some psychoactive medications, topical fluorouracil and isotretinoin).

Radiation damage is cumulative and irreversible. The tissue changes involve both surface epithelium (which show areas of both atrophy and hyperplasia) with surface hyperkeratosis giving the typical blotchy appearance of actinic cheilitis.

N. Savage
Brisbane Oral Medicine and Pathology, Brisbane, Australia

S. Chaw (✉)
School of Dentistry, The University of Queensland, Brisbane, Australia
e-mail: s.chaw@uq.edu.au

R. Balasubramaniam et al. (eds.), *Oral Medicine - A Clinical Guide*,
https://doi.org/10.1007/978-3-031-36797-7_38

Atrophic epithelium allows additional UV exposure of the superficial dermis causing a change from a fibrillar collagenous tissue to amorphous non-fibrillar tissue (*solar elastosis*). This changes the absorptive and reflective capacity of the tissues with reflection of UV towards the already atrophic epidermis. It lessens reparative capacity of the vermilion with chronic fissures, ulcerated areas and surface crusting. There may be an inflammatory infiltrate (*cheilitis*) and epithelial dysplasia, with potential for progression to intraepithelial carcinoma and overt squamous cell carcinoma (SCC).

38.2 Clinical Presentation and Investigations

The appearance of AC reflects genotoxic stresses induced by UV radiation on the premorbid lip. These may be generalised or focal and temporary, stable or progressive depending on the level of individual susceptibility, prior sun exposure and previous history.

Clinical and histological features of AC are presented in Table 38.1.

No grading system exists for AC. Assessment relies on visual and tactile examination, with diagrammatic and photographic records. Advanced visual and tactile changes suggesting neoplasia require urgent referral.

Table 38.1 Clinical and histological features of actinic cheilitis

Clinical features (Fig. 38.1)	Histological features
• Atrophy and blotchy appearance	• Atrophy
• Surface desquamation	• Hyperplasia, hyperkeratosis
• Hyperkeratosis	• Solar elastosis
• Ulceration/crusting	• Inflammation and cheilitis glandularis (red spots)
• Fissures	• Hyperaemia
• Inflammation/erythema	• Dysplasia
• Indistinct vermilion border	
• Extension onto skin surface	

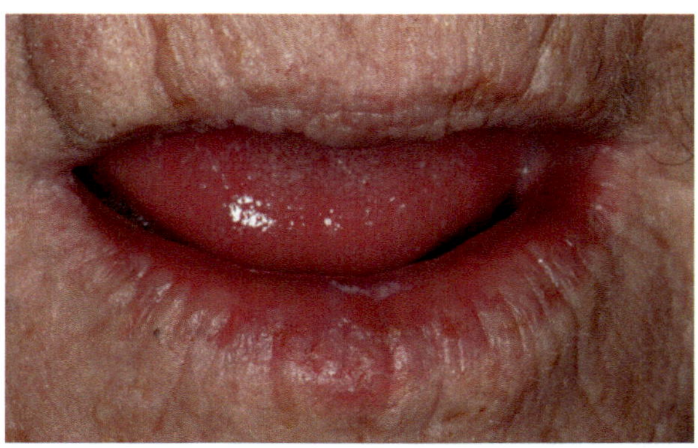

Fig. 38.1 Actinic cheilitis with crusting, atrophy and loss of a sharp vermilion border

Qualitative and quantitative clinical assessment is additional to age, skin type, previous and ongoing sun exposure, occupation, lifestyle, tobacco and alcohol use, and comorbidities.

Biopsy to assess for the presence and severity of dysplasia/SCC is indicated, particularly in the presence of scaling, crusting or ulceration.

38.3 Management and Prognosis

Management of AC includes prevention, risk assessment, active treatment and long-term surveillance. Considerations include clinical indication for immediate treatment, likelihood of disease progression, balanced against patient's age and occupational/social/lifestyle history.

Preventive care is essential and includes limiting UV radiation exposure and implementing sun protection measures (e.g. daily sunscreen, broad-brimmed hat), smoking cessation and minimising alcohol, as these factors can increase the risk of malignant transformation.

Active treatment may include topical therapies such as 5-fluorouracil, imiquimod, trichloroacetic acid chemical peel and photodynamic therapy. Physical treatments may include surgical excision of focal progressive lesions, or complete removal of the external lip (vermilionectomy) via surgery or laser ablation.

AC is an irreversible condition with potential for progression to SCC. It requires close follow-up with general dental practitioners and oral medicine specialists.

Recommended Reading

Poitevin NA, Rodrigues MS, Weigert KL, Macedo CLR, Dos Santos RB. Actinic cheilitis: proposition and reproducibility of a clinical criterion. BDJ Open. 2017;3:17016.

Savage NW, McKay C, Faulkner C. Actinic cheilitis in dental practice. Aust Dent J. 2010;55 Suppl 1:78–84.

Shah AY, Doherty SD, Rosen T. Actinic cheilitis: a treatment review. Int J Dermatol. 2010;49(11):1225–34.

Oral Squamous Cell Carcinoma 39

Antonio Celentano and Giuseppe Pantaleo

39.1 Introduction

Oral cancer includes a group of neoplasms affecting any region of the oral cavity, and more rarely arising from the intraoral salivary glands. Even though the term oropharyngeal cancer and cancers of the external lip are often used interchangeably with oral cancer, these are anatomically different subsets, whose cancer have different aetiology, incidence and prognosis. OSCC is the most common malignant epithelial neoplasm affecting the oral cavity (90% of all oral neoplasms) and remains a major health issue with significant morbidity, mortality and quality of life issues.

The aetiology of OSCC is multifactorial. Commonly demonstrated aetiological factors are smoking, alcohol and betel quid use. The role of oral human papillomavirus (HPV) in OSCC is more nuanced than in the oropharynx, as the presence of transcriptionally active HPV rarely occurs in OSCC. Other factors which may increase the risk of developing OSCC include age, diet and environmental factors. OSCC is generally preceded by oral mucosal precursor lesions collectively known as oral potentially malignant disorders (OPMD), that are very common (4.5% of the population) and have a variable tendency to progress to OSCC.

OSCC is a common and deadly disease affecting more than 750,000 people and with an annual incidence of 300,000 cases globally. There is a substantial global disparity in geographical incidence of OSCC, and this is mainly due to geographic variations of risk factors. Increasing age is a significant OSCC risk factor, while the male to female ratio is on average 2:1. Interestingly, OSCC incidence is increasing among females aged 18–44 years.

A. Celentano (✉)
Melbourne Dental School, The University of Melbourne, Carlton, VIC, Australia
e-mail: antonio.celentano@unimelb.edu.au

G. Pantaleo
Department of Medicine and Surgery, University of Salerno, Fisciano, Italy

R. Balasubramaniam et al. (eds.), *Oral Medicine - A Clinical Guide*,
https://doi.org/10.1007/978-3-031-36797-7_39

39.2 Clinical Presentation and Investigations

OSCC usually manifests as red and white mucosal lesions (Fig. 39.1) which may present with an ulcerated surface or an exophytic growth where papillary or verrucous projections can be also observed. Verrucous carcinoma is a rare clinical variant and forms 3% of all OSCC (Fig. 39.2). About 40% of OSCC begin on the floor of the mouth or on the lateral/ventral surfaces of the tongue. Histologically, OSCC shows invasive tumour islands in the connective tissue arising from dysplastic overlying epithelium. Varying amounts of surface keratinisation may be present.

OSCC is detected visually through a systematic conventional oral examination and evaluation of the neck for palpable nodes. Early presentation is usually asymptomatic. Late symptoms include pain, discomfort, reduced mobility of the tongue and inability to wear dentures. Incisional biopsy is required for diagnosis. Head and neck CT, PET/CT and a chest X-ray are usually performed. The differential diagnosis for OSCC includes other malignant lesions of the oral cavity as well as infection (such as tuberculosis, deep fungal infection) and oral potentially malignant disorders.

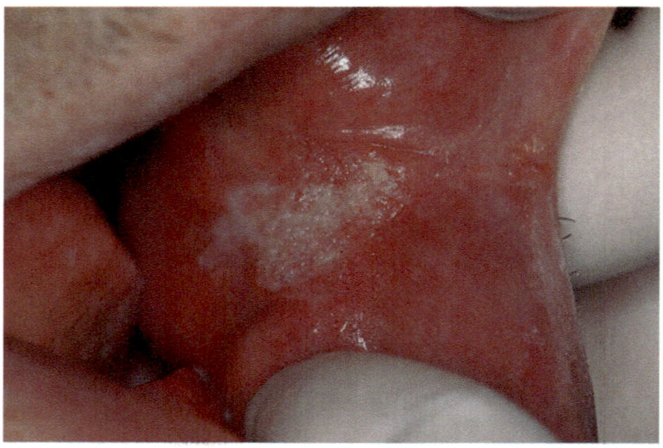

Fig. 39.1 A 56-year-old Caucasian male presented with a mixed red/white nodular fixed lesion on the left buccal mucosa. Histology revealed a moderately differentiated OSCC

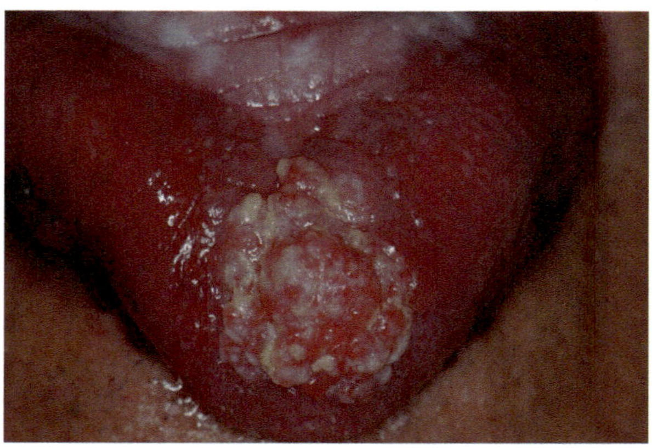

Fig. 39.2 Verrucous carcinoma of the tongue tip in a 72-year-old male smoker patient. Posteriorly, papillomatosis and raised diffuse fixed white patches were also present, which histologically revealed epithelial dysplastic features

39.3 Management and Prognosis

For most oral cavity cancers, surgery is the primary treatment. Adjuvant radiation or chemoradiation may be utilised for cases of suspected or confirmed locoregional or distant spread. However, despite the advances of therapeutic approaches, percentages of morbidity and mortality of OSCC have not improved significantly during the last 30 years. The percentage of 5-year survival for patients with OSCC varies from 40–50%. Locoregional spread to lymph nodes decreases survival rate by about half. Metastases reach the regional lymph nodes first and later the lungs.

Recommended Reading

Machiels JP, René Leemans C, Golusinski W, Grau C, Licitra L, Gregoire V. Squamous cell carcinoma of the oral cavity, larynx, oropharynx and hypopharynx: EHNS-ESMO-ESTRO Clinical Practice Guidelines for diagnosis, treatment and follow-up. Ann Oncol. 2020;31(11):1462–75.

Mody MD, Rocco JW, Yom SS, Haddad RI, Saba NF. Head and neck cancer. Lancet. 2021;398(10318):2289–99.

Thomson PJ. Perspectives on oral squamous cell carcinoma prevention-proliferation, position, progression and prediction. J Oral Pathol Med. 2018;47(9):803–7.

Other Mucosal Malignancies: Oral Melanoma, Oral Lymphomas, and Oral Kaposi Sarcoma

40

Antonio Celentano and Elena Calabria

40.1 Introduction

Oral melanoma (OM) is a rare neoplasia that arises from malignant transformation and clonal expansion of melanocytes in the oral mucosa. Oral lymphomas (OLs) consist in a group of haematolymphoid malignant tumours, developing from an uncontrolled proliferation of B-, T- or NK-cells. Mature B-cell lymphomas are the most common lymphoma subtype found in the oral cavity. Kaposi sarcoma (KS) is a locally aggressive muco-cutaneous vascular malignancy of endothelial origin associated with the gamma-2 Human Herpesvirus 8 (HHV8). Oral KS (OKS) is most frequently affecting patients with acquired immunodeficiency syndrome.

While the pathogenic mechanisms underlying the OM and OL are far to be clarified, OKS's causative factor is HHV8. Only few data are available for OM's mutation patterns which differ from that of the cutaneous form, with less than 20% having BRAFV600E mutations.

Incidences of the above oral malignancies are not well known. Though rare entities, OM accounts for 2% of all mucosal and cutaneous melanomas; OLs account for the 3.5% of oral cavity malignancies. OKS is estimated to develop in almost 20% of HIV-1 positive subjects in their fourth-fifth decades of life.

A. Celentano (✉)
Melbourne Dental School, The University of Melbourne, Carlton, VIC, Australia
e-mail: antonio.celentano@unimelb.edu.au

E. Calabria
School of Dentistry, Magna Graecia University of Catanzaro, Catanzaro, Italy

R. Balasubramaniam et al. (eds.), *Oral Medicine - A Clinical Guide*,
https://doi.org/10.1007/978-3-031-36797-7_40

40.2 Clinical Presentation and Investigations

OM, OLs and OKS are predominantly localised on the hard palate and gingiva (Fig. 40.1a, b). OM presents as an asymmetric black to red or white area with flat to nodular irregular morphology associated with 30% lymph node metastasis at diagnosis. OLs present as a non-tender mass with or without ulceration and is generally diagnosed at early stages. OKS can be clinically unifocal or multifocal and present as pink to bluish-purple or brown lesion resembling a macula that progresses into nodules or exophytic masses. Although OKS may manifest at any stage of the HIV infection, it is more often diagnosed in patients with low CD4+ T-cell count with already disseminated lesions and visceral involvement.

Mucosal biopsy is required for diagnosis of these malignancies. OM's differential diagnosis comprise other pigmented lesions such as benign melanocytic lesions, melanotic macule, oral mucosal nevus, and melanoacanthoma. With respect to OLs and OKS, other benign and malignant oral neoplasms must be ruled out (OSCC, salivary gland carcinomas, secondary metastasis, vascular malformations, pyogenic granuloma, etc.). For all these neoplasms, further investigations, such as magnetic resonance imaging, CT scans, PET scans, may be required for evaluating local and systemic involvement.

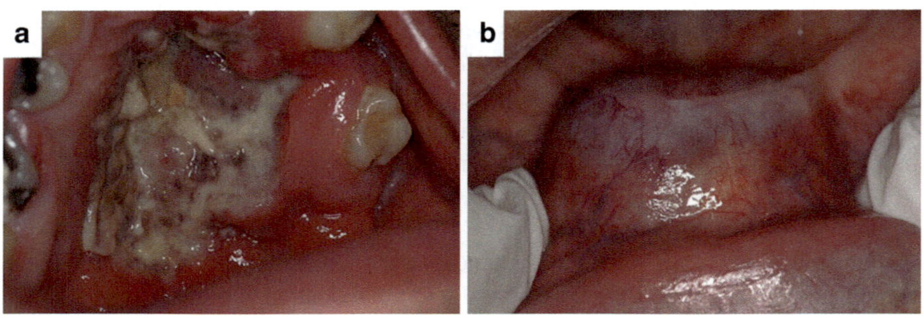

Fig. 40.1 (**a**) Peripheral T/NK-cell lymphoma (nasal type): Case of a 37-year-old immunocompetent man presenting with a deep ulcerative and destructive palatal lesion with an extensive zone of necrosis. Histology revealed T/NK-cell lymphoma. (**b**) Non-Hodgkin's B-cell lymphoma: diffuse enlargement of the edentulous mandibular bone in a 74-year-old Caucasian male. The histopathological/immunohistochemical exam showed diffuse lymphoid proliferation, positivity for cd20, pax5, mum1 e bcl-6, and a ki67 PI of 75%

40.3 Management and Prognosis

Treatments and prognoses may widely vary depending on the histopathology, stage of the diseases, and immune status of the patients. OM is generally treated with surgical excision which often lacks to achieve negative margins due to the lentiginous growth pattern and the multifocal nature, leading to a high local relapse rate. Differently from cutaneous melanoma, first-line therapeutic option for OM still remains chemotherapy. Being diagnosed often at advanced stages, its prognosis remains poor, with a 5-year overall survival rate of 33.1%. Therapeutic strategies for OLs depend on the subtype and disease stage. Also the prognosis can range from an indolent clinical course to rapidly progression and death. OKS treatments rely predominantly on ART (Anti-Retroviral Therapy) and chemotherapy; however, prognosis remains poor with a high fatality rate.

Recommended Reading

Ma Y, Xia R, Ma X, Judson-Torres RL, Zeng H. Mucosal Melanoma: Pathological Evolution, Pathway Dependency and Targeted Therapy. Front Oncol. 2021;11:702287.

Pantanowitz L, Khammissa RA, Lemmer J, Feller L. Oral HIV-associated Kaposi sarcoma. J Oral Pathol Med. 2013;42(3):201–7.

Wagner VP, Rodrigues-Fernandes CI, Carvalho MVR, Dos Santos JN, Barra MB, Hunter KD, et al. Mantle cell lymphoma, malt lymphoma, small lymphocytic lymphoma, and follicular lymphoma of the oral cavity: An update. J Oral Pathol Med. 2021;50(6):622–30.

Pleomorphic Adenoma

41

Norman Firth

41.1 Introduction

Pleomorphic adenoma (PA) is a common benign salivary gland tumour. The lesion is also known as, "benign mixed tumour". The term pleomorphic is appropriate as it refers to both the histogenesis and histopathology of the tumour.

PA is associated with chromosomal translocation which activates one of the two genes, PLAG1 or HMGA2. Their activation results in cell proliferation and abnormal differentiation, which explains the varied tissues seen in PA. Additional chromosomal abnormalities involving oncogenes and tumour suppressor genes are found and may account for the potential malignant transformation seen in long-standing lesions. Radiation increases the risk of developing PA.

PA is common in both the major salivary glands and minor salivary glands. About 75% of parotid tumours and approximately 40% of minor salivary gland tumours are PA. PA can be seen at any age but are most common in middle age. The lesion is slow growing, and there is a slight female predilection. The superficial lobe of the parotid gland in front of the ear is the most common site, and bilateral lesions are rarely encountered.

N. Firth (✉)
School of Dentistry, The University of Queensland, Herston, QLD, Australia

Capital, Coast and Hutt Valley District Health Board, Wellington, New Zealand

Clinipath, Osborne Park, WA, Australia
e-mail: Norman.firth@xtra.co.nz; Nfirth@clinipath.net

R. Balasubramaniam et al. (eds.), *Oral Medicine - A Clinical Guide*,
https://doi.org/10.1007/978-3-031-36797-7_41

41.2 Clinical Presentation and Investigations

The lesion presents as a lump and is painless, well circumscribed, and firm or rubbery in consistency. Lesions presenting in the parotid tail may be associated with eversion of the earlobe. The most common site for intra-oral lesions is the palate, particularly the junction of the hard and soft palate, followed by the upper lip (approximately 25%) and buccal mucosa (approximately 15%). Intra-oral lesions present as a rubbery and firm lump covered by normal coloured oral mucosa without ulceration.

Diagnosis is based on the history, clinical features and investigations including ultrasound, computed tomography and magnetic resonance imaging to assess the extent of the tumour. Fine needle/core needle biopsy or incisional biopsy precedes definitive surgical removal.

PA is circumscribed but often multi-lobular and microscopically has an incomplete capsule. The lesion consists of a mixture of glandular epithelium and myoepithelial cells in a background of variable stromal tissues (mucous, myxomatous and cartilaginous) (Fig. 41.1). The ratio of epithelial, myoepithelial cellular areas compared with stroma is greater in intra-oral lesions than in parotid lesions. Mucin filled duct-like structures may be conspicuous or barely detectable. Myoepithelial cells frequently take on the structure and function of connective tissue cells and may be spindle or stellate in shape or may resemble plasma cells (plasmacytoid cells).

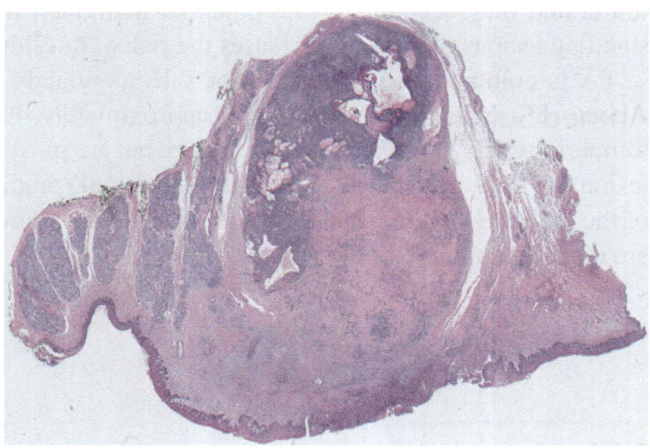

Fig. 41.1 Pleomorphic adenoma presented as a palatal lump in a 69-year-old male. The lesion had an incisional biopsy prior to excision. The lesion has a thin capsule, and the deeper portion of the lesion is much more cellular. A small superficial incisional biopsy in such cases may be mis-diagnosed as it may not be representative of the whole lesion. (H&E)

41.3 Management and Prognosis

Treatment is by excision. Parotid lesions have a reputation for recurrence (15–25%), which may be related to several factors. These include the lack of complete encapsulation and presence of lobules extending through the capsule, the relationship with branches of the facial nerve and myxomatous lesions may be disrupted during excision possibly leading to widespread seeding of residual neoplastic tissue. Multifocal recurrences may be difficult to remove surgically, and radiotherapy is occasionally employed. The recurrence rate for intra-oral lesions is much lower even if lesional tissue extends to surgical margins.

Untreated lesions continue to grow slowly over many years and may attain a large size. These are at risk of malignant transformation particularly in older patients with long standing untreated lesions. Other risk factors for malignant transformation include the presence of multiple recurrences and a primary location in the submandibular gland.

Recommended Reading

El-Naggar AK, Chan JKC, Grandis JR, Takata T, Slootweg P, editors. Tumours of the Salivary Glands. World Health Organization Classification of Head and Neck Tumours. 4th ed. IARC Lyon; 2017. p. 185-186.

Mendenhall WM, Mendenhall CM, Werning JW, Malyapa RS, Mendenhall NP. Salivary gland pleomorphic adenoma. Am J Clin Oncol. 2008;31(1):95–9.

Neville BW, Damm DD, Allen CM, Chi AC. Salivary gland tumors. In: Oral and Maxillofacial Pathology. 4th ed. St. Louis, MO: Elsevier; 2016. p. 440–465.

Odell E. Cawson's Essentials of Oral Pathology and Oral Medicine. 9th ed. Edinburgh: Elsevier; 2017. p. 355–365.

Warthin's Tumour and Other Benign Neoplasms of the Salivary Glands

42

Norman Firth

42.1 Warthin's Tumour (Papillary Cystadenoma Lymphomatosum)

42.1.1 Introduction

Warthin's tumour is a benign tumour of the parotid gland (almost exclusively) characterised by the presence of lymphoid tissue and cystic and oncocytic changes associated with the inflammatory infiltrate (Fig. 42.1).

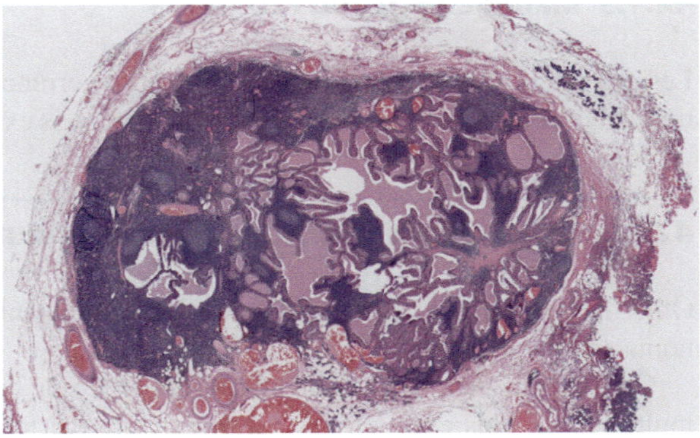

Fig. 42.1 Male aged 71 years with a lump in the right parotid gland showing the characteristic features of Warthin's tumour (H&E)

N. Firth (✉)
School of Dentistry, The University of Queensland, Herston, QLD, Australia

Capital, Coast and Hutt Valley District Health Board, Wellington, New Zealand

Clinipath, Osborne Park, WA, Australia
e-mail: Norman.firth@xtra.co.nz; Nfirth@clinipath.net

R. Balasubramaniam et al. (eds.), *Oral Medicine - A Clinical Guide*,
https://doi.org/10.1007/978-3-031-36797-7_42

Tumours develop from heterotopic salivary ducts trapped within intraparotid or paraparotid lymphoid tissue.

Warthin's tumour is the second most common salivary gland tumours accounting for approximately 10% of all parotid tumours. Lesions present as a slow growing, painless, firm or fluctuant mass usually located near the tail of the parotid gland. Up to 17% may be bilateral. The majority of patients are smokers over 40 years of age, and it has been reported that males are affected more than females, however this may not be the case.

42.1.2 Clinical Presentation and Investigations

Lesions may occur in the oral cavity, larynx or cervical lymph nodes. Lymph node lesions are not metastatic but arise in developmental salivary gland ductal heterotopic tissue found in lymph nodes in this region.

Diagnosis is based on history and clinical findings, and fine needle aspiration (FNA) is accurate in confirming the diagnosis. The tumours have a thin capsule and are partially solid and partially cystic. They have both a lymphoid and ductal epithelial component. The lymphoid tissue contains lymphoid follicles with germinal centres. The epithelial cells are columnar cells with eosinophilic cytoplasm and form multi-folded epithelial lined cystic structures with papillary projections into the cystic spaces. The eosinophilic cytoplasm represents oncocytic change.

42.1.3 Management and Prognosis

Lesions are treated by surgical removal, and recurrence is rare (5%) although new unrelated lesions may develop. Malignant change is extremely rare.

42.2 Other Benign Salivary Gland Tumours

Other benign salivary gland tumours include canalicular adenoma, basal cell adenoma and oncocytoma.

Canalicular adenomas occur in the upper lip or buccal mucosa. Occasionally multifocal lesions encountered. Histologically the lesions consist of strands, trabeculae or duct-like structures of basaloid epithelial cells with a uniform cytological appearance. Myoepithelial cells and stromal changes seen in pleomorphic adenoma are not encountered. Lesions do not recur following simple excision.

Parotid gland basil cell adenomas are encountered in elderly patients and consist of an encapsulated lesion composed of basaloid cells arranged in solid sheets, strands or trabeculae. Stromal changes are not encountered. Lesions do not recur following excision. FNA may prove difficult to differentiate basal cell adenoma from pleomorphic adenoma and often the diagnosis is confirmed after examination of the lesion following superficial parotidectomy.

Oncocytomas are rare tumours of the parotid gland seen in the elderly. The lesion consists of solid cords, nests or sheets of oncocytes (large eosinophilic cells with small compact nuclei). These cells are similar to the eosinophilic cells seen in Warthin's tumour; however, the stroma lacks the lymphoid component required for that diagnosis.

Recommended Reading

El-Naggar AK, Chan JKC, Grandis JR, Takata T, Slootweg P, editors. Tumours of the Salivary Glands. World Health Organization Classification of Head and Neck Tumours. 4th ed. Lyon: IARC; 2017. p. 186–202.

Neville BW, Damm DD, Allen CM, Chi AC. Salivary gland tumors. In: Oral and Maxillofacial Pathology. 4th ed. St. Louis, MO: Elsevier; 2016. p. 440–465.

Odell E. Cawson's Essentials of Oral Pathology and Oral Medicine. 9th ed. Edinburgh: Elsevier; 2017. p. 355–365.

Quer M, Hernandez-Prera JC, Silver CE, Casasayas M, Simo R, Vander Poorten V, et al. Current Trends and Controversies in the Management of Warthin Tumor of the Parotid Gland. Diagnostics (Basel). 2021;13;11(8):1467.

Mucoepidermoid Carcinoma 43

Norman Firth

43.1 Introduction

Mucoepidermoid carcinoma (MEC) has a variable clinical and histopathological presentation, but is characterised by the presence of both and epithelial and a mucous secreting glandular component.

The cause is translocations of METC1 and MAML2 genes causing a new fusion gene promoting neoplastic change.

MEC is the most common malignant salivary gland neoplasm accounting for up to 15% of all salivary gland tumours and 12–29% of malignant salivary gland tumours. MEC occurs in both children and adults. There is a female predilection.

43.2 Clinical Presentation and Investigations

The parotid gland is most commonly affected followed by the submandibular and sublingual glands. Facial nerve involvement may occur. The palate is the most frequent intra-oral site. Lesions are often slow growing and asymptomatic. Intraoral lesions may have an extensive cystic component and appear as a fluctuant bluish mass.

N. Firth (✉)
School of Dentistry, The University of Queensland, Herston, QLD, Australia

Capital, Coast and Hutt Valley District Health Board, Wellington, New Zealand

Clinipath, Osborne Park, WA, Australia
e-mail: Norman.firth@xtra.co.nz; Nfirth@clinipath.net

R. Balasubramaniam et al. (eds.), *Oral Medicine - A Clinical Guide*,
https://doi.org/10.1007/978-3-031-36797-7_43

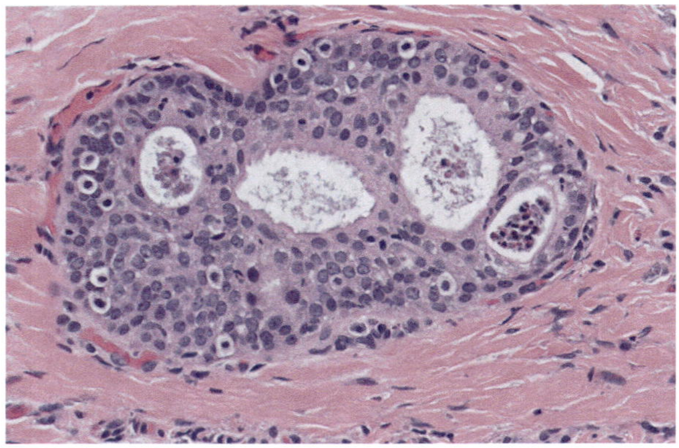

Fig. 43.1 Female aged 59 years with a lump in the palate noted 6 months ago. Photomicrograph showing both an epithelial component together with mucous secretion (H&E)

Fine needle aspirate (FNA) followed by biopsy is essential for diagnosis. Histopathological features are variable, and the neoplasm is graded as low intermediate and high-grade types. Lesions consist of a mixture of mucous secreting cells, squamous epithelial cells, intermediate cells and clear cells. Keratinisation is usually not a feature in the epithelial component. The greater the cystic component the lower the tumour grade (Fig. 43.1). MEC is unencapsulated and shows an infiltrative growth pattern. Higher grade lesions demonstrate cytological atypia and a higher mitotic rate. Low-grade lesions with an extensive cystic component and mucin production are difficult to diagnose on FNA and/or incisional biopsy. In some instances, diagnosis of MEC is aided by detecting the translocation.

43.3 Management and Prognosis

Treatment is surgical removal of the tumour. Radiotherapy is used in selected cases. Chemotherapy is used in cases which are not amenable to surgery or radiotherapy. Approximately 40% of high-grade lesions metastasise and carry a 30% mortality rate.

Recommended Reading

El-Naggar AK, Chan JKC, Grandis JR, Takata T, Slootweg P, editors. Tumours of the Salivary Glands. World Health Organization Classification of Head and Neck Tumours. 4th ed. Lyon: IARC; 2017. p. 186–202.

Neville BW, Damm DD, Allen CM, Chi AC. Salivary gland tumors. In: Oral and Maxillofacial Pathology. 4th ed. St. Louis, MO: Elsevier; 2016. p. 440–465.

Odell E. Cawson's Essentials of Oral Pathology and Oral Medicine. 9th ed. Edinburgh: Elsevier; 2017. p. 355–365.

Rapidis AD, Givalos N, Gakiopoulou H, Stavrianos SD, Faratzis G, Lagogiannis GA, Katsilieris I, Patsouris E. Mucoepidermoid carcinoma of the salivary glands. Review of the literature and clinicopathological analysis of 18 patients. Oral Oncol 2007;43(2):130–6. doi: https://doi.org/10.1016/j.oraloncology.2006.03.001. Epub 2006 Jul 20.

Adenoid Cystic Carcinoma

44

Norman Firth

44.1 Introduction

Adenoid cystic carcinoma (ACC) is nearly as common as mucoepidermoid carcinoma occurring in any salivary gland including about a third of malignant minor salivary glands. ACC has a deceptively benign histological appearance and is composed of cells with bland cytology, but the lesion is locally invasive with a high propensity for local recurrence and distant metastasis. ACC in most cases is caused by activation of the oncogenic transcription factor gene MYB, and over-expression of MYB may be detected by fluorescence in situ hybridisation as an adjunct to diagnosis.

ACC is most commonly (about 40% of cases) seen in the fifth and sixth decades of life. There is a slightly increased female: male ratio.

44.2 Clinical Presentation and Investigations

The lesion presents as a slow growing mass and pain maybe a presenting feature. Swelling may be present, and the surface may be ulcerated. Radiographic evidence of bone destruction may be seen. The lesion has a propensity for neural involvement (perineural spread). Parotid lesions may present with facial nerve involvement or other unusual sensory changes such as pain or paraesthesia/anaesthesia rather than a clearly discernible mass. Magnetic resonance imaging features such as abnormal

N. Firth (✉)
School of Dentistry, The University of Queensland, Herston, QLD, Australia

Capital, Coast and Hutt Valley District Health Board, Wellington, New Zealand

Clinipath, Osborne Park, WA, Australia
e-mail: Norman.firth@xtra.co.nz; Nfirth@clinipath.net

R. Balasubramaniam et al. (eds.), *Oral Medicine - A Clinical Guide*,
https://doi.org/10.1007/978-3-031-36797-7_44

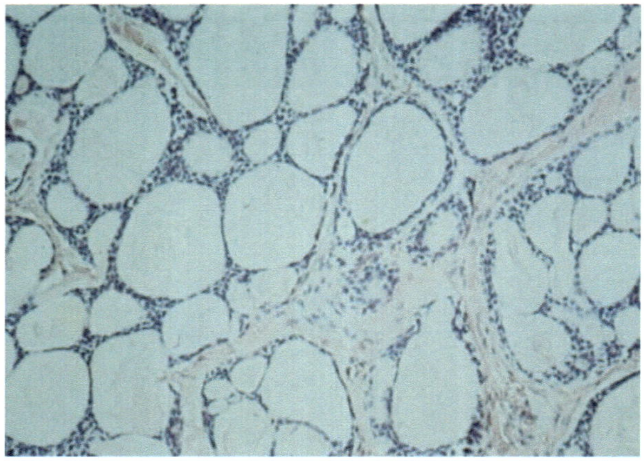

Fig. 44.1 Photomicrograph of an adenoid cystic carcinoma presenting as a buccal mucosal mass in a 65-year-old male (H&E)

density/signal intensity, contrast enhancement or widening of a fossa, foramen or canal in which nerves are located are indicative of neural involvement.

Histopathologically, ACC is characterised by groups of myoepithelial and ductal cells with three distinct patterns cribriform, tubular and solid. Multiple cystic spaces surrounded by darkly staining cells without cytological atypia are seen and form a characteristic cribriform or "Swiss cheese" pattern (Fig. 44.1). The solid or trabecular pattern may however predominate. Perineural invasion is helpful in diagnosis. The lesion is unencapsulated with an infiltrative destructive growth pattern.

44.3 Management and Prognosis

Treatment is surgery, or surgery in combination with radiotherapy. Unfortunately, radiotherapy is sometimes ineffective in radioresistant tumours. The 5-year survival rate is fair; however, the long-term survival rate is very poor. Lymph node metastasis is uncommon; however, a third of patients develop bloodborne metastasis in lung, liver or bone.

Recommended Reading

Dillon PM, Chakraborty S, Moskaluk CA, Joshi PJ, Thomas CY. Adenoid cystic carcinoma: A review of recent advances, molecular targets, and clinical trials. Head Neck. 2016;38(4):620–7.

El-Naggar AK, Chan JKC, Grandis JR, Takata T, Slootweg P, editors. Tumours of the Salivary Glands. World Health Organization Classification of Head and Neck Tumours. 4th ed. Lyon: IARC; 2017. p. 186–202.

Neville BW, Damm DD, Allen CM, Chi CA. Salivary gland tumours. In: Oral maxillofacial pathology 4th ed Elsevier, St Louis, MO 2016. p. 440–465.

Odell E. Cawson's Essentials of Oral Pathology and Oral Medicine. 9th ed. Edinburgh: Elsevier; 2017. p. 355–365.

Other Malignant Salivary Gland Tumours 45

Norman Firth

45.1 Acinic Cell Carcinoma

Acinic cell carcinomas are slow growing but have an unpredictable behaviour and may metastasise. The mean age of occurrence is 50 years, but they occur over a wide age range and are the second most common malignant salivary gland tumour in children. The parotid gland is by far the most common site. They may have a benign appearance including having an encapsulated appearance. They are composed of acinar-like cells, and a variety of histological appearances are encountered.

45.2 Polymorphous Adenocarcinoma

Polymorphous adenocarcinoma is the current name of the lesion formerly called polymorphous low-grade adenocarcinoma. It is a common malignant minor salivary gland tumour generally, but not exclusively seen after the age of 50 years. Most are low grade, but approximately 10% metastasise and some recur due to the infiltrative growth pattern. The histological appearance varies through the lesion (Fig. 45.1), and misdiagnosis particularly with the more aggressive ACC is possible particularly if based on relatively small incisional biopsies. Treatment is by local excision.

N. Firth (✉)
School of Dentistry, The University of Queensland, Herston, QLD, Australia

Capital, Coast and Hutt Valley District Health Board, Wellington, New Zealand

Clinipath, Osborne Park, WA, Australia
e-mail: Norman.firth@xtra.co.nz; Nfirth@clinipath.net

R. Balasubramaniam et al. (eds.), *Oral Medicine - A Clinical Guide*,
https://doi.org/10.1007/978-3-031-36797-7_45

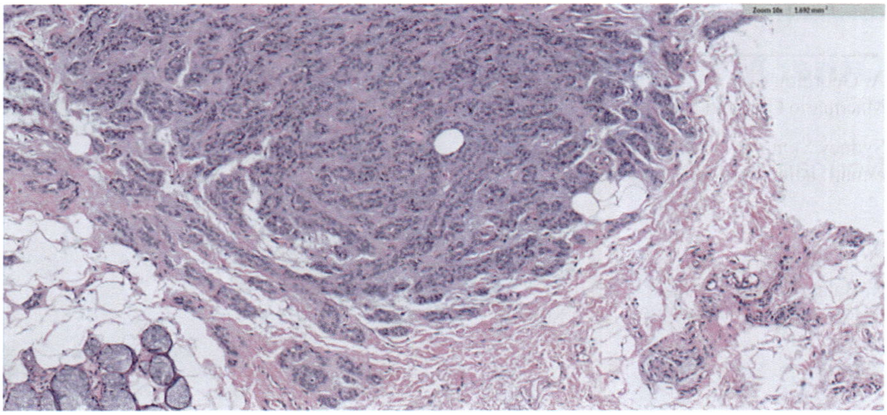

Fig. 45.1 An incisional biopsy of a purplish palatal lump in an 83-year-old female, diagnosed as polymorphous adenocarcinoma. The tumour is composed of polygonal to cuboidal cells set in a fibromyxoid stroma in this region and is not encapsulated (H&E)

45.3 Secretory Carcinoma

Secretory carcinoma was originally called mammary analogue secretory carcinoma in 2010 and renamed in the 2017 WHO classification of salivary gland tumours. It is a generally regarded as a low-grade carcinoma (although up to 20% metastasise) which resembles mammary secretory carcinoma and is caused by a translocation between the ETV6 and NTRK3 genes. It occurs in any of the salivary glands, but most frequently the parotid gland. Histologically, it has solid and ductal areas and spaces containing secretory material.

Salivary duct carcinoma is an aggressive high-grade carcinoma resembling ductal carcinoma of the breast. Growth is usually rapid, with perineural and lymphatic invasion common. Any of the salivary glands may be affected, and there is a male predilection. Large oncocytic cells forming ducts and cystic spaces are present. Androgen expression is strongly expressed in these lesions. Death occurs in more than 50% of cases within 5 years due to metastasis (lung, liver, bone,) and local recurrence is also common.

45.4 Epithelial-Myoepithelial Carcinoma

Epithelial-myoepithelial carcinoma is a low-grade malignancy with a characteristic biphasic growth pattern seen in the major glands of the elderly. Often, they appear to be a benign lesion clinically and histologically, but local recurrence and distant metastases occur in about 10% following local excision.

Recommended Reading

El-Naggar AK, Chan JKC, Grandis JR, Takata T, Slootweg P, editors. Tumours of the Salivary Glands. World Health Organization Classification of Head and Neck Tumours. 4th ed. Lyon: IARC; 2017. p. 166–184.

Jo VY, Krane JF. Clinicopathologic features of secretory carcinoma of salivary gland. AJSP Rev Reports. 2018;23(1):9–24.

Neville BW, Damm DD, Allen CM, Chi AC. Salivary gland tumors. In: Oral and Maxillofacial Pathology. 4th ed. St. Louis, MO: Elsevier; 2016. p. 440–465.

Odell E. Cawson's Essentials of Oral Pathology and Oral Medicine. 9th ed. Edinburgh: Elsevier; 2017. p. 355–365.

Vander Poorten V, Triantafyllou A, Skálová A, Stenman G, Bishop JA, Hauben E, et al. Polymorphous adenocarcinoma of the salivary glands: reappraisal and update. Eur Arch Otorhinolaryngol. 2018;275(7):1681–95.

Differential Diagnosis of Mucosal Swellings

46

Janina Christoforou

Soft and hard tissue swellings or enlargements within the oral cavity present often with histologically normal overlying oral mucosa. There may be a traumatic overlay to the mucosal appearance which may appear as keratosis, erythema or ulceration. The most common of theses 'swellings' include tori or exostoses, fibroepithelial polyps, fibrous epulides and denture-induced hyperplasia. Mucoceles are also very common, but these will be described in Chap. 28 on benign salivary gland conditions. Typically, these oral mucosal swellings are postulated to be initiated by trauma, persistent local irritation or other local environmental factors. The clinical features, diagnosis and treatment of common oral hard and soft tissue swelling are presented in Table 46.1. There are also other causes of swellings which include developmental such as a hamartoma, part of a syndrome, such as neurofibromatosis, or of neoplastic origin which can be benign, such as a schwannoma or malignant, such as a lymphoma.

The benign mucosal swellings described here are generally reactive in nature. Nevertheless, neoplastic lesions may also present similarly and hence the clinician should remain vigilant. Features which may raise suspicion for a malignancy include induration, surface ulceration, an increase in size of the swelling.

J. Christoforou (✉)
UWA Dental School, The University of Western Australia, Nedlands, WA, Australia

Perth Children's Hospital, Nedlands, WA, Australia
e-mail: Janina.Christoforou@health.wa.gov.au

R. Balasubramaniam et al. (eds.), *Oral Medicine - A Clinical Guide*,
https://doi.org/10.1007/978-3-031-36797-7_46

Table 46.1 Features and management of common oral mucosal swellings

	Tori/exostoses	Fibroepithelial polyp	Fibrous epulis	Denture-induced hyperplasia
Aetiology/ pathogenesis	• Genetic background • Environmental factors (masticatory stress, calcium supplementation)	• Trauma or local irritation	• Originates from the periodontal ligament and the periosteum • From chronic local irritation (dental calculus, dental defects)	• Reactive to chronic irritation from a removable prosthesis
Epidemiology	• Tori prevalence = 12–15% • Exostoses less frequent and more prevalent with age	• Prevalence of 26.5% amongst other reactive lesions	• Females > Males • Accounts for 40% of periodontal and gingival pathology • Wide age distribution	• Females > Males • Arises in 43.8% of denture wearers
Clinical presentation	• Palatal tori: Either a single outgrowth or multiple outgrowths from the midline palate • Mandibular tori: Single or multiple projections of bone generally located on the lingual surface of the mandible in the premolar area, superior to the mylohyoid ridge • Exostoses- Single or multiple projections on the buccal or palatal alveolar surface • Asymptomatic, unless traumatised	• A smooth surface and similar to the colour of the adjacent mucosa • Painless, unless traumatised • Can increase in size with persistent stimulation	• An exophytic growth commonly arising below the free gingival margin or the interdental papilla • Smooth surface with a pale pink colour similar to adjacent gingiva • Asymptomatic unless traumatised • Has a variable growth rate	• Asymptomatic, fibrous, hyperplastic swelling(s), seen on the alveolar vestibules corresponding to denture flanges or posterior denture extension on the soft palate • Smooth folds of a pale pink colouration similar to adjacent mucosa
Clinical investigations	• Clinical diagnosis • Further confirmation can be obtained with intraoral or panoramic radiographs, or computer tomography imaging	• Clinical and histopathological evaluation	• Clinical and histopathological evaluation	• Generally, a clinical diagnosis
Differential diagnoses	• If unilateral: osteomas, osteoid osteomas, peripheral ossifying fibromas, osteochondroma, osteosarcoma or osteoblastoma	• Localised: neurofibroma, lipoma or a schwannoma • Multiple: focal epithelial hyperplasia	• Peripheral ossifying fibroma • May occasionally mimic squamous cell carcinoma • If traumatised: angiogranuloma or peripheral giant cell granuloma	• May occasionally mimic a squamous cell carcinoma

Table 46.1 (continued)

	Tori/exostoses	Fibroepithelial polyp	Fibrous epulis	Denture-induced hyperplasia
Treatment and prognosis	• Surgical removal is generally not indicated unless the torus/exostosis interferes with the placement of dental prosthesis or becomes recurrently traumatised • Slow growth may be noted	• Generally excised with narrow margins • Reduced propensity for recurrence should the local irritation be removed	• Excised with curettage of the supra-alveolar connective tissue and debridement of the adjacent dental surfaces making the surface free from calculus and plaque-retaining defects • Recurrence can occur if the potential irritant is not removed	• Surgical removal of the hyperplastic tissue • Rectification of denture fit or new appliance

Recommended Reading

Alhindi NA, Sindi AM, Binmadi NO, Elias WY. A retrospective study of oral and maxillofacial pathology lesions diagnosed at the Faculty of Dentistry, King Abdulaziz University. Clin Cosmet Investig Dent. 2019;11:45–52.

García-García AS, Martínez-González JM, Gómez-Font R, Soto-Rivadeneira A, Oviedo-Roldán L. Current status of the torus palatinus and torus mandibularis. Med Oral Patol Oral Cir Bucal. 2010;15(2):e353–60.

Jainkittivong A, Langlais RP. Buccal and palatal exostoses: prevalence and concurrence with tori. Oral Surg Oral Med Oral Pathol Oral Radiol Endod. 2000;90(1):48–53.

Jones AV, Franklin CD. An analysis of oral and maxillofacial pathology found in adults over a 30-year period. J Oral Pathol Med. 2006;35(7):392–401.

Mandali G, Sener ID, Turker SB, Ulgen H. Factors affecting the distribution and prevalence of oral mucosal lesions in complete denture wearers. Gerodontology. 2011;28(2):97–103.

Mermod M, Hoarau R. Mandibular tori. CMAJ. 2015;187(11):826.

Morrison MD, Tamimi F. Oral tori are associated with local mechanical and systemic factors: a case-control study. J Oral Maxillofac Surg. 2013;71(1):14–22.

Veena K, Jagadishchandra H, Sequria J, Hameed S, Chatra L, Shenai P. An extensive denture-induced hyperplasia of maxilla. Ann Med Health Sci Res. 2013;3(Suppl 1):S7–9.

Differential Diagnosis of Localised Gingival Swellings

47

Janina Christoforou

Localised swellings are a common occurrence on the gingivae. These swellings may be a variation of normal anatomy or arise from embryological remnants. Additionally, the swellings may also result from chronic irritation which leads to a hyperplastic inflammatory response. Table 47.1 summarises the pathogenesis, clinical features, differential diagnosis and treatment of the common localised benign gingival swellings (Fig. 47.1).

Clinicians should be aware of the difference in the types of localised swellings on the gingivae. This awareness will result in accurate diagnosis and the most effective management which also eliminate the need for unnecessary treatment.

J. Christoforou (✉)
UWA Dental School, The University of Western Australia, Nedlands, WA, Australia

Perth Children's Hospital, Nedlands, WA, Australia
e-mail: Janina.Christoforou@health.wa.gov.au

R. Balasubramaniam et al. (eds.), *Oral Medicine - A Clinical Guide*,
https://doi.org/10.1007/978-3-031-36797-7_47

Table 47.1 Features and management of localised gingival swellings

	Angiogranuloma (pyogenic granuloma)/ pregnancy epulis	Peripheral giant cell granuloma	Retrocuspid papillae	Gingival cyst (adults)
Aetiology and pathogenesis	• Low-grade chronic irritation, trauma or hormonal imbalances resulting in an exaggerated proliferation of vascular type of connective tissue • Pregnancy epulis is an angiogranuloma in pregnant women	• Originates from the periosteum or periodontal membrane in response to local irritation • Multinucleated giant cells (derived from osteoclasts) and extravasated red blood cells are histopathologically noted	• A normal anatomical variant • Postulated that the swelling(s) represents the fibrovascular connective tissue covering nutrient vascular canals in the region	• An odontogenic cyst that arises from the rests of the dental lamina
Epidemiology	• Females predilection • 32% of periodontal and gingival pathology • A pregnancy epulis is reported over 5% of pregnancies	• Females predilection • 5–6% of periodontal and gingival pathology	• 67% of children aged 3–5 years showed the presence of retrocuspid papillae	• ~0.3% of all odontogenic cysts • More common in the anterior dentition (mandible greater than maxilla) • Usually present at age 50–60 years
Clinical presentation	• A red/purple, slow-growing swelling which is most commonly found on the attached gingiva, followed by the lateral aspect of the tongue, lower lip and buccal mucosa • Readily bleeds on provocation	• Red/purple, exophytic lesion which is usually noted on the gingivae or the alveolar mucosa • Have an increased propensity to grow, extend locally and at times result in cortical bone erosion and dental root alterations • May represent peripheral extension of central lesion	• A smooth, circumscribed, sessile gingival swelling, approximately 1-4 mm in dimension, which present lingual to the mandibular cuspid/canine • Commonly noted bilaterally, but they can be unilateral	• Presents on the attached gingivae or interdental papillae as a single, small, asymptomatic, bluish swelling • More common in the mandibular incisor, canine and premolar regions • Bony erosion is often noted at the site
Clinical investigations	• Clinical presentation and histopathological evaluation	• Clinical presentation and histopathological evaluation • Intraoral radiographs or computer tomography imaging should be performed to exclude a central giant cell granuloma which extends peripherally	• A clinical diagnosis is generally made based on appearance and location of the swelling(s)	• Clinical presentation and histopathological evaluation

Table 47.1 (continued)

	Angiogranuloma (pyogenic granuloma)/ pregnancy epulis	Peripheral giant cell granuloma	Retrocuspid papillae	Gingival cyst (adults)
Differential diagnosis	• A vascular malformation • Peripheral giant cell granuloma • Vascular tumours: Kaposi's sarcoma or angiosarcoma • Squamous cell carcinoma	• Central giant cell granuloma • Vascular tumours: Kaposi's sarcoma or angiosarcoma • Brown tumour of hyperthyroidism	• If unilateral: fibrous epulis	• A lateral periodontal cyst which has perforated the alveolar cortex • A mucocele of the gingiva
Treatment and prognosis	• Surgical excision of the swelling with the removal of the irritant (subgingival calculus) • Pregnancy epulides most commonly resolve or reduce following delivery	• Complete surgical excision of the swelling with curettage • Risk of recurrence	• Treatment is not required since the swelling(s) appear to regress or disappear with age	• Surgical excision of the swelling • Recurrence is unlikely

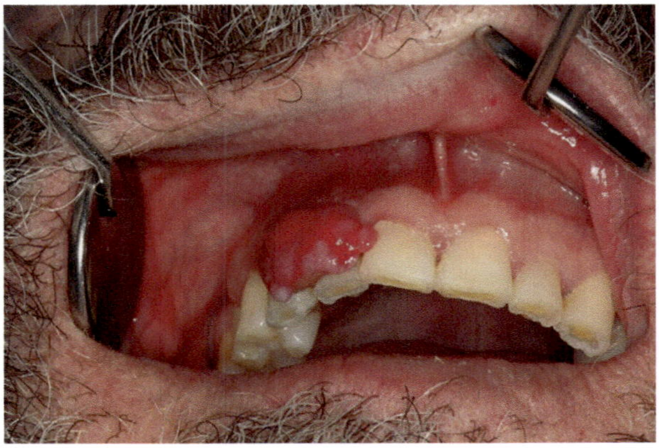

Fig. 47.1 Angiogranuloma of the right labial gingiva

Recommended Reading

Bhari Sharanesha R, Manjunatha BS, Dave B, Mallikarjuna R. Retrocuspid papillae: a study on its prevalence. Res J Pharm, Biol Chem Sci. 2015;6:1632–6.

Bodner L, Peist M, Gatot A, Fliss DM. Growth potential of peripheral giant cell granuloma. Oral Surg Oral Med Oral Pathol Oral Radiol Endod. 1997;83(5):548–51.

Brannon RB, Pousson RR. The retrocuspid papillae: a clinical evaluation of 51 cases. J Dent Hyg. 2003;77(3):180–4.

Capella DL, Zimmermann C, Scotti F, Santos A, Grando L, Maurer M, et al. Gingival cyst of adult with radiographic findings: a rare case. Oral Surg Oral Med Oral Pathol Oral Radiol 2017;124(2):e94–e94.

Chrcanovic BR, Gomez RS. Gingival cyst of the adult, lateral periodontal cyst, and botryoid odontogenic cyst: An updated systematic review. Oral Dis. 2019;25(1):26–33.

Viveiros SK, Pinho RFC, Custódio M, de Sousa SCOM, Santos Pinto DD. A rare odontogenic cyst: Gingival cyst of the adult. A series of 20 new cases from a single center. J Craniomaxillofac Surg. 2019;47(4):647–50.

Levin MP, Cutright DE. The retrocuspid papilla: hypothesis of origin. J Periodontol. 1977;48(8):464–6.

Saravana GH. Oral pyogenic granuloma: a review of 137 cases. Br J Oral Maxillofac Surg. 2009;47(4):318–9.

Savage NW, Daly CG. Gingival enlargements and localized gingival overgrowths. Aust Dent J. 2010;55 Suppl 1:55–60.

Rajashekhara BS, Manjunatha BS, Bhavana D, Rachappa MM. Retrocuspid papillae: a study on its prevalence. Res J Pharm Biol Chem Sci. 2015;6(3):1632–1636.

Differential Diagnosis of Diffuse Gingival Swelling

48

Janina Christoforou

48.1 Introduction

There are a number of aetiologies associated with generalised gingival enlargement. Many are oral manifestations of systemic disease such as granulomatosis with polyangiitis, sarcoidosis, Crohn's disease or even neoplastic processes such as leukaemia. Additionally, gingival overgrowth may be due to hormonal influences such as pregnancy / puberty, malnutrition (vitamin C deficiency) or allergy, presenting as plasma cell gingivitis. Drug-induced gingival overgrowth and hereditary gingival fibromatosis are some other conditions which can lead to progressive gingival overgrowth, to a degree which affects aesthetics, function and complications in oral hygiene maintenance. Table 48.1 summarises the pathogenesis, clinical presentation, differential diagnoses and treatment of drug-induced gingival overgrowth and hereditary gingival fibromatosis.

Diffuse gingival swellings have many aetiologies including genetic predisposition, hormonal alterations, gastrointestinal disorders, immune-mediated multisystem disorders, allergies and neoplastic conditions. It is therefore of utmost importance that a definitive diagnosis is obtained to ensure effective management.

J. Christoforou (✉)
UWA Dental School, The University of Western Australia, Nedlands, WA, Australia

Perth Children's Hospital, Nedlands, WA, Australia
e-mail: Janina.Christoforou@health.wa.gov.au

R. Balasubramaniam et al. (eds.), *Oral Medicine - A Clinical Guide*,
https://doi.org/10.1007/978-3-031-36797-7_48

Table 48.1 Drug-induced gingival overgrowth and hereditary gingival fibromatosis

	Drug-induced gingival overgrowth	Hereditary gingival fibromatosis
Aetiology and pathogenesis	• Arises as a side effect to a number of medications. The most common of these are anticonvulsants (phenytoin), immunosuppressants (cyclosporine A) and various calcium channel blockers (nifedipine, verapamil, diltiazem) • The gingival enlargement often appears within 1–3 months after commencement of the medications • The medications have an effect on the connective tissue, and a bacterial inflammatory component is an essential influencing factor • There may be a decreased synthesis of collagenase and downregulation of genes coding for matrix metalloproteinase, resulting in reduced collagen degradation subsequent collagen accumulation	• A rare, generally autosomal dominant (but can be an autosomal recessive) condition • The fibrous enlargement of the gingiva is seen to be secondary to an increase in submucosal, dense, fibrous connective tissue • The increase in collagen is due to heterogenous genetic alterations of gingival fibroblasts which results in increased synthesis of TGF-β. This promotes elevated collagen and fibronectin production, reduced synthesis and activity of extracellular matrix-degrading metalloproteinases and the transdifferentiation of fibroblasts to myofibroblasts contributing to fibrosis
Epidemiology	• Male predilection • Accounts for less than 1.5% of periodontal and gingival pathology	• Prevalence is 1 per 175,000 individuals • Mainly seen in children and adolescents
Clinical presentation	• The gingival enlargement is generalised, but it presents more severely in the maxillary and mandibular anterior regions with the formation of periodontal pockets/ pseudopockets. See Fig. 48.1	• Characterised by a slowly progressive, firm, fibrous hyperplasia of the gingiva which is usually noted at the time of eruption of the permanent dentition, but it can occur prior • It can affect both the maxillary and mandibular gingivae and the fibrous hyperplasia can result in dental spacing, dental malalignment, retention of primary teeth and gingival coverage of the dental crowns • Usually develops as an isolated disorder with clinical manifestations which may include hypertrichosis, epilepsy, mental retardation and supernumerary teeth

Table 48.1 (continued)

	Drug-induced gingival overgrowth	Hereditary gingival fibromatosis
Clinical investigations	• Evaluation of a patient's medical history to identify a potential implicating medication • Awareness of family history • Blood investigations and histopathological analyses may be warranted	Evaluation of a patient's medical history, family history and clinical presentation
Differential diagnoses	• Hormonal-associated gingival enlargements, hereditary gingival hyperplasia, oral sarcoidosis, Crohn's disease, granulomatosis with polyangiitis, plasma cell gingivitis and leukaemia	• Drug-induced gingival overgrowth, as well as being a feature in syndromes such as Laband, Rutherfurd, Jones and Cross syndromes
Treatment and prognosis	• May respond to local treatment, such as gingival debridement and maintenance of meticulous oral hygiene • Altering the medication if there is minimal local treatment response and difficulty maintaining of oral hygiene • Periodontal resective surgery may be necessary • Risk of recurrence increases as early as 3–6 months following the surgical approach, and this is positively correlated with poor oral hygiene	• When maintaining good oral hygiene becomes difficult, there is the risk of periodontal disease • Dental debridement and good oral hygiene maintenance • Resective surgery to maintain function, aesthetics and oral hygiene is often required

Fig. 48.1 Drug-induced gingival overgrowth from cyclosporine

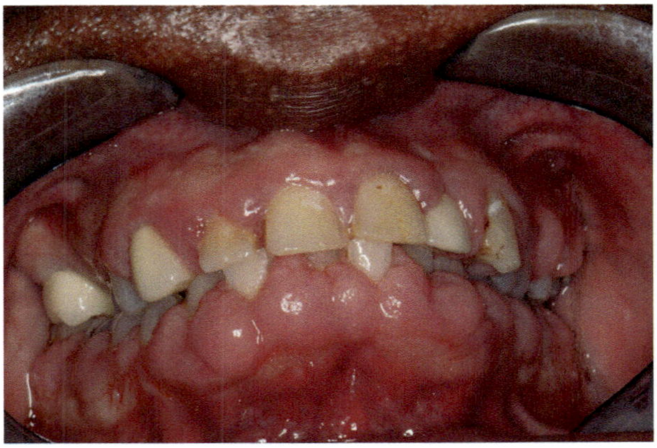

Recommended Reading

Almiñana-Pastor PJ, Buitrago-Vera PJ, Alpiste-Illueca FM, Catalá-Pizarro M. Hereditary gingival fibromatosis: Characteristics and treatment approach. J Clin Exp Dent. 2017;9(4):e599–e602.

Boutiou E, Ziogas IA, Giannis D, Doufexi AE. Hereditary gingival fibromatosis in children: a systematic review of the literature. Clin Oral Investig. 2021;25(6):3599–607.

Coletta RD, Graner E. Hereditary gingival fibromatosis: a systematic review. J Periodontol. 2006;77(5):753–64.

Ilgenli T, Atilla G, Baylas H. Effectiveness of periodontal therapy in patients with drug-induced gingival overgrowth. Long-term results. J Periodontol. 1999;70(9):967–72.

Katz J, Guelmann M, Barak S. Hereditary gingival fibromatosis with distinct dental, skeletal and developmental abnormalities. Pediatr Dent. 2002;24(3):253–6.

Lauritano D, Moreo G, Della Vella F, Palmieri A, Carinci F, Petruzzi M. Biology of drug-induced gingival hyperplasia: in vitro study of the effect of nifedipine on human fibroblasts. Appl Sci 2021;11(7):3287.

Lauritano D, Moreo G, Limongelli L, Palmieri A, Carinci F. Drug-Induced Gingival Overgrowth: The Effect of Cyclosporin A and Mycophenolate Mophetil on Human Gingival Fibroblasts. Biomedicines. 2020;8(7):221.

Livada R, Shiloah J. Calcium channel blocker-induced gingival enlargement. J Hum Hypertens. 2014;28(1):10–4.

Meraw SJ, Sheridan PJ. Medically induced gingival hyperplasia. Mayo Clin Proc. 1998;73(12):1196–9.

Differential Diagnosis of Oral Ulceration 49

Ajith Polonowita and Guangzhao Guan

The term "ulcer" usually refers to a break in the epithelium exposing the underlying connective tissue but whereas the term erosion refers to the loss of some but not all layers of the epithelium. There are many causes of oral ulceration and a wide range of clinical presentations. The estimated worldwide prevalence is of any ulceration is 4%. The differential diagnoses of oral ulceration are presented in Table 49.1.

Differential diagnosis requires a good history, examination and special investigations as required (Fig. 49.1).

Oral ulcers are common, and a systematic approach will lead to the correct diagnosis which will then direct treatment of the underlying cause.

A. Polonowita (✉)
Faculty of Dentistry, The University of Otago, Dunedin, New Zealand

Christchurch Hospital, Canterbury, New Zealand
e-mail: ajith.polonowita@otago.ac.nz

G. Guan
Faculty of Dentistry, The University of Otago, Dunedin, New Zealand

R. Balasubramaniam et al. (eds.), *Oral Medicine - A Clinical Guide*,
https://doi.org/10.1007/978-3-031-36797-7_49

Table 49.1 Differential diagnoses of oral ulceration (diagnostic Sieve 5I 3-Scotland)

Diagnostic category	Subcategory	Examples
Infection	Bacterial, fungal, viral	Bacterial: • Tuberculosis • Syphilis • NUG/NUP(necrotising ulcerative gingivitis/ periodontitis) Viral: • Varicella zoster • Human immunodeficiency virus Fungal: • Candidosis • Deep mycoses
Inflammation	Systemic disorders	Mucocutaneous diseases: • Oral lichen planus (see Chap. 17) Connective tissue diseases: • Lupus (see Chap. 18) Gastrointestinal disorders: • Coeliac disease, Crohn's and ulcerative colitis (see Chaps. 26, 78 and 79)
Immune	Aphthous-like ulcers	Behcet's disease; Sweet's syndrome; periodic fever, aphthous stomatitis, pharyngitis, adenitis (PFAPA)
	Recurrent aphthous stomatitis (RAS)	Major, minor, herpetiform ulceration
	Hypersensitivity	Rarely, it may manifest as erosions or ulcers RAS
Iatrogenic	Fictitious (self-harm) ulceration	
Idiopathic		• Necrotising sialometaplasia
	Medications related	Multiple medications could cause/contribute to oral erosion or ulceration: • Methotrexate, nicorandil and propylthiouracil (see Chap. 53)
Neoplastic	Malignancy	• Oral squamous cell carcinoma • Lymphoma • Leukaemia • Sarcomas • Salivary gland tumours
Trauma	Mechanical, thermal, chemical, radiation	• Aspirin burn, radiotherapy, chemotherapy mucositis

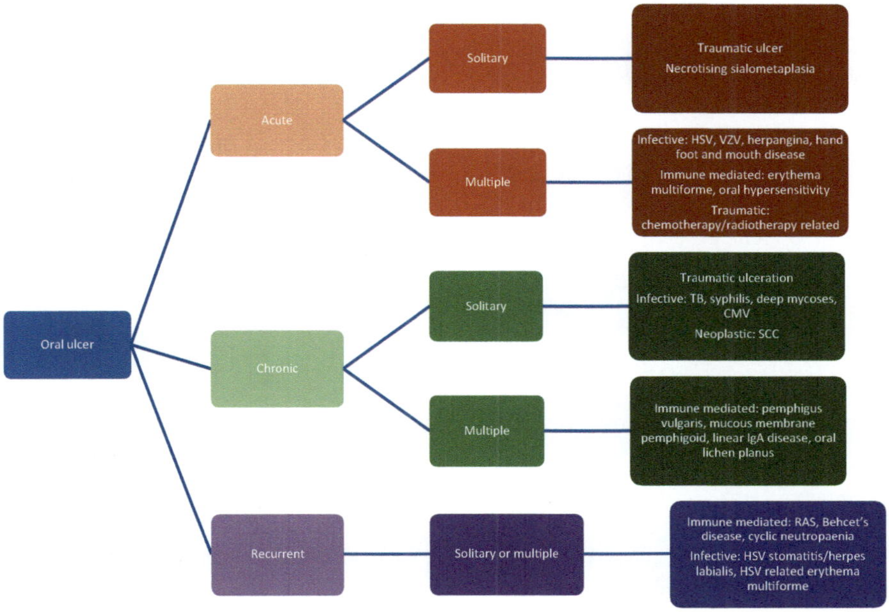

Fig. 49.1 Diagnostic algorithm for oral ulceration. *HSV* herpes simplex virus, *VZV* varicella zoster virus, *TB* tuberculosis, *CMV* cytomegalovirus, *SCC* squamous cell carcinoma, *RAS* recurrent aphthous stomatitis. Clinical algorithm adapted from Lodi G, Varoni E, Robledo-Sierra J, Villa A, and Jontel M. Oral ulcerative lesions. In: Farah CS, Balasubramaniam R, McCullough MJ, editors. Contemporary oral medicine. Cham, Switzerland: Springer; 2019

Recommended Reading

Preeti L, Magesh K, Rajkumar K, Karthik R. Recurrent aphthous stomatitis. J Oral Maxillofac Pathol. 2011;15(3):252–6.

Scully C, Shotts R. ABC of oral health. Mouth ulcers and other causes of orofacial soreness and pain. BMJ. 2000;321(7254):162–5.

Traumatic Oral Ulceration

50

Ajith Polonowita and Guangzhao Guan

50.1 Introduction

Trauma induced ulceration may be caused by mechanical, chemical, electrical, thermal or radiation-related injury, with an approximate prevalence of 1%. A temporal history of commencement of the ulceration combined with examination of the ulcer and contacting structures including teeth, restorations and prostheses can assist to indicate trauma as a potential cause.

50.2 Clinical Presentation and Investigations

Traumatic ulcers are usually solitary, with irregular borders. Patients often report an inciting traumatic event (Fig. 50.1).

Ulceration due to mechanical trauma will be located adjacent to the cause (e.g. fractured or sharp tooth) and may be shallow or deep. These ulcers are generally painful, and the patient will recall the inciting incident such as accidently biting the buccal mucosa while chewing gum.

Chemical traumatic ulceration (e.g. aspirin burn) will often be white in appearance with a wrinkled surface where the chemical has been in contact with the mucosa. Chemical injuries that cause traumatic ulcerations include dental materials (example local anaesthetics) those from self-harm with sulphuric acid.

A. Polonowita (✉)
Faculty of Dentistry, The University of Otago, Dunedin, New Zealand

Christchurch Hospital, Canterbury, New Zealand

G. Guan
Faculty of Dentistry, The University of Otago, Dunedin, New Zealand

R. Balasubramaniam et al. (eds.), *Oral Medicine - A Clinical Guide*,
https://doi.org/10.1007/978-3-031-36797-7_50

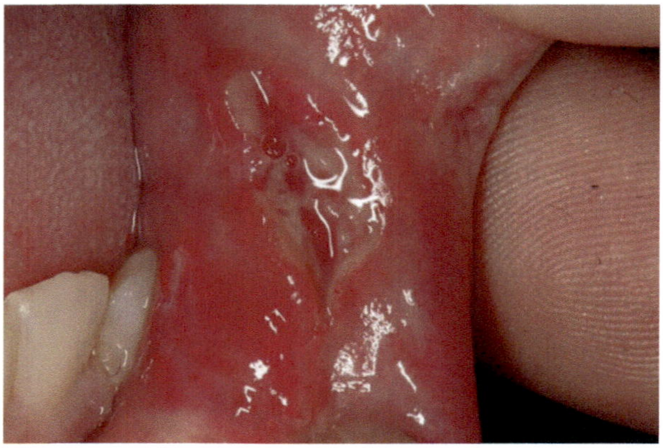

Fig. 50.1 Traumatic ulcer of the left commissure region

Trauma from electrical burns commonly affect the oral commissure and are increasingly caused by exploding e-cigarettes.

Ulcerations from thermal trauma (e.g. pizza burn) commonly appear shallow, with areas of whiteness at the periphery. Usually, these will illicit mild to moderate pain at the site of ulceration.

Ulcerations from factitious causes are included here for completeness. These will not follow a natural history or be obvious or fit into a typical clinical pattern. These ulcerations may be caused by the use of an object such as a knife, pencil and may be associated with compulsion. Social history may be of important relevance.

History, together with the clinical findings usually confirm the traumatic aetiology of the ulceration. Differential diagnoses are discussed in Chap. 49.

The cause of a suspected traumatic ulcer should be addressed (e.g. smoothing a sharp tooth), and the area should be reviewed after 2 weeks. If there is no resolution, further investigations such as blood tests (including full blood count, CRP, iron studies, Vitamin B12, folate, ANA and serum IgE) to exclude systemic causes of oral ulcerations and/or histopathological examination are recommended.

50.3 Management and Prognosis

Identification and cessation or elimination of the traumatic cause are crucial. Supportive measures such as warm salt water gargle every few hours, chlorhexidine 0.2% mouthwash three times daily and an anaesthetic mouthwash (such as cepacaine) for pain relief may be sufficient whilst awaiting re-epithelialisation.

Recommended Reading

Ge S, Liu L, Zhou Q, Lou B, Zhou Z, Lou J, et al. Prevalence of and related risk factors in oral mucosa diseases among residents in the Baoshan District of Shanghai, China. PeerJ. 2020;(8):e8644.

Scully C, Shotts R. ABC of oral health. Mouth ulcers and other causes of orofacial soreness and pain. BMJ. 2000;321(7254):162–5.

Aphthae and Recurrent Aphthous Stomatitis

51

Ajith Polonowita and Guangzhao Guan

51.1 Introduction

The term "aphthous" is derived from the Greek word meaning "ulcer". Aphthae or aphthous ulcers typically present as oval ulcers with a red peripheral halo and a yellow/brown fibrinous centre. Patients commonly report prodromal discomfort at the site prior to observation of an ulcer.

Aphthous ulcers occur in as part of several systemic conditions including haematinic deficiency, Behçet's disease, inflammatory bowel disease (especially related to coeliac disease), MAGIC (mouth and genital ulcers with inflamed cartilage syndrome), Sweet's syndrome, PFAPA (periodic fever, aphthous ulcers, pharyngitis and cervical adenitis syndrome), Reiter's syndrome, cyclic neutropenia. Aphthous-like ulceration may also occur with medications and hypersensitivity reactions.

The terms "recurrent aphthous stomatitis" (RAS) and "recurrent aphthous ulceration" (RAU) are often used interchangeably and refer to an idiopathic recurrence of aphthous ulceration on a spectrum of severity. "Stomatitis" refers to inflammation of the mouth.

51.2 Clinical Presentation and Investigations

RAS is common (reported rates 5–20%) with a multifactorial aetiology. The risk factors for RAS are shown in Table 51.1. Clinically, ulcers appear ovoid or round, with an erythematous halo and most patients report significant pain.

A. Polonowita (✉)
Faculty of Dentistry, The University of Otago, Dunedin, New Zealand

Christchurch Hospital, Canterbury, New Zealand

G. Guan
Faculty of Dentistry, The University of Otago, Dunedin, New Zealand

R. Balasubramaniam et al. (eds.), *Oral Medicine - A Clinical Guide*,
https://doi.org/10.1007/978-3-031-36797-7_51

Table 51.1 Risk factors for recurrent aphthous stomatitis

Younger age
Genetics (e.g. HLA types)
Haematinic deficiencies
Hypersensitivity
Smoking cessation
Immunological disturbances
Stress and anxiety
Female gender
Ethnic heritage (e.g. Irish)
Higher socioeconomic status

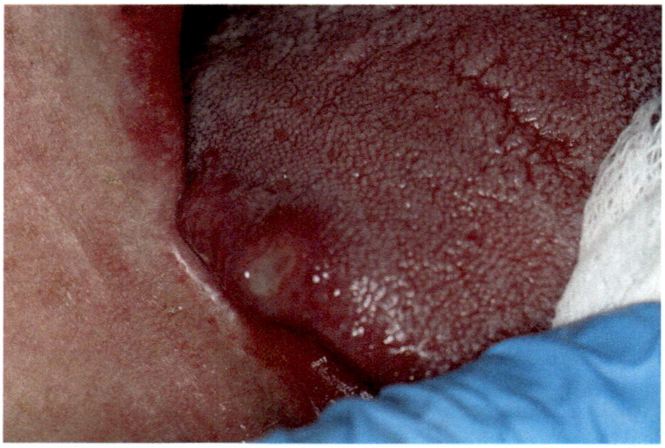

Fig. 51.1 Minor aphthous ulcers affecting the dorsal tongue

Clinically RAS may present as three variants:

1. Minor RAS: Most common variety (80%). These ulcers are usually less than 1 cm in diameter, last up to 2 weeks, and heal without scarring. They often occur as single or multiple ulcers, especially of the non-keratinised oral epithelium (buccal mucosa, lip mucosa, ventral tongue and floor of the mouth) (Fig. 51.1)
2. Major RAS: Also referred to as Sutton's disease, represent about 15% of all RAS. These ulcers are more than 1 cm in diameter and last for more than 4 weeks. They are painful and heal with scarring. They more commonly occur on the lips, soft palate and the region of the pillars of fauces (Fig. 51.2)
3. Herpetiform RAS: Crops of small (2 mm diameter) ulcers that may coalesce to form larger ulcers. They may initially resemble herpetic ulceration, but are not virally induced. Herpetiform RAS last 2–3 weeks and heals without scarring.

RAS may occur in an atypical and more extensive manner in immune-compromised individuals such as patients with human immunodeficiency virus.

The diagnosis of aphthous ulceration is based on the history of the ulcer (the recurrent nature and timeline), clinical features such as appearance (ovoid with

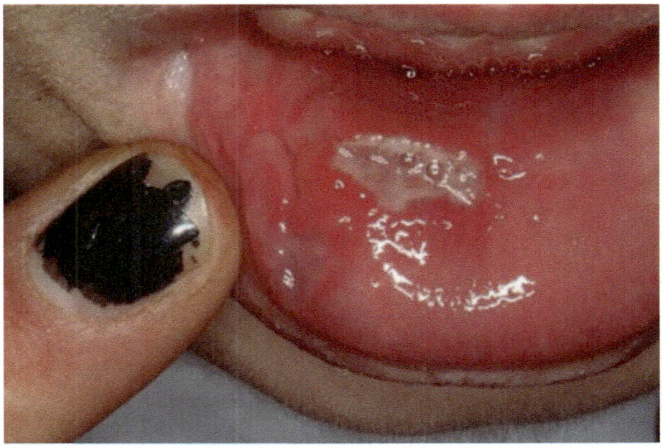

Fig. 51.2 Major aphthous ulcer affecting the lower labial mucosa

erythematous peripheral halo and yellowish fibrinous centre) and location. Investigations may include blood tests (FBE, CRP, serum iron, folate, Vitamin B12 and C, Serum IgE, ANA, HIV, coeliac antibodies, HLA typing) and faecal calprotectin as well as documentation of food triggers. Diagnosis of RAS is by exclusion of underlying systemic disease.

51.3 Management and Prognosis

Management is based on identification and addressing of underlying aetiologies as applicable. Patients may fare better on a toothpaste that does not contain sodium lauryl sulphate (SLS). Sources of trauma such as sharp restorations or removable appliances should be modified.

Management aims to reduce the duration of painful ulceration (symptomatic management) or to reduce the occurrence of ulceration (prevention). Symptomatic care includes topical medications such as anaesthetic mouthwashes (cepacaine without alcohol), antibiotic mouthwashes and topical steroids. Intra-lesional steroids may also be used for intractable ulcerations as well as systemic steroid pulses. Forms of cauterisation such as laser ablation have been also utilised but require surgical attendance. Preventive medications prescribed under specialist care include colchicine, dapsone, immunosuppressive agents (azathioprine, calcineurin inhibitors) and thalidomide.

Recommended Reading

Jurge S, Kuffer R, Scully C, Porter SR. Mucosal disease series. Number VI. Recurrent aphthous stomatitis. Oral Dis. 2006;12(1):1–21.

Scully C, Shotts R. ABC of oral health. Mouth ulcers and other causes of orofacial soreness and pain. BMJ. 2000;321(7254):162–5.

Infection-Related Oral Ulceration

52

Ajith Polonowita and Guangzhao Guan

52.1 Introduction

Alterations in local or systemic immunosurveillance may predispose patients to infections. There is a wide range of viral, bacterial and fungal infections that may manifest orally. The most common are discussed below. Thorough history combined with clinical presentation will steer the investigations, leading to an accurate diagnosis.

52.2 Viral

The most common cause of viral infection in the oral cavity are those associated with the human herpes viruses 1–8. HSV-1 & 2 primary infection is described below. Other virus families including mumps, rubella, human papilloma virus and paramyxovirus can present in the mouth but may not commonly cause oral ulceration. Aphthous-like ulceration and HSV-8 (Kaposi's sarcoma virus) may be seen in individuals who are HIV positive depending on disease control.

A. Polonowita (✉)
Faculty of Dentistry, The University of Otago, Dunedin, New Zealand

Christchurch Hospital, Canterbury, New Zealand

G. Guan
Faculty of Dentistry, The University of Otago, Dunedin, New Zealand

R. Balasubramaniam et al. (eds.), *Oral Medicine - A Clinical Guide*,
https://doi.org/10.1007/978-3-031-36797-7_52

52.2.1 Herpes Simplex Virus 1 and 2 (HSV 1 & 2)

The primary infection presents symptomatically as primary herpetic gingivostomatitis (HSV-1 typically in pre-school age; HSV-2 in teenagers and young adults). There is significant pain and oedema of the gingiva with lymphadenopathy. It may also involve the oral mucosa and oropharynx with widespread fluid-filled vesicles (pinpoint which coalesces to form large ulcers). Fever and malaise can be a feature.

Diagnosis is usually based on clinical history and appearance, however a swab of the fluid within a vesicle can be submitted for PCR testing of viral presence and there may be initial serum antibodies (IgM) to HSV on presentation (with conversion to IgG positivity in the following weeks).

The primary infection symptoms are self-limiting but pain and fever should be managed with antipyretics/analgesics and fluids to prevent dehydration. Antiviral (acyclovir) medication may be indicated in immunocompromised patients.

52.3 Bacterial

52.3.1 Tuberculosis

Tuberculosis is a disease characterized by granulomatous lesions caused by *Mycobacterium tuberculosis* (TB). Oral TB lesions commonly occur on the tongue, although may involve any oral mucosal site, presenting as ulcers, fissures, plaques or vesicles. They usually occur after active pulmonary lesions. The common presentation is a painless ulcer with erythematous margins and sloughing tissue, often accompanied by cervical lymphadenopathy. Chronic TB ulceration may be irregular, deep and painful resembling a malignant ulcer.

History and biopsy show the presence of caseating granulomas, with the presence of acid-fast bacilli on Ziehl–Neelsen staining.

Patients should be referred to a respiratory physician. Usual management involves combination drug therapy of four drugs (isoniazid, rifampicin, pyrazinamide and ethambutol) administered daily for 2 months, followed by 4 months of isoniazid rifampicin, pyrazinamide.

52.3.2 Syphilis

Syphilis is a bacterial infection caused by Treponema pallidum and usually spread by sexual contact.

Oral lesions occur in 30% of patients. Three classical stages are noted: primary (chancre), secondary (snail track ulcers) and tertiary (gumma).

The painless indurated ulcers of primary stage may involve the tongue, gingiva, soft palate and lips. They usually last 3–7 weeks and may mimic other causes of mucosal disease. Painless lymphadenopathy is often present.

Diagnosis is most often based on serology or direct visual inspection of a tissue sample using darkfield microscopy or use of immunohistochemistry. Microscopically there is dense lymphocytic infiltration beyond the lamina propria. Non-caseating granulomas may be present in later stages. Serology involves nontreponemal (venereal disease research laboratory and rapid plasma regain tests) and treponemal tests (treponemal pallidum particle agglutination or fluorescent treponemal antibody absorption test).

Curative management involves benzathine penicillin G (2.4 million-unit IM). Some effects established by the Treponema pallidum before diagnosis are irreversible, particularly in the tertiary stage.

52.4 Fungal

The most common fungal infection of the oral cavity associated with oral ulceration is erythematous form of candidosis most commonly caused by the yeast *C. albicans*. Other fungal infections that may present with oral ulceration include mucormycosis, histoplasmosis, blastomycosis, aspergillosis, cryptococcosis, geotrichosis and coccidioidomycosis (see Chaps. 15 and 16).

Recommended Reading

Ficarra G, Carlos R. Syphilis: the renaissance of an old disease with oral implications. Head Neck Pathol. 2009;3(3):195–206.

Leão JC, Gueiros LA, Porter SR. Oral manifestations of syphilis. Clinics (Sao Paulo). 2006;61(2):161–6.

Santosh ABR, Muddana K. Viral infections of oral cavity. J Family Med Prim Care. 2020;9(1):36–42.

Taylor M, Brizuela M, Raja A. Oral candidiasis. In: StatPearls [Internet]. Treasure Island (FL): StatPearls Publishing; 2023. Available at: https://www.ncbi.nlm.nih.gov/books/NBK545282/. Accessed 10 Oct 2023.

Drug-Induced Ulceration

53

Hadleigh Clark

53.1 Introduction

Oral mucosal ulceration may result as an adverse drug reaction (ADR) to systemic and topical medications. Diagnosis is based on both the chronology of ulcer appearance *after* drug administration or dose escalation and having excluded the condition predating this. In some conditions, e.g. oral lichenoid drug reactions (OLDRs), confirming a temporal relationship may be complicated by long asymptomatic periods. In others, e.g. Stevens-Johnson syndrome (SJS) and toxic epidermal necrolysis (TEN), precipitous and morbid presentation necessitates rigorous enquiry of a drug as causative, given the implicated role of drug exposures in these conditions.

The aetiopathogenesis of drug-induced oral ulceration is complex and includes type A and B reactions. Type A reactions are dose-dependent, based on drug pharmacology. Type B reactions are idiosyncratic and may be immunologic (T-cell or antibody mediated) or non-immunologic. Non-immunologic causes include direct mucosal toxicity, as seen in conventional and targeted anti-cancer therapy. Pharmacogenetics also play a role. In Han Chinese and South-East Asian populations, the HLA-B*1502 polymorphism is associated with an increased risk of SJS and TEN with anticonvulsants.

Many chemotherapeutics, e.g. 5-fluorouracil and cyclophosphamide are stomatotoxic. 20–40% of those undergoing conventional chemotherapy and 60–85% of those undergoing conditioning for haemopoietic stem cell transplant (HSCT) develop oral mucositis. Up to 70% of patients receiving mammalian target of rapamycin inhibitors (mTOR inhibitors), e.g. sirolimus, develop oral aphtheiform ulcers. More recently anti-programmed death receptor-1 (PD-1) immunotherapies, e.g. pembrolizumab, have been implicated with both OLDRs and immunobullous

H. Clark (✉)
Te Toka Tumai Auckland, Te Whatu Ora, Auckland, New Zealand
e-mail: hadleighc@adhb.govt.nz

R. Balasubramaniam et al. (eds.), *Oral Medicine - A Clinical Guide*,
https://doi.org/10.1007/978-3-031-36797-7_53

oral ulceration. The epidemiology of other drug-induced ulcers is often less clearly defined: examples do not always have clear temporal relationship or demonstrate resolution on drug withdrawal or reappearance on medication re-challenge.

53.2 Clinical Presentation and Investigations

Compared to skin, oral mucosa displays a limited array of oral ulcerative inflammatory patterns (aphtheiform, immunobullous, lichenoid, lupus-like) and typically arises within weeks or months after taking the medication (or days in the case of chemotherapy). Fixed-drug eruptions, i.e. those recurring at the same site with drug exposure, may cover a range of presentations. Examples of oral ulcers and the medications implicated are provided in Table 53.1.

Additional to a corroborative medication history, biopsy may support an OLDR by demonstration of deep and polymorphous chronic inflammatory cell infiltrate (eosinophilic/plasmacytic) or perivascular inflammation. HLA allele testing should be considered in at-risk populations prior to starting anticonvulsants, such as carbamazepine and lamotrigine.

Table 53.1 Drug-induced oral mucosal ulceration

Reaction type	Class	Examples
Chemical burns	Analgesic	Aspirin
	Dental	Chlorhexidine, hydrogen peroxide, sodium lauryl sulphate, trichloroacetic acid
	Recreational	Cocaine
Aphtheiform (aphthous-like)	Analgesic	Naproxen, piroxicam, indomethacin
	Antianginal	Nicorandil
	Antihypertensive	Captopril, labetalol, losartan
	Antimicrobial	Co-trimoxazole
	Bisphosphonate	Alendronate, etidronate, risedronate
	mTOR inhibitor	Sirolimus, everolimus
Fixed drug eruptions	Analgesic	Paracetamol, naproxen, etoricoxib, piroxicam
	Antihistamines	Levocetirizine
	Antimicrobial	Co-trimoxazole, fluconazole, clarithromycin, ornidazole, tetracyclines
	Other	Gabapentin, simvastatin
Lichenoid	Analgesic	Aspirin, diclofenac, ibuprofen, indomethacin, naproxen, piroxicam
	Antidiabetic	Glibenclamide, glipizide, metformin, tolbutamide
	Antihypertensive	Atenolol, captopril, enalapril, methyldopa, nifedipine, prazosin, propranolol
	Antimicrobial	Chloroquine, ethambutol, isoniazid, ketoconazole, rifampicin, sulfamethoxazole, tetracyclines
	Diuretic	Bendrofluazide, frusemide, hydrochlorothiazide
	DMARD	Gold, penicillamine, sulfasalazine
	Fusion protein	Imatinib, sunitinib, etanercept, abatacept,
	Monoclonal antibody	Adalimumab, infliximab, nivolumab, pembrolizumab
	Other	Carbamazepine, allopurinol, lithium, pravastatin, simvastatin, levothyroxine

Table 53.1 (continued)

Reaction type	Class	Examples
Lupoid (lupus-like)	Antihypertensive	Captopril, hydralazine
	Antimicrobial	Minocycline, isoniazid
	Others	Carbamazepine, chlorpromazine, penicillamine, sulfasalazine, anti-TNF inhibitors, procainamide
Immunobullous	Erythema multiforme/toxic epidermal necrolysis	Anticonvulsants—carbamazepine, lamotrigine, phenytoin, antibiotics—amoxicillin/clavulanic acid, co-trimoxazole, xanthine oxidase inhibitors—allopurinol, NSAIDs -diclofenac
	Pemphigoid-like	Chlorpromazine, furosemide, spironolactone, sulphonamides
	Pemphigus-like	Diclofenac, penicillamine, rifampicin

53.3 Management and Prognosis

For chemotherapy-related oral mucositis, cryotherapy, palifermin, and low-level laser therapy (photobiomodulation) may mitigate oral ulceration. For other reactions, resolution should occur after discontinuation of a suspected medication, although adjunctive steroids, and other supportive measures, e.g., antiseptics may be required for inflammatory and immunobullous presentations.

Recommended Reading

Teoh L, Moses G, McCullough MJ. A review and guide to drug-associated oral adverse effects-Oral mucosal and lichenoid reactions. Part 2. J Oral Pathol Med. 2019;48(7):637–46.

Yuan A, Woo SB. Adverse Drug Events in the Oral Cavity. Dermatol Clin. 2020;38(4):523–33.

Chronic Ulcerative Stomatitis

54

Hadleigh Clark

54.1 Introduction

Chronic ulcerative stomatitis (CUS) is a rare, poorly understood mucocutaneous disease. It shares significant clinical and histologic overlap with oral lichen planus (OLP). Just over 70 cases have been reported in the literature and debate remains whether it is a distinct diagnosis or a variant of OLP. The hallmark feature is the finding of stratified epithelium-specific antinuclear antibodies (SES-ANA) on direct and indirect immunofluorescence testing.

The target antigen of the autoantibodies in CUS is the ΔNp63α antigen of a 70kD protein member of the p53 tumour suppressor gene family, important in epithelial differentiation and stratification. It remains unclear if the autoantibodies are pathogenic or an epiphenomenon. Circulating antibodies to ΔNp63α can be seen in some patients with OLP on indirect immunofluorescence (IIF) studies. Similarly, a full epithelial thickness SES-ANA staining pattern can be seen on DIF in systemic lupus erythematosus, CREST syndrome and mixed connective tissue disease. Medications have been implicated as a possible aetiology.

Due to the limited cases, population incidence and prevalence data are lacking. CUS almost exclusively affects females (>90% of cases), typically in their sixth and seventh decades and >50% are Caucasian. No paediatric or adolescent cases have been reported.

H. Clark (✉)
Te Toka Tumai Auckland, Te Whatu Ora, Auckland, New Zealand
e-mail: hadleighc@adhb.govt.nz

R. Balasubramaniam et al. (eds.), *Oral Medicine - A Clinical Guide*,
https://doi.org/10.1007/978-3-031-36797-7_54

54.2 Clinical Presentation and Investigations

CUS characteristically presents with erythema, pain and non-scarring, often non-specific appearing ulcers or erosions, with the buccal and lingual mucosae being predominant sites. The oropharynx and soft palate are not known to be involved. Desquamative gingivitis occurs in up to half of cases. In a minority of cases, reticular striae can be seen around ulcer peripheries, as can white papules, patches and plaques, clinically resembling OLP. Although the name suggests a primary oral condition, approximately 20% of patients demonstrate cutaneous papulosquamous eruptions that resemble idiopathic cutaneous lichen planus.

Differential diagnoses include OLP and oral lichenoid lesions (e.g. contact, drug and lupus-like reactions), as well as immunobullous diseases (e.g. mucous membrane pemphigoid), particularly in gingival-limited disease. DIF on perilesional tissue demonstrates a speckled IgG deposition (and rarely IgA, C3) in the keratinocyte nuclei of the basal and parabasal third of the epithelial compartment. IIF and enzyme-linked immunosorbent assay (ELISA) techniques have been described for diagnosis but are not universally available.

54.3 Management and Prognosis

Spontaneous remission of CUS is exceedingly rare, and it demonstrates recalcitrance to topical and systemic steroids strategies employed in OLP. Rather, it demonstrates rapid response to hydroxychloroquine (or its analogue chloroquine), an antimalarial agent with immunomodulatory properties. Systemic cyclosporine has also been used successfully. Given the need to exclude cutaneous involvement and necessary treatment monitoring with hydroxychloroquine (which can lead to retinopathic blindness), referral to an oral medicine specialist, with close liaison with a dermatologist and ophthalmologist is recommended (Figs. 54.1 and 54.2).

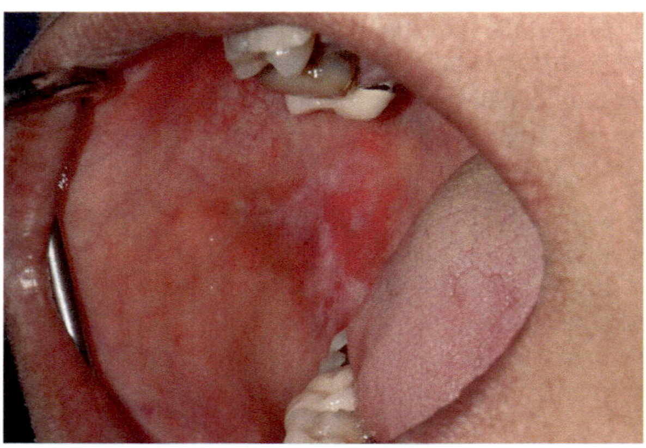

Fig. 54.1 Chronic ulcerative stomatitis involving the right buccal mucosa

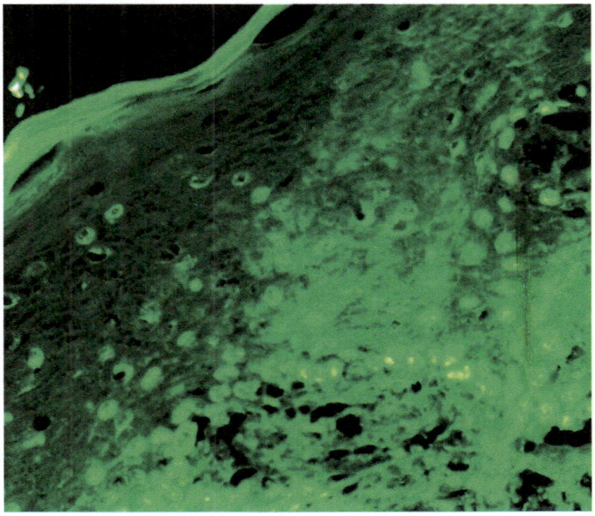

Fig. 54.2 Direct immunofluorescence from buccal mucosal biopsy of above demonstrating speckled IgG deposition of keratinocyte nuclei in the parabasal lower aspects of the epithelium. (Courtesy of Drs. Chris Hallot, Sinead McCarthy and Jessica Buchanan, Te Toka Tumai Auckland, Te Whatu Ora)

Recommended Reading

Qari H, Villasante C, Richert J, Rees T, Kessler H. The diagnostic challenges of separating chronic ulcerative stomatitis from oral lichen planus. Oral Surg Oral Med Oral Pathol Oral Radiol. 2015;120(5):622–7.

Reddy R, Fitzpatrick SG, Bhattacharyya I, Cohen DM, Islam MN. Seventeen New Cases of Chronic Ulcerative Stomatitis with Literature Review. Head Neck Pathol. 2019;13(3):386–96.

Solomon LW. Chronic ulcerative stomatitis. Oral Dis. 2008;14(5):383–9.

55 Traumatic Ulcerative Granuloma with Stromal Eosinophilia

Hadleigh Clark

55.1 Introduction

Traumatic ulcerative granuloma with stromal eosinophilia (TUGSE) is a rare, self-limiting, and benign form of oral ulceration. It has previously been known under a variety of monikers including eosinophilic ulcer and atypical histiocytic granuloma. As it may present with clinical features mimicking malignancy in adults, biopsy is typically warranted. In infants, it is known as Riga-Fede disease.

Trauma from sharp dental margins, ill-fitting dentures and incisional biopsies has been aetiologically implicated. It has been postulated that an influx of microbes, toxins and foreign debris into traumatised tissue leads to a mononuclear and eosinophilic cell infiltrate. Delayed healing is thought to result from a lack of secretion of TGF-α and TGF-β by recruited eosinophils. Sporadic reports of TUGSE-like lesions demonstrating CD30+ T lymphocyte monoclonality suggest that some may represent low-grade lymphoproliferative disorders or T cell lymphomas. However, this remains controversial and many of these also appear to follow a self-limited, benign clinical course.

A bimodal age distribution is described: Riga-Fede occurs in infancy associated with the presence of natal (present at birth) and neonatal (erupt within first 30 days) teeth and a second peak of TUGSE occurs in adults in the fifth to seventh decades. TUGSE may, however, occur at any age and affects genders roughly equally, with a slight male predilection. Both are rare.

H. Clark (✉)
Te Toka Tumai Auckland, Te Whatu Ora, Auckland, New Zealand
e-mail: hadleighc@adhb.govt.nz

R. Balasubramaniam et al. (eds.), *Oral Medicine - A Clinical Guide*,
https://doi.org/10.1007/978-3-031-36797-7_55

55.2 Clinical Presentation and Investigations

Topographic association of a ventral tongue ulcer, in the presence of natal and neonatal teeth suggests Riga-Fede, which may interfere with breast-feeding. TUGSE typically presents as an isolated ulcer of the anteroventral or dorsal tongue (two-thirds of cases). Other affected sites are the buccal, labial, palatal, and vestibular mucosa, and the floor of mouth. Ulcer morphology can appear sinister, with indurated, elevated borders and a yellow, fibrinous base. Ulcers may arise rapidly, be painful and persist for many weeks or months. Cervical lymphadenopathy is occasionally described.

In adults, differential diagnoses include malignancy (e.g. squamous cell carcinoma, lymphoma), major aphthous ulceration, deep fungal or necrotising bacterial infection (e.g. histoplasmosis, primary syphilis, tuberculosis) and in selected patients, lymphoproliferative disorders (LPD) and EBV-associated mucocutaneous ulcer. Where lymphoma is considered, a lymphoma work-up should be considered (e.g. fresh tissue submitted for cell marker studies; bloods FBC, LFTs, LDH, urate, calcium, peripheral flow cytometry, CRP). Microbiologic investigations (e.g. treponemal serology, quantiFERON gold) may be considered based on other suggestive history and exam findings.

55.3 Management and Prognosis

For Riga-Fede, loose or supernumerary teeth should be extracted as they may interfere with breast-feeding and pose an aspiration risk. Sound teeth can be smoothed with addition of restorative material or minor odontoplasty and the ulcer gently cleaned with sterile saline-soaked gauze. For TUGSE, the mainstay of management is addressing injurious trauma and biopsy. Where restoration or odontoplasty cannot be undertaken, a soft occlusal guard can be fashioned. The act of biopsy, even incisional, appears to stimulate healing, with most ulcers spontaneously resolving in about 2–3 weeks. For those persisting following confirmational biopsy, both topical and intralesional corticosteroids have been used successfully. Recurrence is rare. With the finding of lymphocyte monoclonality, lifelong mucosal screening has been advocated.

Recommended Reading

Boffano P, Gallesio C, Campisi P, Roccia F. Traumatic ulcerative granuloma with stromal eosinophilia of the retromolar region. J Craniofac Surg. 2009;20(6):2150–2.

Shen WR, Chang JY, Wu YC, Cheng SJ, Chen HM, Wang YP. Oral traumatic ulcerative granuloma with stromal eosinophilia: A clinicopathological study of 34 cases. J Formos Med Assoc. 2015;114(9):881–5.

Necrotising Sialometaplasia

56

Hadleigh Clark

56.1 Introduction

Necrotising sialometaplasia (NSM) is a rare reactive, self-limiting condition primarily affecting palatal mucosa, where it arises from the minor salivary glands. Recognition is important as it mimics malignant squamous and salivary epithelial neoplasms both clinically and histologically. A subacute non-ulcerating variant is also recognised.

The exact cause of NSM is unknown. The initiating event is believed to be vascular infarction resulting in salivary gland ductal ischaemia. Precipitants include local trauma (physical, thermal, barometric, irritant), surgical manipulation and local anaesthesia. However, often presentations are spontaneous with no inciting event identified. Risk factors contributing to vascular ischaemia include cigarette smoking, cocaine use, alcohol, vomiting (e.g. bulimia nervosa), diabetes mellitus, vascular disease, allergy, infection, ionizing radiation, pressure from dentures or intubation and the presence of synchronous, adjacent minor salivary gland neoplasms.

NSM is rare, representing <1% of oral biopsies; this may be underestimated due to underappreciation of the condition. NSM occurs across all ages from paediatric patients to the elderly, although mean presentation is in the fourth to fifth decades, with a male predominance.

H. Clark (✉)
Te Toka Tumai Auckland, Te Whatu Ora, Auckland, New Zealand
e-mail: hadleighc@adhb.govt.nz

R. Balasubramaniam et al. (eds.), *Oral Medicine - A Clinical Guide*,
https://doi.org/10.1007/978-3-031-36797-7_56

56.2 Clinical Presentation and Investigations

Approximately 70% of cases present as unilateral crateriform ulceration of the palate, which is often painless, but may be deep, extending to underlying bone. Bilateral presentation has been described and may be associated with bulimia nervosa and atherosclerotic disease. NSM can arise in other mucous glandular tissues of the oral cavity and upper aerodigestive tract (e.g. trachea, larynx, nasopharynx). Up to 10% of cases occur in the major salivary glands. Extra-palatal sites include the lips, tongue, buccal mucosa, retromolar area and tonsillar fossa, where involvement occurs more frequently postoperatively, after trauma or radiation therapy. Co-association with salivary neoplasms occurs infrequently but poses diagnostic and therapeutic challenges. The subacute, non-ulcerated form, in contrast, typically arises as an abrupt, painful palatal tissue swelling (Fig. 56.1).

The main differentials for the ulcerative variant are oral or penetrating antral squamous or nasopharyngeal carcinomas, malignant salivary neoplasms (e.g. mucoepidermoid carcinoma) and non-Hodgkin's lymphoma. Infectious diseases such as deep fungal infection and granulomatous palatal destructive lesions (e.g. syphilis, tuberculosis, polyangiitis with granulomatous inflammation) may suggest a need for additional microbiologic (e.g. culture, PCR), serologic (e.g. treponemal serology) and immunologic investigation (e.g. ANCA). The subacute form may mimic benign salivary neoplasia (e.g. pleomorphic adenoma). Biopsy is usually indicated, although for non-ulcerated variants, fine-needle aspiration may be considered to exclude salivary tumours. Imaging (OPG, CT with contrast or MRI) may be considered to exclude bony involvement when clinically indicated.

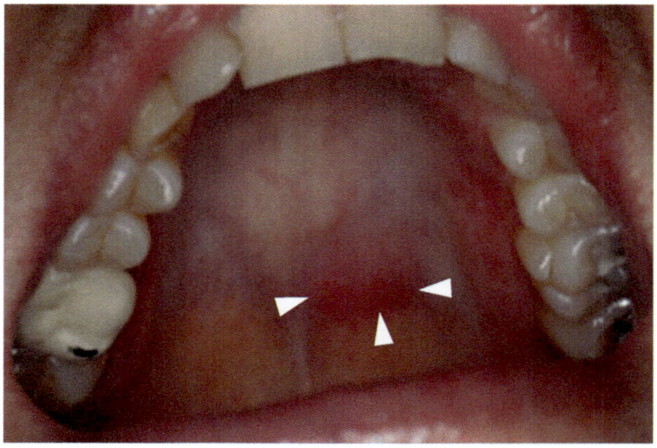

Fig. 56.1 Biopsy confirmed subacute non-ulcerative necrotising sialometaplasia of the left palate. (Courtesy of Dr. Jacinta Vu, Centre for Oral Medicine and Facial Pain)

56.3 Management and Prognosis

Post biopsy, healing usually occurs slowly over 6–10 weeks by secondary intention. Supportive therapy during tissue healing may include simple analgesia, topical antiseptics (e.g. chlorhexidine 0.12%, hydrogen peroxide mouthwash) and soft palatal coverage stents.

Recommended Reading

Neville BW, Damm DD, Allen CM, Chi AC. Oral and Maxillofacial Pathology. 4th ed. Missouri: WB Saunders, Elsevier; 2016. p. 439–440.

Regezi JA, Sciubba JJ, Jordan RCK. Oral Pathology: Clinical Pathologic Correlations. 7th ed. Missouri: Elsevier; 2017. p. 189.

Ulceration Related to Systemic Disorders 57

Hadleigh Clark

57.1 Introduction

Disorders of almost every major organ system can give rise to oral ulcers. Some may present non-specifically, whereas others may indicate the start or progression of a disease and be the first, single, most florid, or most morbid clinical sign. In some conditions, e.g. *Behçet's disease*, oral ulcers are pathognomonic of the disease, being a core diagnostic criterion.

The aetiology of oral ulceration secondary to systemic disorders is broadly related to disruption of mucosal integrity and/or host defence through genetic, haematologic, immunologic, neoplastic or infectious mechanisms. Rare heritable disorders inciting oral ulcers include genodermatoses (e.g. *Hailey-Hailey disease, epidermolysis bullosa*) and those arising from defects in the cellular or humoral arms of immunity (e.g. *cyclic neutropaenia, X-linked hyper IgM syndrome*). Both therapy-related (e.g. *agranulocytosis, neutropaenia*) and pathologic (e.g. *cytopaenias and hypoglobulinaemias secondary to haematologic malignancy, hypocomplementaemia*) causes may equally disrupt host immunity. Many dermatologic (e.g. *pemphigus, pemphigoid*), gastroenterologic (e.g. *dermatitis herpetiformis*) and rheumatologic diseases cause oral ulceration secondary to autoantibody and/or immune complex deposition (e.g. *systemic lupus erythematosus*), while autoinflammatory T-cell mediated responses may underpin both oral dermatoses (e.g. *oral lichen planus*) and ulcers seen in rheumatologic disease (e.g. *reactive arthritis*). Keratinocyte turnover is dependent on haematinic status, and deficiencies in folate, vitamin B12 and iron may predispose to ulcers. Micronutrient deficiencies can arise through impaired absorption due to the disease itself (e.g. *coeliac disease, Crohn's disease*), inadequate intake (e.g. *veganism*) or autoimmune cause (e.g. *pernicious*

H. Clark (✉)
Te Toka Tumai Auckland, Te Whatu Ora, Auckland, New Zealand
e-mail: hadleighc@adhb.govt.nz

R. Balasubramaniam et al. (eds.), *Oral Medicine - A Clinical Guide*,
https://doi.org/10.1007/978-3-031-36797-7_57

anaemia). Inflammatory bowel disorders (e.g. *Crohn's disease, ulcerative colitis*) are also associated with granulomatous inflammatory ulcers.

Aphthous ulceration may affect 1 in 5 individuals during their lifetime and can be found in 97–100% with Behçet's disease, and up to 40% of those with coeliac disease. 10% of those with Crohn's disease develop oral ulcers. Mucosal ulceration is omnipresent in acquired and paraneoplastic immunobullous disorders and is frequently encountered in inflammatory dermatoses.

57.2 Clinical Presentation and Investigations

A high index of suspicion for a systemic cause should be held for bilaterally distributed oral ulcers and characteristic ulcer morphology patterns, such as reticulate striations (e.g. *oral lichen planus*), linear ulceration with associated mucosal tags (e.g. *Crohn's disease*) or dermatomal distribution (e.g. *herpes zoster*). In paediatric patients, viral enanthems (e.g. *herpetic gingivostomatitis, varicella, measles, herpangina, hand-foot-mouth*) frequently present with vesiculoulcerative lesions associated with malaise, lymphadenitis and fever. Bacterial (e.g. *tuberculosis*), fungal (e.g. *histoplasmosis*) and viral (e.g. cytomegalovirus) ulcers are more frequent in immunocompromised individuals with tumeriform or locally destructive ulcers.

A review of systems and targeted clinical system examination will determine the most appropriate investigations. As a minimum, dermatologic, rheumatologic, ophthalmic, gastrointestinal, haematologic, ENT, immune and infectious disease enquiry will aid this process. The full blood count is probably the most diagnostically yielding initial test as it may suggest (a) haematinic deficiency relating to erythrocyte assessment, (b) infection or lymphoreticular malignancy relating to white cell assessment, and (c) a number of autoimmune disorders are associated with eosinophilia.

57.3 Management and Prognosis

Following diagnosis, patient management benefits from integrated and specialised care with the appropriate medical team. The latter is often guided by the primary organ system suspected as causative in oral ulcer development.

Recommended Reading

Akintoye SO, Greenberg MS. Recurrent aphthous stomatitis. Dent Clin North Am. 2014;58(2):281–97.

Mays JW, Sarmadi M, Moutsopoulos NM. Oral manifestations of systemic autoimmune and inflammatory diseases: diagnosis and clinical management. J Evid Based Dent Pract. 2012;12(3 Suppl):265–82.

Porter SR, Mercadante V, Fedele S. Oral manifestations of systemic disease. Br Dent J. 2017;223(9):683–91.

Differential Diagnosis of Pigmented Lesions

58

Anastasia Georgiou

58.1 Introduction

"Pigment" refers to any substance that gives colour to tissue. Oral pigmentation refers to deposition of pigment, either of endogenous or exogeneous origin, within the oral mucosa. In the oral cavity, pigmentation generally refers to grey, blue, brown and/or black colour changes. The differential diagnosis of pigmented lesions requires a thorough medical, family and social history. This should be followed by examination to assess clinical parameters of the pigmentation and to determine if further investigations are required. Oral pigmented lesions range from innate to abnormal discolouration and benign to malignant. They may also signal underlying inherited and/or systemic disease.

The most common endogenous sources of pigmentation are melanin, haemoglobin and haemosiderin. Melanin is produced by melanocytes which reside in the basal epithelial layer of oral mucosa. The number of melanocytes is similar between light- and dark-skinned individuals. The difference in physiologic pigmentation relates to increased melanocytic activity. Oral melanotic lesions can be due to an increased production of melanin from various stimuli or may represent benign or malignant melanocytic growth and proliferation.

Exogenous sources of pigmentation are often a result of traumatic or iatrogenic deposition of foreign material directly into the mucosa. Amalgam is the most common source of exogeneous pigmentation with traumatic implantation at time of tooth extraction, restoration placement, polishing or removal. Other exogeneous materials include graphite and ornamental ink. Chronic exposure to heavy metals can result in discolouration of the skin and oral mucosae.

Oral pigmentation is extremely common with global prevalence estimated to be approximately 20%. Physiologic pigmentation is the most common diffuse/multifocal pigmentation followed by smoker's melanosis. The most common focal

A. Georgiou (✉)
Sydney Skin, Newtown, NSW, Australia

Macquarie Oral and Maxillofacial Specialists, Sydney, NSW, Australia

R. Balasubramaniam et al. (eds.), *Oral Medicine - A Clinical Guide*,
https://doi.org/10.1007/978-3-031-36797-7_58

pigmented lesions are amalgam tattoos and focal melanotic macules. Benign pigmentation is far more common than oral malignant melanoma, which comprise of less than 1% of all malignant melanoma.

58.2 Clinical Presentation and Investigations

Oral pigmented lesions can present as focal (Fig. 58.1), multifocal (Fig. 58.2) or diffuse (Fig. 58.3). The colour of pigmentation can give an indication of whether origin is endogenous or exogeneous (Table 58.1). The intensity of the colour depends on the amount pigment and spatial location within the oral mucosa (superficial compared with deep). The ABCD checklist (asymmetry, border irregularities, colour variegation and diameter >6 mm) used for cutaneous melanoma may be of value in oral melanoma with location of the palate increasing rate of suspicion. In general, pigmented lesions are painless.

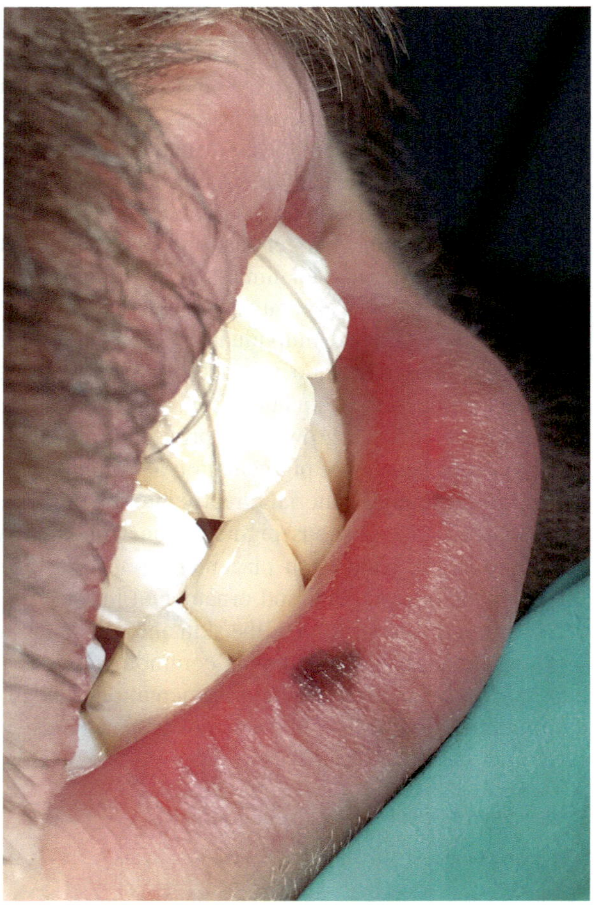

Fig. 58.1 Presentation of focal pigmentation with an oral melanotic macule

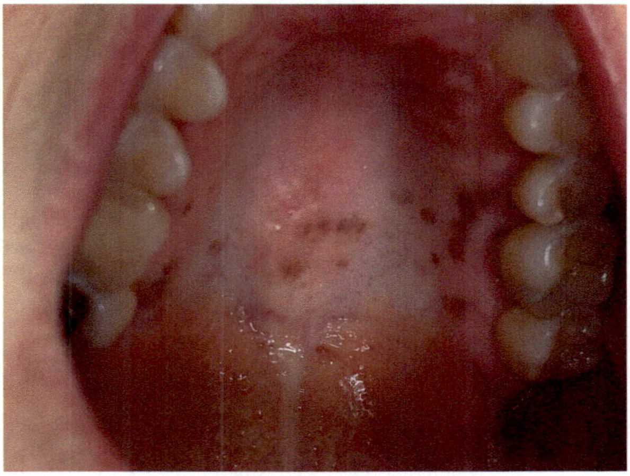

Fig. 58.2 Multifocal pigmentation of the hard palate

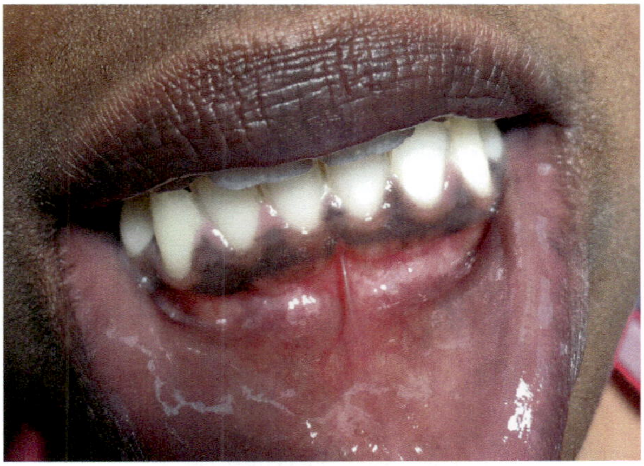

Fig. 58.3 Diffuse pigmentation as seen with physiologic pigmentation

A thorough and detailed family and medical history is essential in the diagnostic work up of pigmented lesions and for identifying the source of pigmentation and underlying aetiology. Laboratory tests may be required when oral pigmentation is thought to be related to an underlying systemic condition. Diascopy is valuable to differentiate vascular lesions, and radiographs can assist in diagnosis of amalgam tattoos. Dermoscopy, epiluminescence microscopy, is routinely used when investigating cutaneous pigmented lesions but may also be used for oral pigmented lesions when access permits. Biopsy with histological examination remains the gold standard for diagnosis of pigmented lesions.

Table 58.1 Sources of oral pigmentation

Origin	Source	Colour	Focal	Diffuse/Multifocal
Endogenous	Melanin	Brown (predominantly) Blue/Black	• Ephelis • Melanotic Macule • Oral melano-acanthoma • Melanocytic naevus • Malignant melanoma	• Physiologic pigmentation • Drug-induced melanosis • Smoker's melanosis • Post inflammatory hyperpigmentation • Addison's disease/ adrenal insufficiency • Cushing's syndrome • Peutz-Jeghers syndrome • Laugier-Hunziker syndrome
	Haemoglobin	Red/Blue/Purple	• Varix/venous ectasia • Vascular malformations • Kaposi's sarcoma	• Hereditary haemorrhagic telangiectasia • CREST syndrome
	Haemosiderin	Blue-grey to brown	• Haematoma • Ecchymosis • Purpura • Petechiae	• Haemochromatosis • Ecchymosis • Purpura • Petechiae
	Bilirubin	Yellow		• Jaundice
Exogeneous	Amalgam	Grey/Black	• Amalgam tattoo	
	Graphite	Grey/Black	• Graphite tattoo	
	Ink	Black Variable	• Ornamental tattoos	• Ornamental tattoos
	Heavy metals	Grey/Black		• Heavy metal pigmentation • Chrysiasis
	Chromogenic bacteria	Green/Brown/ Black		• Hairy tongue
	Medications	Blue/Grey		• Drug-induced pigmentation

58.3 Management and Prognosis

Treatment of pigmented lesions is dependent on establishing a diagnosis and identifying the underlying cause. Benign focal pigmented lesions do not need to be treated, but sometimes are removed for aesthetic reasons. Management of any associated underlying systemic cause of multifocal pigmented lesions may require multidisciplinary care. Oral malignant melanoma requires radical surgical excision. The prognosis of oral malignant melanoma is poor, with a 5-year survival rate of 10–25%.

Recommended Reading

Alawi F. Pigmented lesions of the oral cavity: an update. Dent Clin North Am. 2013;57(4):699–710.

Fernandes D, Ferrisse TM, Navarro CM, Massucato EM, Onofre MA, Bufalino A. Pigmented lesions on the mucosa: a wide range of diagnoses. Oral Surg Oral Med Oral Pathol Oral Radiol. 2015;119(4):374–8.

Stooper ET, Alawi A. Pigmented Lesions of the Oral Mucosa. In: Farah CS, Balasubramaniam R, McCullough MJ, editors. Contemporary Oral Medicine. Springer; 2019. p. 1175–1205.

Common Causes of Oral Pigmentation

59

Arabelle Clayden and Anastasia Georgiou

59.1 Introduction

Pigmentations are common in the oral cavity and range from physiologic pigmentation to oral manifestations of systemic disease including malignancies. They can broadly be attributed to either the deposition of endogenous or exogenous pigments. The most common endogenous sources are haemoglobin, haemosiderin or melanin. Exogenous pigmentation is usually associated with a traumatic event resulting in the deposition of the pigment directly into the tissues. Oral pigmentations can range from solitary macules to diffuse mucosal involvement.

Oral pigmentation can be due to the augmentation of melanin production (e.g. Addison's disease), an increased number of melanocytes (e.g. melanoma) or the deposition of exogenous material into the tissues (e.g. amalgam tattoo). Some chromogenic bacteria can also produce discolouration of the dorsum of the tongue (black hairy tongue).

59.2 Clinical Presentation and Investigations

Physiologic pigmentation is the most common cause of multifocal or diffuse oral pigmentation. It is common in people of African descent (racial). It is often first seen in childhood in individuals with Fitzpatrick Skin Types 3–6. It can be

A. Clayden
Dent8 Dental, Taringa, QLD, Australia

A. Georgiou (✉)
Sydney Skin, Newtown, NSW, Australia

Macquarie Oral and Maxillofacial Specialists, Sydney, NSW, Australia
e-mail: admin@sydneyskin.com

R. Balasubramaniam et al. (eds.), *Oral Medicine - A Clinical Guide*,
https://doi.org/10.1007/978-3-031-36797-7_59

widespread, mostly affecting the gingivae with variation between individuals with similar skin types (Fig. 59.1).

Amalgam tattoo is a common cause of localised oral pigmentation with 1–3% of the population affected. The asymptomatic grey/black lesions are typically seen close to amalgam restorations (Fig. 59.2).

Many drugs are known to cause oral pigmentation, particularly antimalarials, oral contraceptives and some cytotoxic drugs. They will cause localised or diffuse pigment changes with no nodularity in the tissues.

Common sites involved are gingivae, palatal mucosa and buccal mucosa.

Black hairy tongue is a harmless asymptomatic condition that presents with elongation of the filiform papillae on the tongue dorsum with colonisation of the papillae with chromogenic bacteria giving superficial black discolouration of the

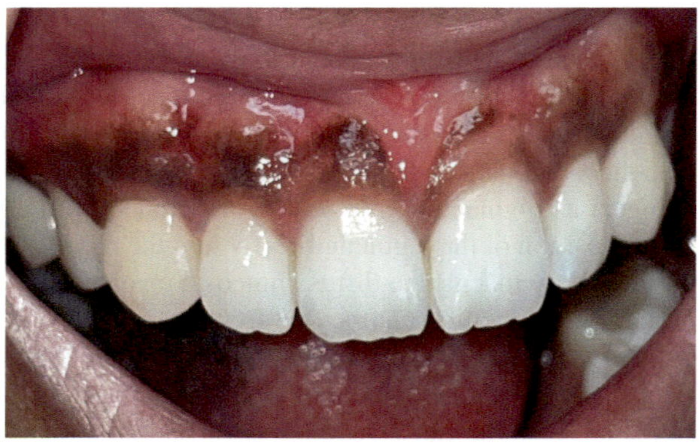

Fig. 59.1 Physiological pigmentation appearing as a dark brown generalised colouration of the buccal gingival tissues

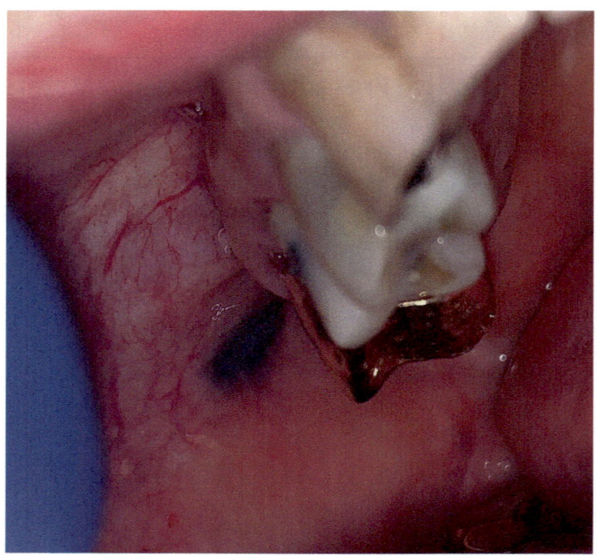

Fig. 59.2 Amalgam tattoo presenting as a localised dark blue-black lesion on the right posterior buccal mucosa in proximity to heavily restored teeth

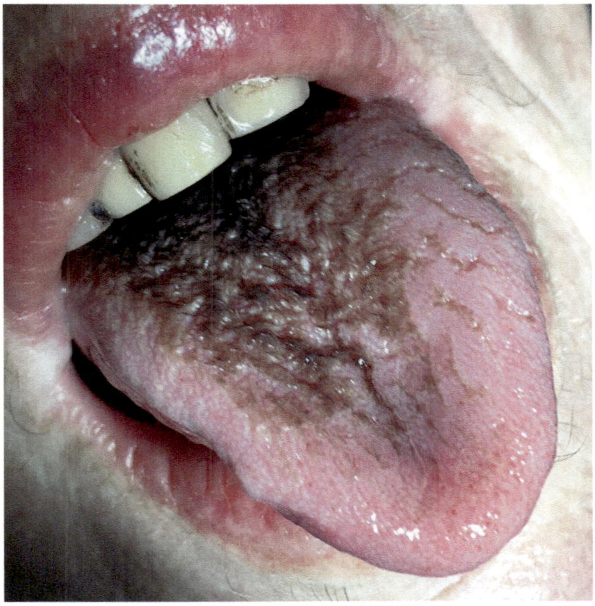

Fig. 59.3 Black hairy tongue presenting as dark pigmentation and elongation of the filiform papillae on the dorsal surface of the tongue

posterior and mid-third of the tongue (Fig. 59.3). It affects older individuals and males more than females.

Any oral pigmentation with an uncertain aetiology needs a biopsy. This will determine if the pigmentation is due to increased melanin production, an increase in melanocytes or exogenous pigment being deposited into the tissues. A medication history and correlation with other systemic diseases known to cause pigmentation is valuable.

59.3 Management and Prognosis

Treatment depends on whether the increased pigmentation is pathological. Physiologic pigmentation does not require treatment. Black hairy tongue improves with mechanical cleaning of the tongue but does have a propensity to recur. Assessment and management by a physician for systemic disorders causing oral pigmentation are essential.

Recommended Reading

Aguirre A, Alawi F, Tapia JL. Pigmented Lesions of the Oral Mucosa. In: Glick M, Greenberg MS, Lockhart PB, Challacombe SJ, editors. Burket's oral medicine. 13th edition. Hoboken, NJ: Wiley Blackwell; 2021. p 139–69.

Diz Dios P, Scully C, Paes de Almeida O, Bagn JV, Mosqueda Taylor A. Oral Medicine and Pathology at a Glance. Hoboken: John Wiley & Sons, Incorporated; 2016. p 26–9.

Sreeja C, Ramakrishnan K, Vijayalakshmi D, Devi M, Aesha I, Vijayabanu B. Oral pigmentation: A review. J Pharm Bioallied Sci. 2015;7(Suppl 2):S403–8.

Melanotic Macules, Naevi, and Melanoma

60

Arabelle Clayden and Anastasia Georgiou

60.1 Introduction

The range of benign melanocytic lesions of the oral mucosa is varied and includes mucosal melanotic macules and melanocytic naevi. These are usually solitary lesions but if multiple may be a sign of an associated syndrome such as Peutz-Jeghers or Laugier–Hunziker syndromes. Oral malignant melanomas are rare malignant tumours with a poor prognosis.

Oral melanotic macules are the most common benign pigmented oral lesions. They are caused by a focal increase in melanin pigmentation in the basal cell layer with no increase in melanocyte numbers or altered morphology.

Naevi are less common on the oral mucosa than their counterparts on the skin. They are due to melanocyte growth and proliferation, and it is not established if they develop from neural crest cells or immature forms of melanocytes. The location of the naevus cells determines whether they are considered junctional, intradermal/intramucosal or compound. The naevus cells initially are located at the basal layer as a small cluster of cells but over time they proliferate as small nests of cells down into the connective tissue. As they mature, they lose their association with the epithelial layer. It is thought that melanocytic naevi of the oral mucosa do not have an increased risk for oral malignant melanoma.

Multiple oral pigmented lesions can be seen in hereditary genetic disorders where there is an alteration in melanogenesis. The most common of these is Laugier–Hunziker syndrome, an innocuous disorder of unknown aetiology causing multiple areas of pigmentation on the buccal mucosa, lips, nails and skin. There is

A. Clayden
Dent8 Dental, Taringa, QLD, Australia

A. Georgiou (✉)
Sydney Skin, Newtown, NSW, Australia

Macquarie Oral and Maxillofacial Specialists, Sydney, NSW, Australia
e-mail: admin@sydneyskin.com

R. Balasubramaniam et al. (eds.), *Oral Medicine - A Clinical Guide*,
https://doi.org/10.1007/978-3-031-36797-7_60

no association with systemic diseases. The second most common, Peutz-Jeghers syndrome, is an autosomal dominant condition due to mutation of the SKT11 gene. As well as the pigmentary changes to the oral mucosa and skin, these patients have multiple gastrointestinal polyps with an increased risk of malignant transformation.

Malignant oral mucosal melanoma is a malignant neoplasm of the melanocyte and is a distinct neoplasm from cutaneous melanoma. They are unrelated to UV exposure. They represent about 1% of all melanomas. The aetiology of these neoplasms is unknown, and BRAF mutations are rare. The histological micro-staging system used in cutaneous melanoma cannot be applied to oral mucosal melanoma.

60.2 Clinical Presentation and Investigations

Melanotic macules are more common in females and often occur on the lower lip or gingivae (Fig. 60.1). They measure less than 1 cm and are well-circumscribed brown macules. They are asymptomatic but maybe of cosmetic concern to the patient if they are visible.

Naevi are rare, asymptomatic, small well-circumscribed macules or slightly raised papules of a grey-black or brown colour, and the most common site is on the hard palate (Figs. 60.2 and 60.3).

Oral malignant melanomas are rare but aggressive tumours with no distinctive clinical features. The palate or maxillary gingivae are the most common sites, and most patients are older than 50 years.

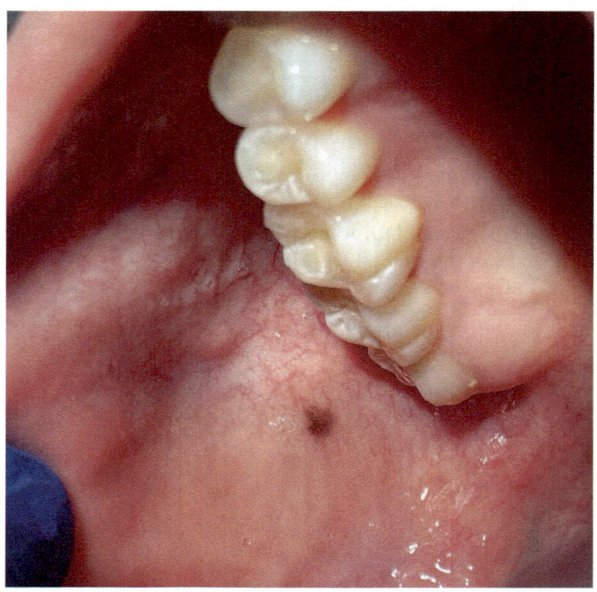

Fig. 60.1 Oral melanotic macule of the buccal mucosa

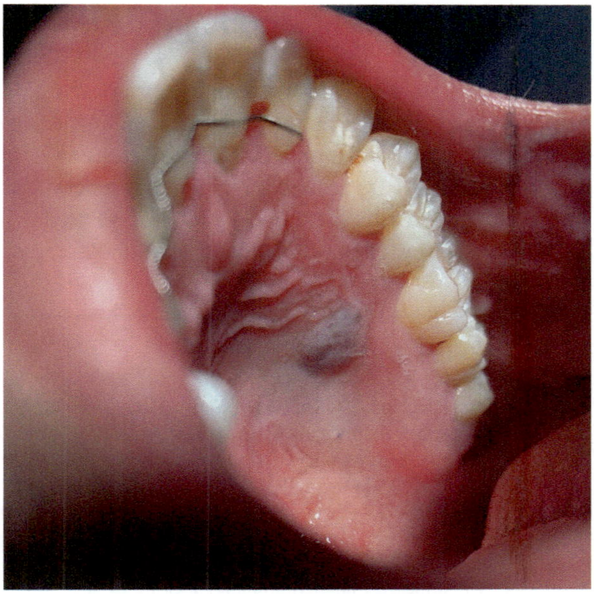

Fig. 60.2 Raised pigmented lesion of the palate shown to be an intramucosal melanocytic naevus on histopathology

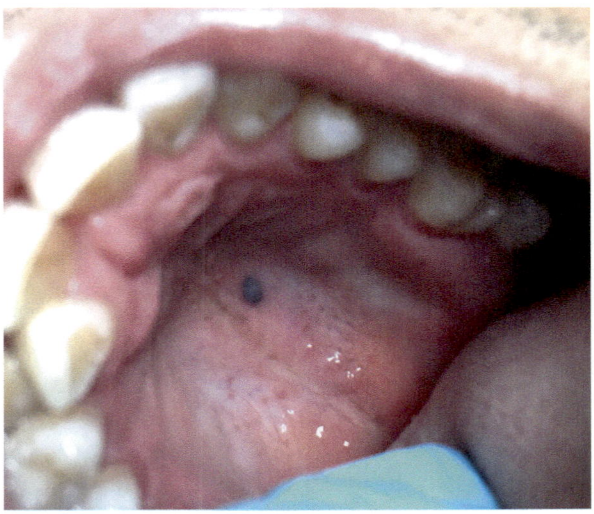

Fig. 60.3 Uncommon blue naevus involving the hard palate, diagnosis confirmed by biopsy

There have been reports of higher incidence in Japanese and African/black populations. They may present as a macular or nodular lesions with variable amounts of pigmentation and are often asymptomatic.

Any oral pigmentation that cannot be confidently diagnosed as benign on clinical grounds requires a mandatory biopsy to exclude malignant melanoma. Clinical features alone are not diagnostic for oral pigmented lesions. If the pigmented lesions are associated with a syndrome, then these patients require appropriate multidisciplinary care.

60.3 Management and Prognosis

The benign melanocytic lesions can just be observed over time but any lesions demonstrating progressive change or a degree of suspicion should be biopsied.

Oral malignant melanomas are aggressive tumours with a poor prognosis. Five-year survival rates are around 15%. There is no consensus or set guidelines for their management. Treatment requires radical resection with wide margins, combined with radiotherapy, chemotherapy and immunotherapy.

Recommended Reading

Ferreira LDS, Calderipe CB, Maass JB, Carrard VC, Martins MD, Abreu LG, et al. Oral pigmented lesions in syndromic individuals: A systematic review. Oral Dis. 2022;28(3):531–40.

Meleti M, Leemans CR, Mooi WJ, Vescovi P, van der Waal I. Oral malignant melanoma: a review of the literature. Oral Oncol. 2007;43(2):116–21.

Meleti M, Mooi WJ, Casparie MK, van der Waal I. Melanocytic nevi of the oral mucosa - no evidence of increased risk for oral malignant melanoma: an analysis of 119 cases. Oral Oncol. 2007;43(10):976–81.

Müller S. Melanin-associated pigmented lesions of the oral mucosa: presentation, differential diagnosis, and treatment. Dermatol Ther. 2010;23(3):220–9.

Oral Mucosal Pigmentation Secondary to Systemic Disorders and Medications

61

Hadleigh Clark

61.1 Introduction

Diffuse and/or multifocal oral or perioral pigmentation is commonly encountered secondary to medications and systemic diseases. When oral pigmentation is observed alongside other cutaneous or mucosal pigmentation, confirming a drug history and a lack of other systemic signs or symptoms is critical. Genetic disorders may present with oral hyperpigmentation in childhood and adolescence; in adults such hyperpigmentation may suggest an underlying acquired endocrinopathy.

Drug-related pigmentation is thought to arise through four major routes: stimulation of melanogenesis (e.g. hydroxychloroquine), medication deposition (e.g. chlorpromazine), de novo pigment synthesis (e.g. amiodarone–lipofuscin) or iron deposition via blood vessel damage and haemolysis (e.g. minocycline). While *hypercarotenaemia* and *haemochromatosis* may also lead to generalised oral pigmentation, most systemic causes relate to disordered melanogenesis. Melanocytes are primarily under the influence of melanocyte stimulating hormone (MSH). A major source of MSH is the anterior pituitary, where it is cleaved from the prohormone pro-opiomelanocortin (POMC), along with adrenocorticotrophin (ACTH). In primary adrenal insufficiency (*Addison's disease*) and in the setting of secretory pituitary adenomas (*Cushing's disease*), hypothalamus-derived corticotrophin-releasing hormone (CRH) acts on POMC to increase ACTH *and* MSH levels, leading to hyperpigmentation. Both thyroxine and oestrogen (*in melasma*) also influence melanogenesis. Germline mutations account for both pigmentation in familial lentiginosis syndromes such as *Peutz-Jegher's syndrome* (PJS), in addition to hamartomas and neoplasm. *Laugier–Hunziker syndrome* (LHS) is thought to be idiopathic and is a diagnosis of exclusion.

H. Clark (✉)
Te Toka Tumai Auckland, Te Whatu Ora, Auckland, New Zealand
e-mail: hadleighc@adhb.govt.nz

R. Balasubramaniam et al. (eds.), *Oral Medicine - A Clinical Guide*,
https://doi.org/10.1007/978-3-031-36797-7_61

Medications account for 10–20% of all acquired hyperpigmentation cases, with antimicrobial (e.g. minocycline, zidovudine), antimalarial (e.g. minocycline), anti-hypertensive (e.g. amlodipine) and anticancer (e.g. imatinib, capecitabine) agents implicated. PJS is rare (1 in 8000–200,000 live births), and Addison's and Cushing disease are more so (incidence 0.6 per 100,000 and 0.6 per one million person-years, respectively). LHS usually affects middle-aged, mainly Caucasian women sporadically, although a few familial cases have been reported.

61.2 Clinical Presentation and Investigations

Melasma may affect the cheeks and forehead of a typically pregnant female, but may occur with oral contraceptive use. In chronic Addison's disease, the skin may appear generally bronzed and there may be complaint of malaise, anorexia, weight loss and joint pain. In Cushing's disease, central adiposity, moon facies and buffalo hump may be seen. In both conditions, patchy hyperpigmentation of the tongue, gingivae, buccal mucosae and hard palate present as blue-black or brown macules and can have a spotty or streaked pattern. Labial and perioral pigmented macules are an early clinical manifestation of PJS and may also be seen in LHS along with acral involvement. The palatal mucosa appears to be a predisposed site for many drug-induced hyperpigmentations.

A biopsy should be undertaken for any unifocal macule demonstrating features concerning for melanoma (Asymmetry, Border irregularity, Colour variegation, Diameter >6mm, Evolution or Elevation). In suspected Addison's disease, a blood pressure may reveal hypotension, and blood testing includes a full blood count (eosinophilia, lymphocytosis), electrolytes (hyponatraemia, hypokalaemia), cortisol and an ACTH stimulation test. Cushing disease requires initial investigation with cortisol levels or a dexamethasone suppression test. Referral to an endocrinologist is required for further work-up and management; brain imaging is required to confirm a pituitary adenoma in the setting of Cushing disease. Patients with suspected PJS should be referred to a gastroenterologist for endoscopic evaluation of the gastrointestinal tract for intestinal polyposis.

61.3 Management and Prognosis

Patients with PJS require lifelong cancer surveillance due to an increased risk of bowel malignancy; pigmentation will persist but can be improved with laser therapy. Hyperpigmentation demonstrates variable regression on drug withdrawal or management of underlying systemic disorders, although aesthetically concerning areas can also be treated by various types of laser therapy.

Recommended Reading

Binmadi NO, Bawazir M, Alhindi N, Mawardi H, Mansour G, Alhamed S, et al. Medication-Induced Oral Hyperpigmentation: A Systematic Review. *Patient Prefer Adherence*. 2020;14:1961–8.
Ko E, Panchal N. Pigmented Lesions. Dermatol Clin. 2020;38(4):485–94.

Differential Diagnosis of White Oral Mucosal Lesions

62

Lara DeAngelis and Nicole Heaphy

62.1 Introduction

White oral mucosal lesions can represent a range of heterogenous conditions as well as a variation of the normal oral mucosa that can include both malignant and benign pathology. White oral lesions can be solitary or multiple and thorough history, examination and special tests are required to achieve the definitive diagnosis. Below we present a limited list of most common differential diagnoses (Table 62.1).

L. DeAngelis
Melbourne Dental School, The University of Melbourne, Carlton, VIC, Australia

N. Heaphy (✉)
Royal Dental Hospital of Melbourne, Carlton, VIC, Australia
e-mail: Nicole.Heaphy@dhsv.org.au

R. Balasubramaniam et al. (eds.), *Oral Medicine - A Clinical Guide*,
https://doi.org/10.1007/978-3-031-36797-7_62

Table 62.1 Differential diagnoses of common white oral mucosal lesions

Category	Condition	Aetiology	Clinical presentation	Management
Developmental				
	Fordyce's spots	Ectopic sebaceous glands present in the oral mucosa	Multiple yellow-white papules typically found on the buccal mucosa	• Variation of normal anatomy • Education and reassurance
Reactive				
	Frictional keratosis (including benign alveolar ridge keratosis)	Repeated mechanical trauma to the oral mucosa	Shaggy, white keratosis that tends to be poorly demarcated	• Clinical diagnosis • Education regarding aetiology and habit reversal if necessary • Biopsy may be required to exclude other conditions that may have a similar presentation • Condition is benign
	Leukoedema	Keratinocyte oedema	Diffuse, pale white lesions present on the buccal mucosa that disappear when the mucosa is stretched	• Clinical diagnosis • Education and reassurance
	Nicotinic stomatitis	Repeated thermal trauma to the oral mucosa from smoking	Diffuse palatal keratosis with erythematous papules representing the inflamed openings on minor salivary glands	• Clinical diagnosis • Smoking cessation
	Actinic cheilitis	Chronic exposure of UV radiation to the lower lip vermilion	Loss of the vermilion-skin interface with diffuse keratosis, fissuring and dry patches	• Education on protection from sun including sun cream and long-brim hat use • Biopsy is required to exclude dysplastic or malignant changes with regular review required

	Submucous fibrosis	The result of chewing areca nut which has carcinogenic properties	Pale mucosa with fibrous banding that can result in limited opening	• Biopsy required • Areca nut cessation
	Hairy tongue	Elongated filiform papillae predominantly on the posterior dorsum of the tongue	White, yellow, or black hairy plaques	• Clinical diagnosis • Mechanical debridement and cessation of smoking
Infective				
	Chronic hyperplastic candidosis	Chronic lesion associated with *Candida* spp.	White plaques typically present bilaterally in the post-commissural area which cannot be wiped off	• Biopsy indicated to rule out dysplasia • Patients require management with topical and/or systemic antifungal therapy
	Pseudomembranous candidosis	Typically, acute but can be chronic. Due to infection with *Candida* spp. usually in patients who are immunocompromised	Creamy white plaques that can be removed to reveal and erythematous mucosa	• Cytology may assist with making a diagnosis • Patients require management with topical and/or systemic antifungal therapy
	Oral hairy leukoplakia	Secondary infection with Epstein–Barr virus in immunocompromised individuals	Homogenous white plaques typically present on the lateral surfaces of the tongue	• Biopsy with immunohistochemistry required
Immunological				
	Oral lichen planus[a]	T cell-mediated chronic inflammatory condition affecting skin and mucosa	White striations, plaques and/or ulcerations presenting on an erythematous background	• Biopsy to confirm diagnosis with ongoing review required • Symptomatic management only
	Oral lichenoid lesion[a]	T cell-mediated chronic inflammatory condition	Solitary white striations, plaques and/or ulcerations presenting on an erythematous background	• Biopsy to confirm diagnosis with ongoing review required • Symptomatic management only

(continued)

Table 62.1 (continued)

Category	Condition	Aetiology	Clinical presentation	Management
	Lupus erythematosus[a]—discoid and systemic	Autoimmune disease with multifactorial aetiology	Intra-oral lesions present as mixed red and white lesions mimicking oral lichen planus	• Biopsy and serology to confirm diagnosis with ongoing review required • Symptomatic management only
	Graft versus host disease[a]	Chronic immune-mediated inflammatory condition as a result of undergoing a haemopoietic stem cell transplant	Intra-oral lesions present as mixed red and white lesions mimicking oral lichen planus	• Biopsy to confirm diagnosis with ongoing review required • Symptomatic management only
Neoplasia and Oral potentially malignant conditions				
	Squamous cell carcinoma[a]	Multiple genetic and environmental factors implicated	Mixed red and white exophytic, indurated swelling with or without ulceration	• Biopsy required; referral to head and neck team
	Leukoplakia[a] with or without dysplasia	Multiple genetic and environmental factors implicated	Usually solitary white plaque that may be homogenous or non-homogenous in appearance. In some cases multiple lesions may be present	• Biopsy required; monitoring or referral for excision or other forms of removal
	Proliferative verrucous leukoplakia	Potentially malignant disorder presenting with multiple white patches progressing over time	Multiple white patches most commonly on the gingival tissues/alveolar process	• Biopsy and screening at regular intervals required
Idiopathic				
	Geographic tongue[a]	Benign, chronic inflammatory condition of the tongue of unknown cause	Relapsing/remitting erythematous lesions on the tongue which show loss of filiform papillae bordered by a keratotic rim	• Education and reassurance

[a] Can be mixed red and white

Recommended Reading

Jessri M, Mawardi H, Farah CS, Woo SB. White and Red Lesions of the Oral Mucosa. In: Farah CS, Balasubramaniam R, McCullough MJ, editors. Contemporary Oral Medicine: A Comprehensive Approach to Clinical Practice. Cham: Springer International Publishing; 2019. p. 1207–1248.

Warnakulasuriya S, Kujan O, Aguirre-Urizar JM, Bagan JV, González-Moles MÁ, Kerr AR, et al. Oral potentially malignant disorders: A consensus report from an international seminar on nomenclature and classification, convened by the WHO Collaborating Centre for Oral Cancer. Oral Dis. 2021;27(8):1862–80.

Differential Diagnosis of Red Oral Mucosal Lesions

63

Lara DeAngelis and Nicole Heaphy

63.1 Introduction

Erythematous or red appearing mucosal lesions can represent a variation of the normal oral mucosa as well as a range of heterogenous conditions that can include both benign and malignant pathology. The appearance of erythema can be focal or multifocal and may be a result of mucosal epithelial thinning, erosion, or ulceration or an increase in vascularity or inflammatory infiltration. Here we present more common and typical presenting causes of erythematous lesions (Table 63.1).

L. DeAngelis
Melbourne Dental School, The University of Melbourne, Carlton, VIC, Australia

N. Heaphy (✉)
Royal Dental Hospital of Melbourne, Carlton, VIC, Australia
e-mail: Nicole.Heaphy@dhsv.org.au

R. Balasubramaniam et al. (eds.), *Oral Medicine - A Clinical Guide*,
https://doi.org/10.1007/978-3-031-36797-7_63

Table 63.1 Differential diagnoses of common red mucosal lesions

Category	Condition	Aetiology	Clinical presentation	Investigations and treatment
Developmental				
	Benign vascular malformation	Capillary/venous/lymphatic high/low flow simple or combined	Soft red, blue or purple raised lesion which may or may not blanch on pressure	Monitor unless aesthetic concerns or bleeding upon trauma
Infective				
	Erythematous candidosis	May be acute or chronic and is due to *Candida* spp. infection. Usually associated with the use of broad-spectrum antibiotics or corticosteroids	Erythematous, painful lesions in the mouth that may affect any area in the mouth and when present on the tongue will present with atrophy	Management includes use of topical and/or systemic antifungal therapy as well as management of any pre-disposing factors
	Denture associated erythematous stomatitis	Overgrowth of oral commensal organisms including *Candida* spp., poor denture hygiene, nocturnal denture wear and ill-fitting dentures	Erythema that usually present on the palate and involves the entire area covered by the denture. Typically lesions are asymptomatic	Management is aimed at treating the biofilm present under the denture
	Angular cheilitis	*Candida* associated lesion which may also be due to co-infection with *Staph. aureus*	Painful erythematous, crusted lesions occurring at the angles of the mouth	Management includes use of topical antifungals
	Median rhomboid glossitis	Aetiology of this lesion is uncertain with some papers attributing this to *Candida* spp. and others disputing this fact	Erythematous, atrophic rhomboid shaped area in the mid-dorsum of the tongue. Kissing lesion may be present on the palate	Management includes use of topical antifungals
Immunological				
	Plasma cell gingivostomatitis	Type I hypersensitivity reaction	Diffuse gingival erythema and oedema with pain and sensitivity	Biopsy and patch testing

	Granulomatosis with polyangiitis	Small to medium blood vessel antineutrophil-cytoplasmic antibody associated vasculitis	Presents as granulomatous inflammation on the gingival tissues known as strawberry gingivitis	Biopsy and multidisciplinary management required
	Oral lichen planus[a]	T cell-mediated chronic inflammatory condition affecting skin and mucosa	White striations, plaques and/or ulcerations presenting on an erythematous background	Biopsy to confirm diagnosis with ongoing review required. Symptomatic management only
	Oral lichenoid lesion[a]	T cell-mediated chronic inflammatory condition	Solitary white striations, plaques and/or ulcerations presenting on an erythematous background	Biopsy to confirm diagnosis with ongoing review required. Symptomatic management only
	Lupus erythematosus[a]—discoid and systemic	Autoimmune disease with multifactorial aetiology	Intra-oral lesions present as mixed red and white lesions mimicking oral lichen planus	Biopsy and serology to confirm diagnosis with ongoing review required. Symptomatic management only
	Graft versus host disease[a]	Chronic immune-mediated inflammatory condition as a result of undergoing a haemopoietic stem cell transplant	Intra-oral lesions present as mixed red and white lesions mimicking oral lichen planus	Biopsy to confirm diagnosis with ongoing review required. Symptomatic management only
	Mucous membrane pemphigoid	Autoimmune vesiculobullous condition	Thick-walled blisters, mucosal oral mucosa, erythema and desquamative gingivitis	Biopsy. Management with topical and/or systemic steroids and steroid sparing agents. Review with dermatology and ophthalmology required
	Pemphigus vulgaris	Autoimmune vesiculobullous condition	Thin-walled blisters, mucosal oral mucosa, erythema and desquamative gingivitis	Biopsy. Management with topical and/or systemic steroids and steroid sparing agents. Review with dermatology and ophthalmology required

(continued)

Table 63.1 (continued)

Category	Condition	Aetiology	Clinical presentation	Investigations and treatment
Neoplasia and oral Potentially Malignant Conditions				
	Erythroplakia	Multiple genetic and environmental factors implicated	Typically solitary, red, velvety lesion	Biopsy is required due to high risk of dysplasia and/or malignancy. Appropriate management will depend on histopathology results and include may include review, surgical excision and/or referral to a multidisciplinary head and neck unit
	Squamous cell carcinoma[a]	Multiple genetic and environmental factors implicated	Mixed red and white exophytic, indurated swelling with or without ulceration	Biopsy is required. Appropriate management will depend on histopathology results and may include surgical excision and/or referral to a multidisciplinary head and neck unit
	Kaposi sarcoma	Endothelial cell cancer associated with human herpes virus 8 and immunodeficiency	Purple/blue/red macule/ nodule	Biopsy is required. Appropriate management will depend on histopathology results and may include surgical excision and/or referral to a multidisciplinary head and neck unit
Idiopathic				
	Geographic tongue[a]	Benign, chronic inflammatory condition of the tongue of unknown cause	Relapsing/remitting erythematous lesions on the tongue which show loss of filiform papillae bordered by a keratotic rim	Reassurance
	Angina bullosa haemorrhagica	Idiopathic however could be related to trauma or long-term use of topical steroids	Formation of blood blisters on the oral mucosa	Symptomatic management only

[a] Can be mixed red and white

Recommended Reading

Jessri M, Mawardi H, Farah CS, Woo S-B. White and Red Lesions of the Oral Mucosa. In: Farah CS, Balasubramaniam R, McCullough MJ, editors. Contemporary Oral Medicine: A Comprehensive Approach to Clinical Practice. Cham: Springer International Publishing; 2019. p. 1207–1248.

Warnakulasuriya S, Kujan O, Aguirre-Urizar JM, Bagan JV, González-Moles MÁ, Kerr AR, et al. Oral potentially malignant disorders: A consensus report from an international seminar on nomenclature and classification, convened by the WHO Collaborating Centre for Oral Cancer. Oral Dis. 2021;27(8):1862–80.

Part III

Oral Considerations in Systemic Disease

Anaemia and Iron Deficiency

64

Ben Karim, Mark Schifter, and Suma Sukumar

64.1 Introduction

The term "anaemia" originates from the Greek root words for "lack of blood" and refers to a decrease in red blood cell numbers or the loss in quantity or quality of haemoglobin. Anaemia may be quantitative (due to loss or decrease in production) or qualitative (Table 64.1).

Iron levels are critical to the formation of haemoglobin, and an iron deficiency is therefore the most common reason for an anaemia to occur. WHO data indicates global prevalence of iron deficiency at approximately 29% of women of reproductive age and as high as 39% in children under 5 years of age. Causes of iron deficiency are varied (Table 64.2).

It should be noted that iron deficiency is not the only nutritional deficiency associated with a decrease in red blood cell formation as several other vitamins and minerals (e.g. vitamin B12 and folate) also play vital roles.

B. Karim (✉) · S. Sukumar
Sydney Dental School, The University of Sydney, Camperdown, NSW, Australia

Westmead Centre for Oral Health, Westmead Hospital, Westmead, NSW, Australia
e-mail: Ben.karim@health.nsw.gov.au

M. Schifter
Sydney Dental School, The University of Sydney, Camperdown, NSW, Australia

Westmead Centre for Oral Health, Westmead Hospital, Westmead, NSW, Australia

The Skin Hospital, Darlinghurst, NSW, Australia

R. Balasubramaniam et al. (eds.), *Oral Medicine - A Clinical Guide*,
https://doi.org/10.1007/978-3-031-36797-7_64

Table 64.1 Causes of anaemia

Dilutional anaemia	
"Spurious Anaemia"	• Increased plasma volume (e.g. due to I.V. fluid excess)
Decreased production	
Genetic disorders	• Defective DNA repair (e.g. Fanconi's anaemia) • Telomerase disorders (e.g. Dyskeratosis congenita) • Ribosomal disorders (e.g. Diamond-Black Anaemia) • Maturation disorders (e.g. Thalassaemias)
Nutritional deficiencies	• Haemoglobin synthesis (e.g. iron deficiency) • DNA synthesis (e.g. Vitamin B12, folate)
Bone marrow dysfunction	• Aplastic anaemia • Haematopoietic malignancy • Space occupying lesions
Increased loss	
Acute	• Trauma • Infection
Chronic	• Gastrointestinal disorders (e.g. coeliac disease, Crohn's disease, ulcerative colitis, ulceration, malignancy) • Menorrhagia • Increased sequestration (e.g. hypersplenism)
Qualitative deficiency	
Genetic	• Thalassaemia

Table 64.2 Causes of iron deficiency

Causes	
Dietary insufficiency	• Lack of iron containing food such as red meat, cereals, legumes, dark green leafy vegetable such as spinach, kale, chard
Poor absorption	• Consumption of inhibitors of iron absorption—tea, calcium, antacids, intestinal mucosal disorders such as coeliac disease, inflammatory bowel disease
Blood loss	• Gastrointestinal blood loss due to ulceration or malignancy • Menorrhagia • Bleeding disorders such as von Willebrand disorder

64.2 Iron Absorption and Testing

Haematinic studies are used to determine the nature of the deficiency or malfunction that leads to decreased in iron levels or function. Interpretation of these test results requires an understanding of the mechanisms by which iron is absorbed, transported, and stored (Table 64.3). A lack of iron supply to the bone marrow results in abnormalities of red blood cell formation volume and colour.

Table 64.3 Common iron studies

Test	Description
Ferritin	A protein produced in response to increased iron and functions as an iron store and is therefore seen as an indicator of iron ***storage***. A low ferritin result indicates the body has low iron stores, a high ferritin level can occur during systemic inflammation, or with certain genetic conditions (haemochromatosis).
Transferrin	A protein produced in the liver. High transferrin levels could be an indicator of iron deficiency as the liver increases the production of transferrin to bind more iron. Transferrin is therefore seen as an indicator of iron ***amount***.
TIBC—Total Iron Binding Capacity	The ability for the blood to bind and transport iron and reflects transferrin levels. The transferrin molecule has several binding sites, which are filled under normal circumstance. An increase in empty binding sites (and therefore an increase in TIBC) indicates an iron deficiency. TIBC is a reflection of transferrin and can be therefore seen as an indicator of the ability to ***transport*** iron.
FBC/CBC	Complete blood count—gross breakdown of the cellular content of blood which could be useful in detecting a quantitative pathology. Especially important in the assessment of anaemia is the number of reticulocytes—newly produced red blood cells. The reticulocyte count reflects bone marrow functionality. An increase in the reticulocyte amount or percentage above normal could be an indication that the body is attempting to compensate for recent blood loss. Conversely, a decrease in reticulocytes could be an indication of a pathology involving red blood cell formation.
Haematocrit	The percentage of whole blood that is made up of red blood cells with an implication of blood viscosity. A low haematocrit may indicate RBC loss or a failure in RBC formation.
Haemoglobin	An iron-based protein within RBCs that binds oxygen and therefore allows the oxygen transport function of blood. A low haemoglobin level may indicate RBC loss or a failure in RBC formation.
MCH/MCHC	Mean cell haemoglobin which indicates the average amount of haemoglobin per cell, and the mean concentration of haemoglobin per cell with altered levels indicating a qualitative pathology.
MCV	Mean cell volume. The average size of the red cells. Normal red blood cell numbers but an altered MCV could indicate a qualitative pathology—either *microcytosis* in which the average cell volume is low, or *macrocytosis* in which the average volume is high.
Folate	Essential for the synthesis of DNA and therefore intrinsically involved in cell division and growth.
Vitamin B12	Acts as a coenzyme facilitating the function of numerous vital enzymes, including those involved the release of chemical energy during metabolism and DNA synthesis.

64.3 Oral Manifestations of Iron Deficiency

The systemic presentation of iron deficiency is generally associated with the reduced oxygen availability to tissues and manifests as fatigue, pallor, and reduced exercise capacity. Oral manifestations and local tissue changes are related to this, but include mechanisms of altered immune response, and impairment of proliferation, maturation, and repair. It is also noted that iron deficiency is commonly associated with deficiencies in other essential nutrients—most notably folate and Vitamin B12, and mucosal changes due to deficiencies in these are difficult to separate.

64.3.1 Atrophic Glossitis

The classic oral mucosal change associated with iron deficiency as well as B12 deficiency is the described "red, beefy" tongue of atrophic glossitis. It has been recorded as occurring in as much as 40–60% of iron deficient persons. It manifests as complete or patchy atrophy of the dorsal papillae, leaving the tongue with smooth, erythematous appearance. Management is usually by the correction of the deficiency.

64.3.2 Plummer-Vinson Syndrome

A rare, pre-malignant condition, in which patients present with glossitis, oral burning sensation, and dysphagia. Barium swallow radiography shows archetypal oesophageal webbing. The glossitis and sensation disturbances usually correct with iron supplementation; however, surgical dilation may be required to correct the established oesophageal stricture.

64.3.3 Fungal Infection

Iron deficiency has been associated with increased rates of oral fungal infections and *Candida*-related conditions, including angular cheilitis, pseudomembranous candidosis, median rhomboid glossitis, and papillary hyperplastic candidosis. The aetiopathogenesis is thought to be associated with epithelial abnormalities and altered immune response.

Recommended Reading

Australian Government Department of Health: Anaemia. https://www.health.gov.au/resources/pregnancy-care-guidelines/part-f-routine-maternal-health-tests/anaemia. Retrieved April 2022.

Finch C. The Role of iron in hemoglobin synthesis. National academy of sciences and National Research Council (US) Division of Medical Science. Conference on Hemoglobin 2–3 May 1957.

Johnson-Wimbley TD, Graham DY. Diagnosis and management of iron deficiency anemia in the 21st century. Therap Adv Gastroenterol. 2011;4(3):177–84.

Lu SY. Perception of iron deficiency from oral mucosa alterations that show a high prevalence of Candida infection. J Formos Med Assoc. 2016;115(8):619–27.

Ogun AS, Adeyinka A. Biochemistry, Transferrin. In: StatPearls [Internet]. Treasure Island (FL): StatPearls Publishing; 2023. Available from: https://www.ncbi.nlm.nih.gov/books/NBK532928/. Accessed 20 Nov 2023.

Pasricha SR, Flecknoe-Brown SC, Allen KJ, Gibson PR, McMahon LP, Olynyk JK, et al. Diagnosis and management of iron deficiency anaemia: a clinical update. Med J Aust. 2010;193(9):525–32.

Jani PG. Plummer Vinson syndrome: case report. East Afr Med J. 2001;78(6):332–3.

Jo, VY, Krane, JF. Clinicopathologic features of secretory carcinoma of salivary gland. AJSP Rev Reports 2018;23(1):9–24.

World Health Organisation: Anaemia in Woman and Children [Internet]. Geneva: WHO; 2023 [cited 2023 Jul 7]. Available from: https://www.who.int/data/gho/data/themes/topics/anaemia_in_women_and_children.

Bleeding Diathesis and Anti-thrombotic Drugs

65

Suma Sukumar, Ben Karim, and Mark Schifter

65.1 Introduction

Bleeding diatheses (defined as group of systemic disorders of haemostasis) together with anti-thrombotic drugs (including both antiplatelet and anticoagulant medications) cause an abnormal susceptibility to bleed that is excessive in volume and/or protracted in duration and collectively present a clinical challenge in dental practice. Disorders of platelets, coagulation and of the vessel wall are summarised in Tables 65.1, 65.2, and 65.3, respectively.

S. Sukumar (✉)
Westmead Centre for Oral Health, Westmead Hospital, Sydney, NSW, Australia

Sydney Dental School, The University of Sydney, Camperdown, NSW, Australia
e-mail: Suma.Sukumar@health.nsw.gov.au

B. Karim
Westmead Centre for Oral Health, Westmead Hospital, Sydney, NSW, Australia

M. Schifter
Westmead Centre for Oral Health, Westmead Hospital, Sydney, NSW, Australia

Sydney Dental School, The University of Sydney, Camperdown, NSW, Australia

The Skin Hospital, Darlinghurst, NSW, Australia

R. Balasubramaniam et al. (eds.), *Oral Medicine - A Clinical Guide*,
https://doi.org/10.1007/978-3-031-36797-7_65

Table 65.1 Disorders of platelets

Abnormalities in number	Abnormalities in function
Decreased platelet number—thrombocytopenia	*Inherited*
Decreased marrow production	Disorders of adhesion: • Von Willebrand disease • Bernard-Soulier syndrome
Increased splenic sequestration	Disorders of granule release: • Platelet storage pool disorder
Increased platelet destruction	Disorders of aggregation: • Glanzmann thrombasthenia • Afibrinogenaemia
Increased platelet number—thrombocytosis	*Acquired*
Iron deficiency	Intended medication effect
Inflammation	Unintended medication effect
Infection	Uraemia
Malignancy	Cardiopulmonary bypass
Myeloproliferative states	

Table 65.2 Disorders of coagulation

Anticoagulant	Procoagulant
Inherited	*Inherited*
Haemophilia A—deficiency of Factor VIII	Factor V Leiden
Haemophilia B—deficiency of Factor IX	Deficiencies in protein C and S and antithrombin III
Other factor deficiencies, e.g. Factor XI	Hyperhomocysteinaemia
Acquired	*Acquired*
Liver disease	Hyperestrogenic states • Pregnancy • Oral contraceptive
Vitamin K deficiency	Antiphospholipid antibodies
Medications: Warfarin Heparin Direct-acting anticoagulants Other drugs that may cause prolonged bleeding include non-steroidal anti-inflammatory medications, selective serotonin uptake inhibitors, and a number of complementary medicines	Increased hepatic protein synthesis

Table 65.3 Disorders of the vessel wall

Inherited	Acquired
Marfan's syndrome	Scurvy
Ehlers-Danlos syndrome	Cushing's syndrome
Hereditary haemorrhagic telangiectasia	Henoch-Schonlein purpura
	Senile purpura
	Multiple myeloma
	Waldenstrom's macroglobulinaemia
	Cryoglobulinaemia

65.2 Dental Considerations

The key to the successful dental management of patients at risk of bleeding is thorough evaluation and risk assessment to determine whether a patient can safely tolerate a planned procedure with the necessary level of haemostatic support in a suitable clinical setting. Risk assessment is generally undertaken by evaluation of the bleeding risk of the patient and the bleeding risk of the planned procedure, taking into consideration the risk of the patient having a thromboembolic event if their anticoagulant/antiplatelet regimen is altered. Treatment planning should involve the patient as well as their medical team, particularly if the patient has a bleeding diathesis. In certain cases, specialist referral may be appropriate.

Recommended Reading

Brennan Y, Gu Y, Schifter M, Crowther H, Favaloro EJ, Curnow J. Dental extractions on direct oral anticoagulants vs. warfarin: The DENTST study. Res Pract Thromb Haemost. 2020;4(2):278–84.

Casari C, Bergmeier W. Acquired platelet disorders. Thromb Res. 2016;141(suppl 2):S73–S75.

Lippi G, Franchini M, Montagnana M, Favaloro EJ. Inherited disorders of blood coagulation. Ann Med. 2012;44(5):405–18.

Scottish Dental Clinical Effectiveness Program. Management of Dental Patients Taking Anticoagulants or Antiplatelet Drugs. Dental Clinical Guidance. 2nd ed. 2022. Sections 2–10. p. 4–28.

Oral and Dental Expert Group. Therapeutic Guidelines: Oral and Dental. Dental Management of Patients with Medical Conditions. Version 3. Melbourne: Therapeutic Guidelines Limited; 2019. p. 153–162.

Ramasamy I. Inherited bleeding disorders: disorders of platelet adhesion and aggregation. Crit Rev Oncol Hematol. 2004;49(1):1–35

Schardt-Sacco D. Update on coagulopathies. Oral Surg Oral Med Oral Pathol Oral Radiol Endod. 2000;90(5):559–63.

Scott A, Gibson J, Crighton A. The management of dental patients taking new generation oral anticoagulants. Prim Dent J. 2014;3(4):54–8.

Haematological Malignancies: Leukaemia, Lymphoma and Multiple Myeloma

66

Mark Schifter, Ben Karim, and Suma Sukumar

66.1 Introduction

Leukaemia, lymphoma, and multiple myeloma are interrelated spectrum of cancers of the myeloid and lymphoid cells and tissues. They may present acutely, where immature blast cells dominate, or chronically, where mature cell forms dominate and may be of myeloid cell or lymphoid lineage.

66.2 The Leukaemias

In leukaemia, there is a persistent clonal population of poorly differentiated, immature precursor cells ("*blasts*"), that invade the bone marrow space, replacing the normal haematopoietic stem cells, eventually entering the circulation. The main subtypes of leukaemia are listed in Table 66.1.

M. Schifter
Westmead Centre for Oral Health, Westmead Hospital, Westmead, NSW, Australia

The Skin Hospital, Darlinghurst, NSW, Australia

Sydney Dental School, The University of Sydney, Camperdown, NSW, Australia

B. Karim · S. Sukumar (✉)
Westmead Centre for Oral Health, Westmead Hospital, Westmead, NSW, Australia

Sydney Dental School, The University of Sydney, Camperdown, NSW, Australia
e-mail: suma.sukumar@health.nsw.gov.au

R. Balasubramaniam et al. (eds.), *Oral Medicine - A Clinical Guide*,
https://doi.org/10.1007/978-3-031-36797-7_66

Table 66.1 Main subtypes of leukaemia

Acute Myeloid Leukaemia (AML)	• Monocytic leukaemia • Granulocytic leukaemia • Megakaryocytic leukaemia • Erythroid leukaemia • "Secondary AML" (AML following previous cytotoxic chemotherapy)
Acute Lymphoblastic Leukaemia (ALL)	• B-cell/T-cell • Pre-B-cell/pre-T-cell
Chronic Myeloid Leukaemia (CML—myeloproliferative disorders)	• Philadelphia positive • Primary myelofibrosis • Essential thrombocytopaenia • Polycythaemia rubra vera • Myelodysplasia
Chronic Lymphocytic Leukaemia (CLL)	• Can present in leukemic phase or as localised disease • 5–10% undergo "Richter's" transformation to lymphoma

66.2.1 Clinical Presentation and Investigations

Leukamias present with symptoms from inadequate haematopoiesis with resultant cytopenia, due to infiltration and proliferation of the leukaemic blast cells in the marrow, as well as in the blood and other tissues. The oral manifestations and complications of leukaemia and its treatment are varied (Table 66.2). Gingival infiltrates consisting of blast cells may be one of the presenting features of acute leukaemia (Fig. 66.1).

Oral signs and symptoms of leukaemia are predominantly related to anaemia (e.g. mucosal pallor), thrombocytopenia (e.g. petechiae of the soft palate (Fig. 66.2)), neutropenia (e.g. odontogenic infection and mucositis), and lymphopenia (e.g. opportunistic viral infection). Diagnosis is made following blood tests, peripheral blood smear, and bone marrow biopsy. Cytogenetic studies are used to further subtype acute leukaemia.

Table 66.2 Oral manifestations and complications of leukaemia and its treatment

Cytopenia and clinical findings	Oral manifestations
Anaemia	
• Pallor • Declining oxygen carrying capacity	• Pallor of oral mucosa
Thrombocytopenia	
• Impaired haemostasis	• Petechiae on soft palate • Oral haemorrhagic bulla • Major bleed into oral soft tissues
Lymphopenia	
• Opportunistic infections	Viral infections • Herpes stomatitis • Herpes zoster • Post-herpetic neuralgia of trigeminal nerve
	Fungal infections • Candidosis • Fungal sinusitis • Aspergillus • Mucormycosis
Neutropenia	
• Bacterial infections	Odontogenic infections • Periapical periodontitis • Periodontal abscess • Severe pericoronitis
• Impaired healing • Mucositis	Mucositis • Wide-spread oral ulceration of the soft palate, buccal, and floor of mouth

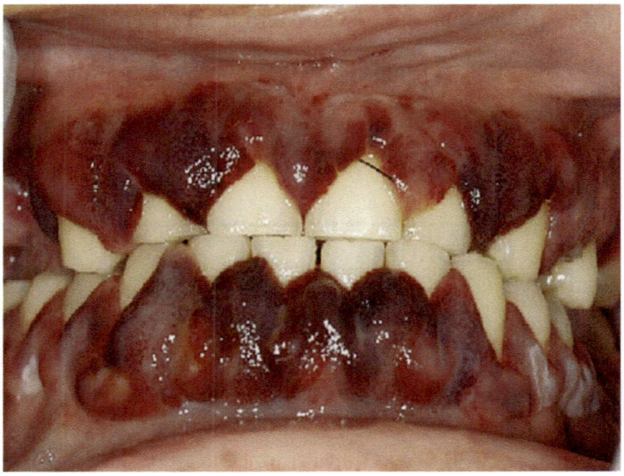

Fig. 66.1 Gingival infiltration of blast cells in acute myelomonocytic leukaemia

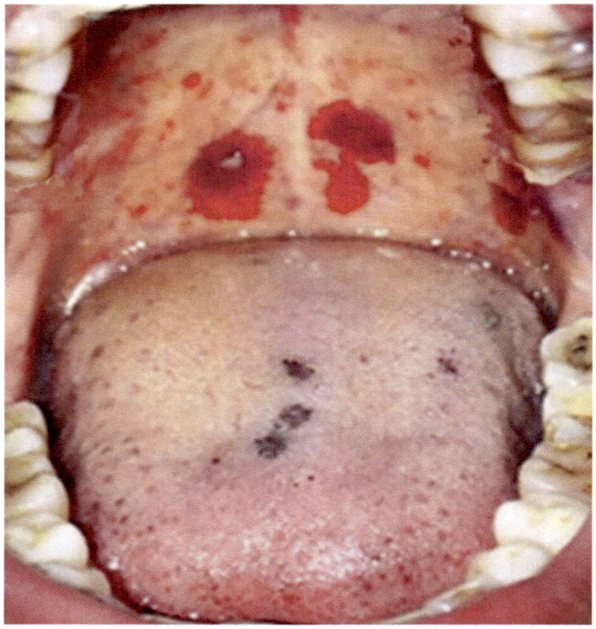

Fig. 66.2 Petechiae and purpura in a patient with severe thrombocytopaenia

66.2.2 Management and Prognosis

Initial treatment entails molecular and genetic subtyping and assessment of the patients' fitness for treatment. Treatment for specific subtypes of leukaemia will entail set protocols of cytotoxic chemotherapy over various cycles, usually consisting of 3–4 different agents to prevent chemotherapy resistance. Chemotherapy may be followed by allogenic haematopoietic stem cell transplantation (HSCT).

66.3 Lymphoma

Lymphoma is a malignancy of lymphocytes, presenting as single or multiple soft tissues masses (Fig. 66.3), with little involvement of the circulating blood and variable involvement of the bone marrow. Some patients present with "B symptoms" of unexplained weight loss, fever, and sweats. There are over 60 different subtypes of lymphoma, reflective that lymphoma may develop from the clonal proliferation of lymphocytes at any stage of their development (Table 66.3).

Diagnosis relies on histological, immunological, and molecular analysis. PET-CT imaging is used for disease staging and monitoring treatment response.

The disease stage, responsiveness to chemotherapy, and the medical status of the patients determine treatment intent. Chemotherapy is the mainstay of treatment, although localised involved field radiotherapy up to 30 Gy is also used. As in

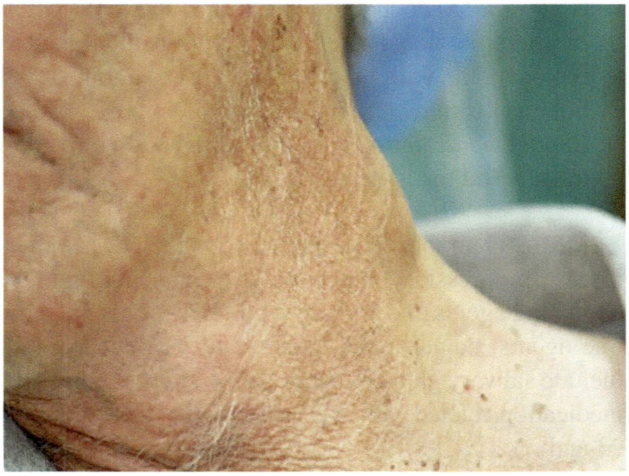

Fig. 66.3 Hodgkin's lymphoma presenting as a mass in the left neck

Table 66.3 Categorisation of lymphoma

Hodgins's Lymphoma (HL)[a]	• **10% of all lymphoma presentations**
Non-Hodgkin's Lymphoma (NHL)[b]	• **90% of all lymphoma presentations**
• B-lymphocyte origin	• 80% (of the 90% of NHL presentations)
• T-lymphocyte origin	• 20% (of the 90% of NHL presentations)
• NK-cell and/or undefined cell origin	• <1% (of the 90% of NHL presentations)

[a] Characterised by finding of Reed-Sternberg cell or its variants

[b] Classified by the absence of Reed-Sternberg cell, according to the cell of origin—T or B lineage and the stage of lymphocytic maturation, i.e. precursor or mature

leukaemia, a combination of cytotoxic chemotherapy agents is used, with the same consequent toxicities, especially myelosuppression. Patients who fail initial treatment may receive a second round of therapy, and possibly HSCT.

66.4 Multiple Myeloma

Multiple myeloma (MM) is a common malignancy of plasma cells, accounting for 1% of all malignancies with a median age of presentation of 60 years. Its main presenting features are attributable to the role and function of plasma cells:

- ***Bone Destruction***: Commonly "punched-out" lytic lesions evident on plain films of the skull and mandible. Sequelae include pathological fractures and severe hypercalcaemia.
- ***Paraproteinemia***: Excessive amounts of protein fragments, mainly from the antibody-derived light chains of the plasma cell clone, with an overall reduction in antibody production.

- ***Cytopenia***: Secondary to infiltration and replacement of the bone marrow by the proliferating plasma cell clone, with resultant anaemia, thrombocytopenia, and lymphopenia with neutropenia. This, combined with reduced antibody production results in recurrent bacterial infections.
- ***Renal Impairment***: From the combination of paraprotein deposition in the renal tubules and the hypercalcaemia.

MM is incurable. Therapeutic aims include eliminating myeloma cells, limiting disease progression, managing pain, and prolonging survival. A combination of alkylating agents and high dose corticosteroids is used initially followed by second-line agents, namely thalidomide and its derivatives. Complications of treatment, include salivary gland exocrinopathy and oral mucosal graft versus host disease and medication-related osteonecrosis of the jaws (MRONJ), are discussed in Chap. 23 and 73.

Recommended Reading

Murphy MF, Pasi KJ, Mead A. Haematological Disease. In: Kumar P, Clark M, editors. Kumar and Clark's Clinical Medicine. 7th ed. Elsevier; 2017. p. 515–582.

Gallagher CJ, Smith M, Sharmash J. Malignant Disease. In: Kumar P, Clark M, editors. Kumar and Clark's Clinical Medicine. 7th ed. Elsevier; 2017. p. 583–644.

Fernandes KS, Gallottini M, Castro T, Amato MF, Lago JS, Braz-Silva PH. Gingival leukemic infiltration as the first manifestation of acute myeloid leukemia. Spec Care Dentist. 2018;38(3):160–2.

Francisconi CF, Caldas RJ, Oliveira Martins LJ, Fischer Rubira CM, da Silva Santos PS. Leukemic Oral Manifestations and their Management. Asian Pac J Cancer Prev. 2016;17(3):911–5.

Silva TD, Ferreira CB, Leite GB, de Menezes Pontes JR, Antunes HS. Oral manifestations of lymphoma: a systematic review. Ecancermedicalscience. 2016;10:665.

Chronic Graft-Versus-Host Disease

67

Katrusha Hull

67.1 Introduction

Haematopoietic stem cell transplantation (HSCT) has evolved as a curative therapy for haematological malignancies, bone marrow failure, immune deficiencies and some solid tumours. Transplant recipients are showing greater survival outcomes and fewer complications; however, significant treatment-related morbidities are still common with the oral cavity frequently affected.

67.2 Early Adverse Effects of Haematopoietic Stem Cell Transplantation

Early adverse effects of transplant include mucositis and both local and systemic infections.

Graft-versus-host disease (GvHD) is one of the most common transplant-associated complications. Second only to infection it is, directly or indirectly, the major cause of early death in patients who are otherwise cured of their original disease.

GvHD is the result of a battle between graft (donated stem cells) and the host (the patient's own residual cells) where immune cells recognise the host as foreign (non-self). The transplanted and now proliferating T-lymphocytes mount an immune response against the host to "reject" the residual host cells. The range of clinical outcomes depend on the body site/system involved and degree of subsequent injury.

K. Hull (✉)
Peter MacCallum Cancer Centre, Parkville, VIC, Australia

Monash Medical Centre, Clayton, VIC, Australia

Royal Dental Hospital of Melbourne, Carlton, VIC, Australia
e-mail: katrusha.hull@dhsv.org.au

R. Balasubramaniam et al. (eds.), *Oral Medicine - A Clinical Guide*,
https://doi.org/10.1007/978-3-031-36797-7_67

67.3 Graft-Versus-Host Disease

GvHD is broadly categorised as an acute or chronic process (Fig. 67.1). The 2005 NIH Consensus Conference redefined acute and chronic GvHD as distinct clinical syndromes without a time restriction (Fig. 67.1).

Acute GvHD typically affects the skin, gastrointestinal tract and liver, and most commonly arises in the early-post transplant period. Patients classically present with a maculopapular rash which is often painful or pruritic. Abdominal pain and diarrhoea plus rising serum bilirubin reflect liver involvement. Oral involvement is uncommon and non-specific but may include erythema, with or without ulceration, of the oral mucosa and lips.

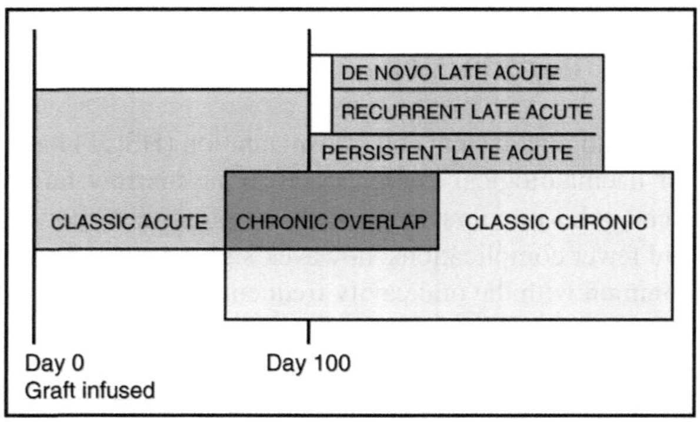

Fig. 67.1 Classification of Graft-Versus-Host Disease

67.4 Late-Effects of Haematopoietic Stem Cell Transplantation

Chronic GvHD (cGvHD) is a systemic disease with a wide range of signs and symptoms with 30–60% incidence. Single organs can be affected; however, multiple sites are nearly always involved. The eyes, skin (Fig. 67.2), mouth, gastrointestinal tract and liver are the most frequently affected organs with their diagnostic features summarised in Table 67.1.

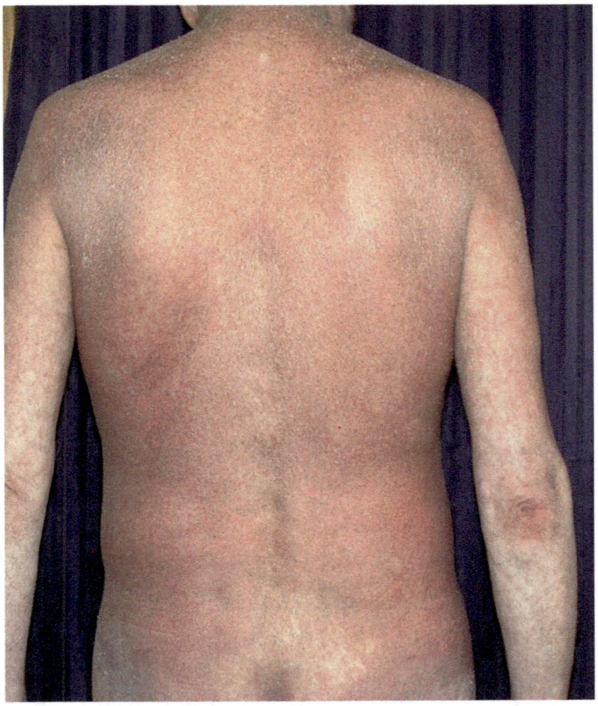

Fig. 67.2 Cutaneous cGvHD

Table 67.1 Diagnostic organ-specific presentation of chronic graft-versus-host disease

Organ	Diagnostic features	Clinical feature
Skin	Poikiloderma	Atrophy, pigmentary changes, telangiectasia
	Lichen planus-like features	Erythematous/violaceous plaques or papules
	Deep sclerotic features	Smooth, waxy, indurated skin with reduced joint mobility
	Morphea-like features	Superficial sclerotic features, i.e. localised patches of moveable leather-like skin
Eyes	New keratoconjunctivitis sicca symptoms where other causes are excluded (classified as "distinctive" only)[a]	Ranges from mild dry eye symptoms to severe symptoms which impact vision and activities of daily living
Gastrointestinal	Oesophageal webs, strictures or concentric rings	Diagnosed on endoscopy or barium contrast radiograph. Symptoms are not specific to cGvHD but may include nausea, vomiting, weight loss and anorexia
Lungs	Bronchiolitis obliterans:	New onset obstructive lung defect presenting as dyspnoea on exertion, cough or wheezing diagnosed on pulmonary function tests or biopsy
Liver	No diagnostic features but can present with acute hepatitis or progressive cholestatic features	Steeply rising serum ALT or slowly progressive cholestatic disorder with elevated alkaline phosphatase and gamma-glutamyl transpeptidase concentrations followed by jaundice

[a] Distinctive features: manifestations which are not ordinarily found in acute GvHD but are not considered sufficient in isolation to establish an unequivocal diagnosis of cGvHD

67.5 Clinical Presentation

67.5.1 Oral Chronic Graft-Versus-Host Disease

cGvHD frequently affects the oral cavity commonly arising within the first year and persisting for many years. Oral mucosal cGvHD is discussed in Chap. 23.

67.5.2 Salivary Gland

Patients present with a dry, glazed oral mucosa with a complaint of xerostomia and difficulty tolerating dry foods. Saliva may be more mucous presenting as opaque ropy saliva with mucous strings. Multiple recurring superficial mucoceles are common (Fig. 67.3).

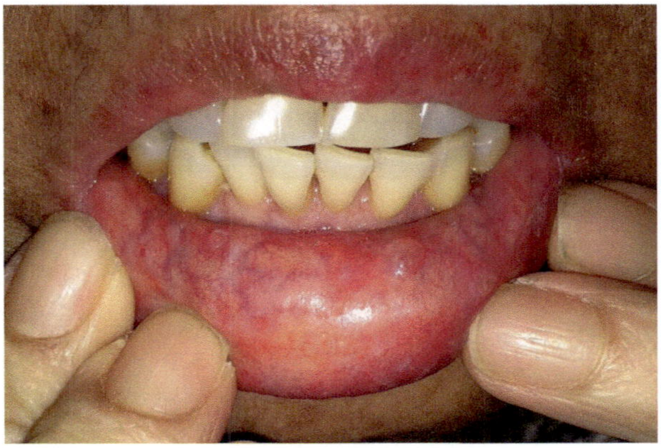

Fig. 67.3 Superficial mucoceles in cGvHD

67.5.3 Musculoskeletal

cGvHD can affect all layers of the skin, the epidermis, dermis and subcutaneous structures. Sclerodermatous cGvHD is a distinct form of cutaneous cGvHD with predominant involvement of skin, subcutaneous tissue and fascia. When the orofacial region is involved, loss of elasticity of the peri-oral tissues and lips may lead to reduced oral aperture.

67.6 Management and Prognosis

The treatment for cGvHD depends on organs affected and disease severity. Moderate to severe symptoms, or multi-organ involvement, often requires systemic therapy with corticosteroids the mainstay. Non-pharmacological modalities may include extracorporeal photopheresis and phototherapy. cGvHD is one of the major causes of late transplant-related mortality.

Recommended Reading

Jagasia MH, Greinix HT, Arora M, Williams KM, Wolff D, Cowen EW, et al. National Institutes of Health Consensus Development Project on Criteria for Clinical Trials in Chronic Graft-versus-Host Disease: I. The 2014 Diagnosis and Staging Working Group report. Biol Blood Marrow Transplant. 2015;21(3):389–401.e1.

Lee SJ. Classification systems for chronic graft-versus-host disease. Blood. 2017;129(1):30–7.

McDonald GB, Sandmaier BM, Mielcarek M, Sorror M, Pergam SA, Cheng GS, et al. Survival, Nonrelapse Mortality, and Relapse-Related Mortality After Allogeneic Hematopoietic Cell Transplantation: Comparing 2003-2007 Versus 2013-2017 Cohorts. Ann Intern Med. 2020;172(4):229–39.

Vermilion Disorders

68

Tami Yap, J. S. Kern, and Laura Scardamaglia

68.1 Introduction

A wide variety of contributory aetiologies can present as chronic, recurrent or widespread mucosal and submucosal changes to the vermilion including frictional trauma, infective causes, UV-related changes, immune-mediated disease, contact irritation and allergic reactions, as well as granulomatous disease. A brief selection of more common clinical presentations and associated diagnoses is described below.

T. Yap (✉)
Melbourne Dental School, The University of Melbourne, Melbourne, VIC, Australia

Royal Melbourne Hospital, Parkville, VIC, Australia
e-mail: tspyap@unimelb.edu.au

J. S. Kern
Royal Melbourne Hospital, Parkville, VIC, Australia

Melbourne Medical School, The University of Melbourne, Parkville, VIC, Australia

L. Scardamaglia
Royal Melbourne Hospital, Parkville, VIC, Australia

Melbourne Medical School, The University of Melbourne, Parkville, VIC, Australia

Western Health, St Albans, VIC, Australia

R. Balasubramaniam et al. (eds.), *Oral Medicine - A Clinical Guide*,
https://doi.org/10.1007/978-3-031-36797-7_68

68.2 Recurrent Ulceration

A common cause of recurrent ulceration of the vermilion is HSV-1 reactivation, termed herpes labialis when it occurs within the vermilion or at the vermilion border. This is often associated with a prodrome of tingling, followed by erythema, and the appearance of a crop of vesicles which coalesce into an ulceration that becomes crusted. This usually resolves within a week. For the majority of individuals, HSV-1 reactivation can be managed symptomatically, with only a minority requiring suppressive therapy typically in the setting of immunocompromise. Ulcerations associated with oral dermatoses may present on the lip and are differentiated from HSV-1 reactivation by their persistence and their tendency to not form a crust.

68.3 Acute Widespread Denudation

Ulceration associated with erythema multiforme, Stevens-Johnson's syndrome or fixed drug reactions may present acutely as widespread ulceration of the vermilion, most commonly with other intraoral mucosal lesions and may be with or without cutaneous or other mucosal site involvement. Bullae formation may or may not be described as preceding. These acute immune-mediated lesions often affect both the upper and lower vermilion. These clinical presentations erupt rapidly and are associated with severe symptoms. HSV-1 reactivation is an important association for erythema multiforme. Management is aimed at symptomatic relief and trigger identification with many individuals requiring inpatient care.

68.4 Persistent Keratosis, Erythema or Ulceration

Persistent ulceration may represent a chronic immunobullous disorder such as pemphigus vulgaris or mucous membrane pemphigoid. Oral lichen planus and oral lichenoid lesions including lupus can also affect the vermilion (Fig. 68.1) including

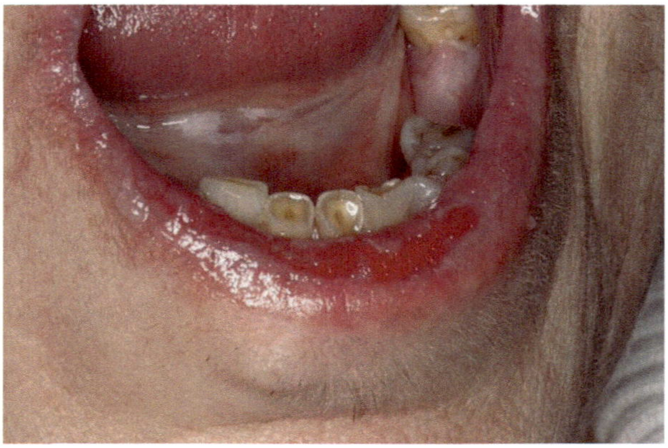

Fig. 68.1 Oral lichenoid inflammation of the lower vermilion including ulceration. Intraoral plaque-like changes are noted on the gingiva and floor of the mouth

typical striae, plaques and ulceration. Management of lichenoid inflammation on the vermilion can be challenging due to expected UV exposure. An important differential diagnosis of persistent crusting and erosions is squamous cell carcinoma. Eczematous cheilitis including irritant and allergic contact reactions to cosmetics and dentifrices should be considered in presentations of erythema, dryness and scaling that may also extend peri-orally. Persistent fissuring ulceration of the vermilion and angular cheilitis may associate acutely with *Staphylococcus aureus*, or represent chronic *candida-associated* conditions in the setting of haematinic deficiency or other disposing factors.

68.5 Chronic UV-Related Changes

Actinic cheilitis presents as changes to the vermilion common in fair-skinned adults with a history of chronic UV exposure and with a history of smoking also being a factor. The affected lip is often dry, scaly, fragile, cracked, with thickened plaques and papules, possibly pallor of leukoplakia, or crusting and erosions. There is loss of definition of the vermilion border. There may be dysplasia on histopathological examination. Squamous cell carcinoma of the lip frequently arise from UV-induced established actinic changes. Actinic changes are prevented by sun protection measures. Established changes are often treated with cryotherapy, laser ablation, surgical excision or topical therapies.

68.6 Persistent Lip Enlargement

Granulomatous cheilitis is a feature of a number of diseases including orofacial granulomatosis, Crohn's disease and sarcoidosis. The history is often described as multiple episodes of lip enlargement with decreasing resolution. Management is dependent upon the identification or ruling out of underlying systemic disease. In particular, granulomatous cheilitis may be the first sign of Crohn's disease.

68.7 Conclusion

The vermilion can present with pathologies which may affect the oral mucosae as well as cutaneous epithelium. Consideration of expected UV exposure of the vermilion, particularly to the lower lip, is important.

Recommended Reading

Celentano A, Tovaru S, Yap T, Adamo D, Aria M, Mignogna MD. Oral erythema multiforme: trends and clinical findings of a large retrospective European case series. Oral Surg Oral Med Oral Pathol Oral Radiol. 2015;120(6):707–16.

Grave B, McCullough M, Wiesenfeld D. Orofacial granulomatosis–a 20-year review. Oral Dis. 2009;15(1):46–51.

Yap T, Kern JS, Wood B, Scardamaglia L. Cutaneous Pathology of the Head and Neck. In: Farah C, Balasubramaniam R, McCullough M, editors. Contemporary Oral Medicine. Cham: Springer; 2019. p 704–60.

Sleep Disordered Breathing

69

Michael Stubbs and Anam Khan

69.1 Introduction

Sleep disordered breathing (SDB) is characterised by frequent episodes of obstructed breathing during sleep, often associated with oxygen desaturation, sleep fragmentation, snoring and daytime sleepiness.

The global prevalence of SDB, especially obstructive sleep apnoea (OSA), is high. Over 936 million individuals (aged 30–69 years) globally have OSA. Due to its high prevalence and adverse long-term outcomes, OSA represents a major burden with an estimated total economic cost of $18.9 billion in Australia in 2019–2020.

OSA is characterised by recurrent collapse of the pharyngeal airway during sleep, causing intermittent reduction in airflow (i.e. hypopnoea) or cessation in airflow (i.e. apnoea) (Fig. 69.1). Based on the current American Academy of Sleep Medicine (AASM 2020, Version 2.6) scoring criteria, a hypopnoea in adults is based on a $\geq$30% reduction in airflow for $\geq$10 seconds associated with an arousal or a $\geq$3% oxygen desaturation; and an obstructive apnoea is based on a $\geq$90% reduction in airflow for $\geq$10 seconds. Central sleep apnoea (CSA) is complete cessation of airflow and intrathoracic pressure, caused by brain inactivity. SDB has significant negative health implications, both intermediate and long term (Table 69.1).

M. Stubbs (✉)
School of Dentistry, Latrobe University, Melbourne, VIC, Australia
e-mail: M.Stubbs@latrobe.edu.au

A. Khan
UWA Dental School, The University of Western Australia, Nedlands, WA, Australia

R. Balasubramaniam et al. (eds.), *Oral Medicine - A Clinical Guide*,
https://doi.org/10.1007/978-3-031-36797-7_69

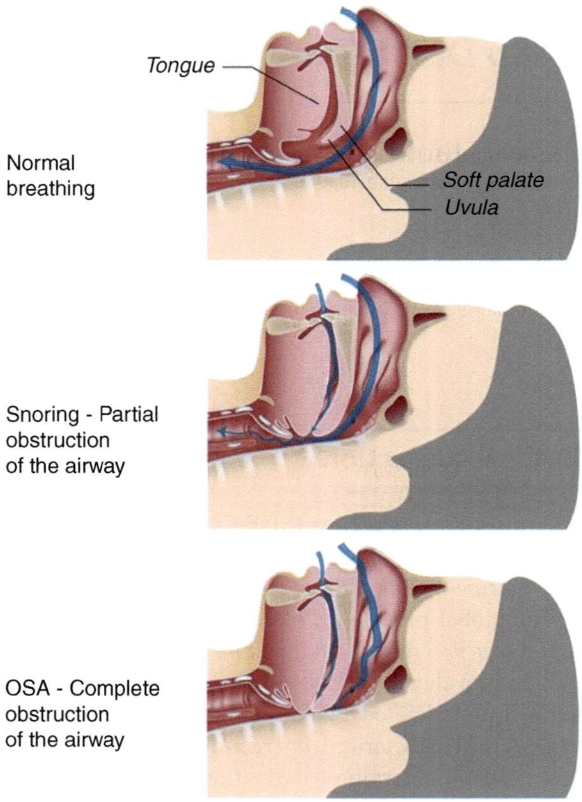

Fig. 69.1 Airway physiology in partial and complete obstruction of the airway in sleep. In the presence of an anatomically compromised, collapsible airway, the sleep-induced loss of compensatory tonic input to the upper airway dilator muscle motor neurons leads to collapse of the pharyngeal airway. (Image Credit: Alila Medical Media)

Table 69.1 Intermediate and long-term health implications of sleep disordered breathing

Intermediate health implications	Long-term comorbidities
• Sleep fragmentation • Intermittent hypoxia • Fluctuations in heart rhythm • Increased blood pressure • Fluctuations in intrathoracic pressure	• Hypertension • Type II diabetes mellitus • Cardiovascular morbidities • Increased risk of stroke • Cognitive impairment • Mood disorders (depression and anxiety) • Premature death

69.2 Clinical Presentation and Investigations

Patient's history of snoring and daytime sleepiness are common indicators for dentists to investigate further. The Epworth Sleepiness Scale (ESS), STOP-BANG (for males) or OSA 50 (for females) questionnaires are useful screening tools. Comorbidities such as high body mass index (BMI), hypertension, type 2 diabetes mellitus, cardiovascular disease or cerebrovascular event necessitate further investigation.

The gold-standard diagnostic tool for SDB is an in-laboratory (monitored) overnight polysomnography (PSG). The primary PSG metric used to define OSA severity is the apnoea-hypopnoea index (AHI), a measure of the number of obstructive breathing events per hour of sleep (Table 69.2). Respiratory disturbance index (RDI) includes AHI as well as the respiratory effort-related arousals (RERAs) that can also be used to determine the severity of the SDB. Upper airway resistance syndrome is diagnosed in the case of absence of AHI but increased RERAs leading to brief arousals and daytime somnolence. The severity of the AHI score does not always correspond to severity of symptoms experienced by the patient.

Table 69.2 The severity of diagnosed obstructive sleep apnoea is based on the apnoea-hypopnoea index (AHI) or respiratory disturbance index (RDI)

AHI/RDI	Severity of OSA
<5 events	No OSA
>5–15 events	Mild OSA
>15–30 events	Moderate OSA
>30 events	Severe OSA

69.3 Management and Prognosis

There are a number of treatment options available. Treatment for SDB should be prescribed by a sleep physician and includes continuous positive air pressure, mandibular advancement appliances or upper airway surgery. Factors for treatment selection include aetiology, disease severity, comorbidities and patient preference (Table 69.3). In addition to the standard treatment options, lifestyle changes and adjunctive therapies such as positional therapy, weight loss, alcohol and sedative avoidance may be effective.

Table 69.3 Clinical considerations and adverse effects of common treatments for sleep disordered breathing

Treatment options	Common indications	Variations	Treatment challenges
Continuous Positive Airway Pressure (CPAP)	• Moderate to severe OSA	• A continuous airflow via facial or nasal mask • Auto-titrating or fixed pressure CPAP	• Difficulty tolerating mask • Claustrophobia • Compliance: 50–60%
Mandibular Advancement Appliance (MAA)	• Mild to moderate OSA • CPAP intolerant patient	• Custom fabricated • Titratable • Many designs	• Long term occlusal changes • Teeth movement • Transient jaw pain
Upper Airway Surgery	• CPAP and MAA intolerant patient	• Bimaxillary advancement surgery • Tonsillectomy • Uvulopalatopharyngoplasty • Tongue reduction • Hypoglossal nerve stimulator	• Post-operative pain • Invasive • Irreversible

Recommended Reading

Baldwin CM, Griffith KA, Nieto FJ, O'Connor GT, Walsleben JA, Redline S. The association of sleep-disordered breathing and sleep symptoms with quality of life in the Sleep Heart Health Study. Sleep. 2001;24(1):96–105.

Berry RB, Quan SF, Abreu AR, Bibbs ML, DelRosso L, Harding SM, et al. The AASM manual for the scoring of sleep and associated events: rules, terminology and technical specifications, version 2.6. Darien, Illinois: American Academy of Sleep Medicine; 2020.

Joosten SA, Edwards BA, Wellman A, Turton A, Skuza EM, Berger PJ, et al. The Effect of Body Position on Physiological Factors that Contribute to Obstructive Sleep Apnea. Sleep. 2015;38(9):1469–78.

Peppard PE, Szklo-Coxe M, Hla KM, Young T. Longitudinal association of sleep-related breathing disorder and depression. Arch Intern Med. 2006;166(16):1709–15.

Peppard PE, Young T, Palta M, Skatrud J. Prospective study of the association between sleep-disordered breathing and hypertension. N Engl J Med. 2000;342(19):1378–84.

Shamsuzzaman AS, Gersh BJ, Somers VK. Obstructive sleep apnea: implications for cardiac and vascular disease. JAMA. 2003;290(14):1906–14.

Shrivastava D. Impact of sleep-disordered breathing treatment on upper airway anatomy and physiology. Sleep Med. 2014;15(7):733–41.

Simou E, Britton J, Leonardi-Bee J. Alcohol and the risk of sleep apnoea: a systematic review and meta-analysis. Sleep Med. 2018;42:38–46.

Streatfeild J, Smith J, Mansfield D, Pezzullo L, Hillman D. The social and economic cost of sleep disorders. Sleep. 2021;44(11):zsab132.

Young T, Finn L, Peppard PE, Szklo-Coxe M, Austin D, Nieto FJ, et al. Sleep disordered breathing and mortality: eighteen-year follow-up of the Wisconsin sleep cohort. Sleep. 2008;31(8):1071–8.

Oral Appliance Therapy for Sleep Disordered Breathing

70

Michael Stubbs and Anam Khan

70.1 Introduction

Mandibular advancement appliance (MAA) reduces the upper airway collapsibility during sleep by advancing the mandible. Mandibular advancement increases airway volume at the velopharynx where the airway space is mostly enlarged laterally by traction of the soft tissue connections between the pharynx and ramus. MAAs are fabricated with biocompatible materials and include a titration mechanism to facilitate gradual mandibular advancement (Fig. 70.1a, b). There is no difference in the efficacy of various titratable MAA based on the design or the material used. The increase in pharyngeal space with the MAA in situ improves the pharyngeal airflow during sleep and is an effective treatment for snoring, upper airway resistance syndrome, mild to moderate obstructive sleep apnoea (OSA) and severe OSA, where continuous positive airway pressure (CPAP) is not tolerated. Although CPAP is more efficacious than MAA, the higher efficacy of CPAP may be offset by greater MAA compliance in some cases.

According to the clinical practice guidelines for oral appliance therapy for OSA and snoring published in 2015, an oral appliance must be presented as a treatment modality by a sleep physician for snoring and OSA. If prescribed, a qualified dentist must deliver a custom fabricated, titratable appliance. The patient must be followed up by their sleep physician and dentist to monitor the efficacy and side effects of oral appliance therapy.

M. Stubbs (✉)
School of Dentistry, Latrobe University, Melbourne, VIC, Australia
e-mail: M.Stubbs@latrobe.edu.au

A. Khan
UWA Dental School, The University of Western Australia, Nedlands, WA, Australia

R. Balasubramaniam et al. (eds.), *Oral Medicine - A Clinical Guide*,
https://doi.org/10.1007/978-3-031-36797-7_70

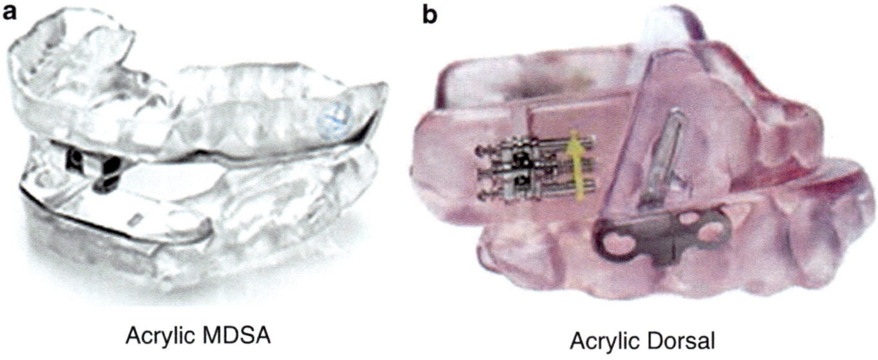

Fig. 70.1 Mandibular advancement appliances (MAAs) are fabricated with biocompatible materials, including titration mechanisms to facilitate gradual mandibular advancement. Here are two examples of MAAs (**a**) Acrylic MDSA, (**b**) Acrylic Dorsal (Image courtesy of Somnomed®). Variations of these designs in 3D printed nylon are also available

Fig. 70.2 Tongue retaining device. (Image courtesy of Zensleep Pty Ltd)

In cases where MAA is not suitable, such as inadequate teeth for retention and if the CPAP is not tolerated, a tongue retaining device may be considered. This is a prefabricated, over-the-counter silicone device that holds the tongue in a forward position during sleep.

70.2 Principles of Mandibular Advancement Appliance Therapy

The benefit of a custom fabricated MAA is both in its comfort, and the ability to gradually advance the mandible over time. Most MAAs are made of acrylic or 3D printed nylon material, with some appliances having a soft inner lining or Adams clasps for retention. The titration mechanism allows gradual mandibular protrusion

Table 70.1 Objective and subjective benefits of mandibular advancement appliance therapy

Objective benefits	Subjective benefits
• Reduction of apnoea-hypopnoea index • Improvement of arousal index, oxygen saturation, rapid eye movement sleep duration and sleep fragmentation • Reduction of Epworth Sleepiness Scale • Improvement in systolic and diastolic blood pressure	• Improved sleep quality • Reduction of snoring • Reduction in daytime sleepiness and lethargy • Improvement in concentration, mood, motivation and quality of life • Reduction of the risk for motor vehicle and work-related accidents due to tiredness

in 0.1–1 mm increment. Determining the range of protrusion and the starting position of the MAA can be measured using a George gauge. The starting position is ideally 50–60% of the total protrusive position. Excessive vertical dimension results excessive posterior rotation of the mandible, reducing the efficacy of the MAA. The ideal mandibular position is subjectively determined by snoring cessation, and improvement in daytime symptoms of tiredness, lethargy, and sleepiness. Once the optimum titration is achieved on the MAA, a treatment polysomnography may be necessary to objectively assess the efficacy of the appliance. The effectiveness of MAA is based on objective and subjective benefits to patients (Table 70.1).

70.3 Side Effects of Mandibular Advancement Appliance Therapy

Informed consent regarding potential side effects is necessary before commencing treatment. These include the arousal or exacerbation of temporomandibular disorders, occlusal and skeletal changes, dry mouth or excess salivation, and soft tissue irritation. Most side effects are transient but patients require monitoring to ensure ongoing compliance with MAA therapy.

Recommended Reading

Andrén A, Hedberg P, Walker-Engström ML, Wahlén P, Tegelberg A. Effects of treatment with oral appliance on 24-h blood pressure in patients with obstructive sleep apnea and hypertension: a randomized clinical trial. Sleep Breath. 2013;17(2):705–12.

Araie T, Okuno K, Ono Minagi H, Sakai T. Dental and skeletal changes associated with long-term oral appliance use for obstructive sleep apnea: A systematic review and meta-analysis. Sleep Med Rev. 2018;41:161–172.

Brown EC, Cheng S, McKenzie DK, Butler JE, Gandevia SC, Bilston LE. Respiratory Movement of Upper Airway Tissue in Obstructive Sleep Apnea. Sleep. 2013;36(7):1069–1076.

Doff MH, Finnema KJ, Hoekema A, Wijkstra PJ, de Bont LG, Stegenga B. Long-term oral appliance therapy in obstructive sleep apnea syndrome: a controlled study on dental side effects. Clin Oral Investig. 2013;17(2):475–82.

Kazemeini E, Braem MJ, Moorkens G, Balina S, Kastoer C, Op de Beeck S, Vanderveken OM, Dieltjens M. Scoring of Hypersomnolence and Fatigue in Patients With Obstructive Sleep

Apnea Treated With a Titratable Custom-Made Mandibular Advancement Device. J Clin Sleep Med. 2019;15(4):623–628.

Marco-Pitarch R, García-Selva M, Plaza-Espín A, Puertas-Cuesta J, Agustín-Panadero R, Fernández-Julián E, Marco-Algarra J, Fons-Font A. Dimensional analysis of the upper airway in obstructive sleep apnoea syndrome patients treated with mandibular advancement device: A bi- and three-dimensional evaluation. J Oral Rehabil. 2021;48(8):927–936.

Mayoral P, Lagravère MO, Míguez-Contreras M, Garcia M. Antero-posterior mandibular position at different vertical levels for mandibular advancing device design. BMC Oral Health. 2019;19(1):85.

Phillips CL, Grunstein RR, Darendeliler MA, Mihailidou AS, Srinivasan VK, Yee BJ, Marks GB, Cistulli PA. Health outcomes of continuous positive airway pressure versus oral appliance treatment for obstructive sleep apnea: a randomized controlled trial. Am J Respir Crit Care Med. 2013;187(8):879–87.

Ramar K, Dort LC, Katz SG, Lettieri CJ, Harrod CG, Thomas SM, et al. Clinical Practice Guideline for the Treatment of Obstructive Sleep Apnea and Snoring with Oral Appliance Therapy: An Update for 2015. J Clin Sleep Med. 2015;11(7):773–827.

Diabetes Mellitus

71

Mark Schifter, Ben Karim, and Suma Sukumar

71.1 Introduction

Diabetes mellitus (DM), generally referred to as "diabetes", is a common metabolic disease characterised by persistently elevated blood glucose levels (hyperglycaemia) resulting from defects in insulin secretion, insulin action or both. Persistent hyperglycaemia can lead to acute and chronic complications, the latter manifesting as damage, dysfunction and failure of various organs, particularly the eyes, kidneys, nerves, heart and blood vessels.

M. Schifter
Westmead Centre for Oral Health, Westmead Hospital, Westmead, NSW, Australia

The Skin Hospital, Darlinghurst, NSW, Australia

Sydney Dental School, The University of Sydney, Camperdown, NSW, Australia

B. Karim · S. Sukumar (✉)
Westmead Centre for Oral Health, Westmead Hospital, Westmead, NSW, Australia

Sydney Dental School, The University of Sydney, Camperdown, NSW, Australia
e-mail: suma.sukumar@health.nsw.gov.au

R. Balasubramaniam et al. (eds.), *Oral Medicine - A Clinical Guide*,
https://doi.org/10.1007/978-3-031-36797-7_71

Glucose homeostasis is a tightly regulated process reliant on the opposing effects of the hormones, insulin and glucagon, in response to blood glucose levels (BGL). High BGL result in the release of insulin, whereas low blood sugar levels in the release of glucagon (and vice versa).

Insulin deficiency may be absolute, resulting from total loss of production, as seen in Type 1 DM, or relative, due to impaired production and/or the function as seen in Type 2 DM. Treatment is tailored to the specific subtype of disease (see Table 71.1). Glycosylated haemoglobin (HbA1c) is used for monitoring disease control which assesses BGL over the previous 120 days.

Table 71.1 Subtypes of diabetes mellitus

Subtype	Epidemiology	Pathophysiology and risk factors	Treatment
Type 1 (T1DM)	• 10% of all diabetics • Onset typically in children and young adults	• Immune-mediated destruction of pancreatic beta islet cells *Risk factors:* • Genetics • Environment: viruses	• Insulin analogues • Insulin pump • Transplantation of pancreatic islet cell or whole pancreas
Type 2 (T2DM)	• 90% of all diabetics • Worldwide: 1 in 11 • Australia 5% • (25% ≥65 years or older)	• Failure in insulin production and/or insulin resistance *Risk factors:* • Body mass index • Diet • Sedentary lifestyle • Genetics	• Diet and exercise • Oral hypoglycaemic medications/agents • Insulin analogues (for refractory DM) • Bariatric surgery
Gestational	• 5–10% of pregnant women (risk for postpartum T2DM)	• Hormone associated Insulin resistance *Risk factors:* • BMI • Genetics • Polycystic ovarian syndrome	• Insulin analogues
Secondary/ Iatrogenic	• DM prevalence in acromegaly: 20–35% • DM prevalence in hypercortisolism: <10% • DM incidence with systemic corticosteroids: 34–56%	• Disease or therapy related insulin resistance *Risk factors:* • Corticosteroid regime • Older age • Body mass index	• Similar to T2DM

BSL blood sugar level (blood glucose level), *HbA1c* glycosylated haemoglobin, *GAD* glutamic acid decarboxylase, *OGTT* oral glucose tolerance test, *BGL* blood glucose levels, *Ivi* intravenous injection

71.2 Complications

Many of the complications related to advanced or poorly controlled DM results from binding of advanced glycation end products to receptors expressed on inflammatory cells, endothelium and vascular smooth muscle leading to vascular insufficiency, tissue hypoxia and subsequently necrosis. Macrovascular disease can cause atherosclerosis, resulting in angina, heart attack and stroke. On the other hand, microvascular disease encompasses diabetic retinopathy, nephropathy and neuropathy.

Given the high prevalence of DM worldwide, dental practitioners must be knowledgeable in recognising common orofacial manifestations of DM. Table 71.2 outlines the more common orofacial manifestations of DM.

Table 71.2 Orofacial manifestations of diabetes mellitus

Clinical condition	Aetiology/pathophysiology
Sialosis including parotid sialoadenopathy	• Autonomic neuropathy • Attempted hypersalivation to combat xerostomia
Angular cheilitis	• Infection (impaired immunity): *Candida* species with coinfection by staphylococcal bacteria of the perioral skin
Hyposalivation	• Autonomic neuropathy • Hyposalivation from dehydration secondary to polyuria
Candidosis	• Infection secondary to impaired immunity
Oropharyngeal candidosis	• Predisposing factors:
Atrophic candidosis	– Hyposalivation
Denture stomatitis	– Increased glucose in saliva
Median rhomboid glossitis	
Periodontal disease	• Infection secondary to impaired immunity • Accumulation of dental plaque • Impaired neutrophil response: – Defective chemotaxis and impaired phagocytosis
Poor oral hygiene	• Impaired manual dexterity secondary to neuropathy affecting hands

Recommended Reading

Australian Institute of Health and Welfare. Diabetes: Australian facts [Internet]. Canberra: Australian Institute of Health and Welfare, 2023 [cited 2023 Dec. 3]. Available from: https://www.aihw.gov.au/reports/diabetes/diabetes.

Schifter M, McLean M, Sukumar S. Disorders of the Endocrine System and of Metabolism. In: Glick M, Greenberg MS, Lockhart, Challacombe SJ, editors. Burket's Oral Medicine. 13th ed. Wiley-Blackwell; 2021. p. 817–902.

Zheng Y, Ley SH, Hu FB. Global aetiology and epidemiology of type 2 diabetes mellitus and its complications. Nat Rev Endocrinol. 2018;14(2):88–98.

Endocrinopathies

72

Mark Schifter, Ben Karim, and Suma Sukumar

72.1 Introduction

The endocrine system is comprised of hormone-secreting glands that control the body's homeostasis, metabolism, growth and development. Endocrinopathies are a result of hormonal excess, deficiency or changes in functional efficacy. Deficiency is usually due to endocrine gland destruction from autoimmune or neoplastic disease and treatment. Hormone resistance occurs when hormones are ineffective in triggering normal physiological responses. Several endocrinopathies can have orofacial manifestations and are discussed in this chapter. A more detailed discussion of diabetes mellitus can be found in Chap. 71.

M. Schifter
Westmead Centre for Oral Health, Westmead Hospital, Westmead, NSW, Australia

The Skin Hospital, Darlinghurst, NSW, Australia

Sydney Dental School, The University of Sydney, Camperdown, NSW, Australia

B. Karim · S. Sukumar (✉)
Westmead Centre for Oral Health, Westmead Hospital, Westmead, NSW, Australia

Sydney Dental School, The University of Sydney, Camperdown, NSW, Australia
e-mail: suma.sukumar@health.nsw.gov.au

R. Balasubramaniam et al. (eds.), *Oral Medicine - A Clinical Guide*,
https://doi.org/10.1007/978-3-031-36797-7_72

72.2 The Pituitary Gland

The pituitary gland is considered the control centre of the endocrine system. Pathological changes of the pituitary can affect singly or multiple trophic hormone release with resulting changes to the function of the effector gland endocrine glands. Pituitary tumours are the most common cause of hypopituitarism that on treatment will require the patient to be provided with life-long hormone replacement.

Abnormalities of growth and stature are due to deficiencies or excess of growth hormone (GH) which is released from the pituitary. GH deficiency can result in short stature, and if recognised, exogenous GH can be given to improve growth height and velocity.

Gigantism is due to GH excess, before the bony epiphyseal growth plates have fused. Acromegaly is due to GH excess after the bony epiphyseal growth plates have fused. Orofacial feature of acromegaly includes coarse facies, macrognathia, macroglossia, enlarging diastemas and enlargement of the nose and frontal bones. Treatment is either surgical or medical management of the causative pituitary adenoma.

72.3 The Parathyroid Glands

Parathyroid hormone (PTH) is secreted by the four parathyroid glands in response to low serum calcium. High serum calcium and/or Vitamin D_3 suppress the production of PTH.

Primary hyperparathyroidism is due to excessive PTH production/secretion from the parathyroid glands, generally due to parathyroid adenoma. Secondary hyperparathyroidism is due to parathyroid hyperplasia as a consequence of chronic hypercalcaemia from renal failure, inadequate vitamin D metabolism or inadequate calcium intake or absorption This can lead to tertiary hyperparathyroidism, with excessive PTH secretion from loss of inhibitory control of the parathyroid glands. The consequences are osteopenia and osteoporosis.

Hyperparathyroidism can cause non-neoplastic lesions in the mandible and maxilla, called brown tumours. These occur in approximately 1.5% of patients with secondary hyperparathyroidism and are more common in females than in males. They are generally asymptomatic, presenting radiographically as a well demarcated uni- or multilocular radiolucency. If the brown tumour enlarges to breach the bone cortex, there may be localised expansion of the jaw with a slow-growing mass noted intra-orally. Brown tumours are managed surgically, together with correction of the hyperparathyroidism.

72.4 The Adrenal Glands

Cushing's syndrome (resulting from use of exogenous glucocorticoids) and Cushing's disease (due to endogenous glucocorticoid secretion by the adrenal glands due to excess ACTH from a pituitary adenoma) may result in easy bruising

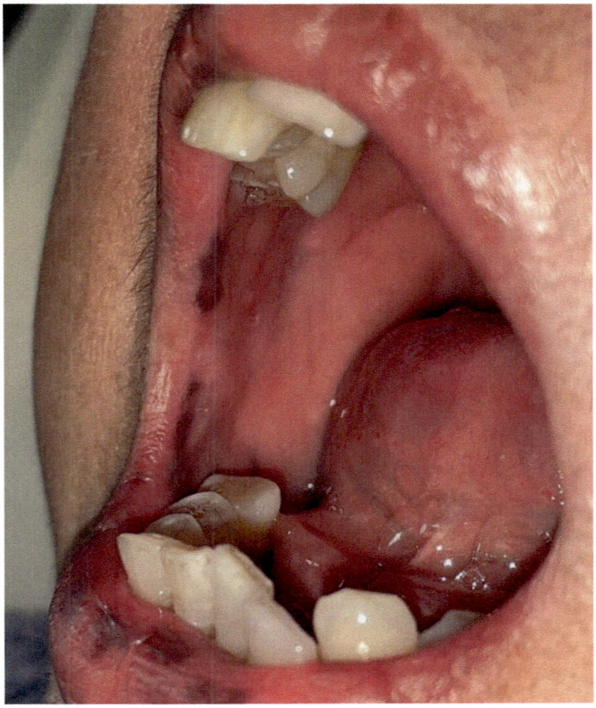

Fig. 72.1 Multifocal pigmentation of the right buccal mucosa and lower lip in a patient with Addison's disease

of the skin, fat deposition of the central upper back, rounded facies, proximal myopathy, diabetes mellitus, hypertension, gastric reflux and psychiatric effects of marked anxiety and/or depression.

Addison's disease is caused by primary adrenal deficiency. A common clinical presentation is mucocutaneous hyperpigmentation, affecting the skin, oral mucosa (Fig. 72.1), conjunctiva and genital mucosa. Orally, this may present as asymptomatic brown macular pigmented lesions on any mucosal surface. Oral pigmentation may be one of the first signs of Addison's disease.

Recommended Reading

Atreja G, Atreja SH, Jain N, Sukhija U. Oral manifestations in growth hormone disorders. Indian J Endocrinol Metab. 2012;16(3):381–3.

Sarkar SB, Sarkar S, Ghosh S, Bandyopadhyay S. Addison's disease. Contemp Clin Dent. 2012;3(4):484–6.

Schifter M, McLean M, Sukumar S. Disorders of the Endocrine System and of Metabolism. In: Glick M, Greenberg MS, Lockhart, Challacombe SJ, editors. Burket's Oral Medicine. 13th ed. Wiley-Blackwell; 2021. p. 817–902.

Kesidi S, Bhayya H. Oral manifestations of endocrine disorders. Int J Appl Dent Sci. 2019;5:476–478.

Osteoporosis and Medication-Related Osteonecrosis of the Jaws

73

Antonio Celentano and Elena Calabria

73.1 Introduction

Osteoporosis is a systemic skeletal disease, characterized by bone fragility and susceptibility to fracture. In Australia, 29% of women and 10% of men aged 75 and over have osteoporosis. The treatment of osteoporosis is aimed at the fracture prevention and is generally based on the use of anti-resorptive agents, such as bisphosphonates and Denosumab. These agents are also used for oncological conditions.

A rare anti-resorptive adverse event is the development of medication-related osteonecrosis of the jaws (MRONJ). MRONJ is defined as the presence of exposed bone or bone that can be probed through an intra-oral or extra-oral fistula(e) in the maxillofacial region that has persisted for more than 8 weeks, in patients with current or previous treatment with anti-resorptive therapy alone or in combination with immune modulators or antiangiogenic medications, and no history of radiation therapy to the jaws or metastatic disease to the jaws.

Suppression of bone remodelling is the most widely accepted pathogenic hypothesis of MRONJ. Anti-resorptive drugs inhibit osteoclastic activity leading to an increase of bone mineral density and to a decrease of the risk of bone fractures. As a counterpart, the alteration of bone turnover impacts bone healing processes, and the newly formed bone may become ischemic and necrotic as a consequence of insufficient blood supply. Pathogenesis remains multifactorial and there are other

A. Celentano (✉)
Melbourne Dental School, The University of Melbourne, Carlton, VIC, Australia
e-mail: antonio.celentano@unimelb.edu.au

E. Calabria
School of Dentistry, University Magna Grecia of Catanzaro, Catanzaro, Italy

R. Balasubramaniam et al. (eds.), *Oral Medicine - A Clinical Guide*,
https://doi.org/10.1007/978-3-031-36797-7_73

leading hypotheses including inflammation or infection, angiogenesis inhibition, innate or acquired immune dysfunction as well as genetic predisposition.

Anti-resorptive therapy increases the risk of MRONJ depending on the type of drug, potency and dosage, route of administration and length of treatment. Indeed, the risk is considerably higher in the oncological (<5%) than in the osteoporotic patients (<0.05%). Other factors include concomitant medications, underlying diseases, dental infections and surgical procedures.

73.2 Clinical Presentation and Investigations

MRONJ includes a non-exposed variant (classified as Stage 0), which include up to 25% of cases, and is characterised by the presence of unexplained symptoms mimicking odontalgia or unexplained loosening of teeth. The exposed variant presents clinically as an area of exposed and necrotic bone or a fistula that connects to bone in the absence (Stage 1) or evidence of infection and symptoms (Stage 2). Stage 3 presentations include exposed necrotic bone extending beyond the alveolar bone, pathologic fractures, extra-oral fistulae, oral-antral or oral-nasal communications, or osteolysis extending to the inferior border of the mandible or sinus floor. Radiographic presentations vary in different stages of MRONJ, from minor bony changes such as sclerosis and periodontal ligament widening to areas of osteolysis with or without sclerotic areas (Fig. 73.1).

In addition to history and examination, computer tomography can be accurate in detecting disease extension and margins. Differential diagnoses include osteoradionecrosis, bone metastasis, odontogenic and non-odontogenic neoplasms.

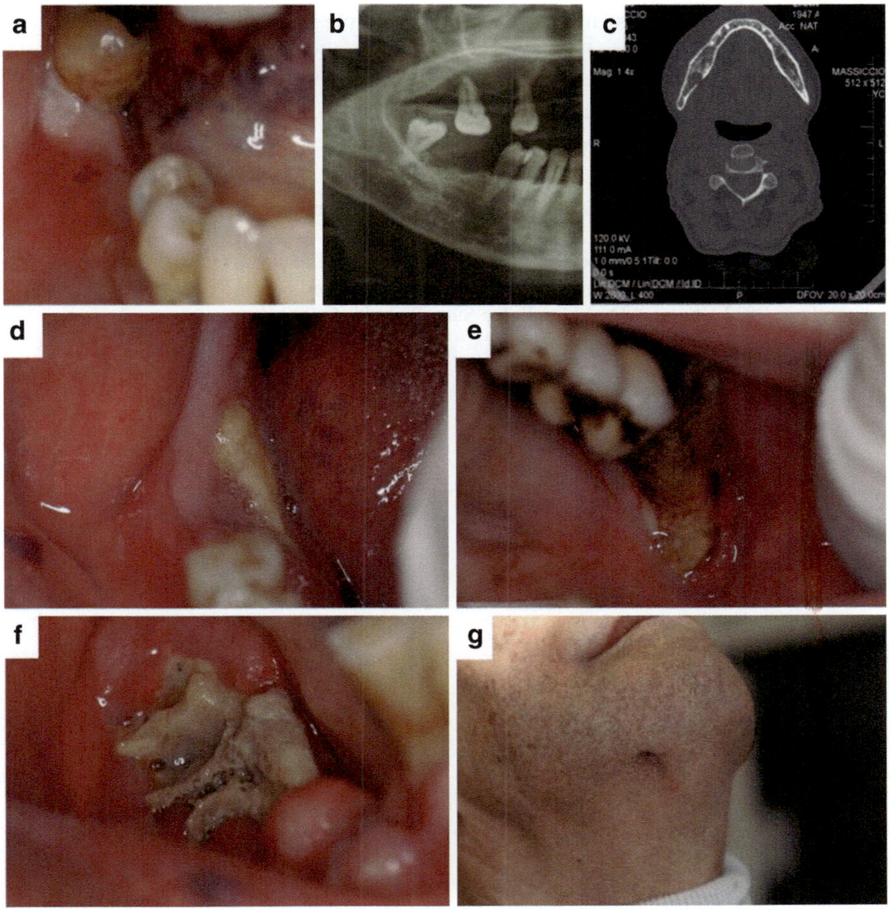

Fig. 73.1 Medication-related osteonecrosis of the jaws (MRONJ): Stage 0 (**a**), non-exposed bone in asymptomatic patient with positive radiographic findings (OPG (**b**) and CT scan (**c**)); stage 1 (**d**), showing exposed and necrotic bone in asymptomatic patient with no evidence of infection; stage 2 (**e**); stage 3 (**f**); and extra-oral view of patient in (**f**) with extra-oral fistula (**g**)

73.3 Management and Prognosis

MRONJ management includes non-surgical and surgical approaches. The goal of both are the prioritisation and support of continued bone health and fractures prevention in osteoporotic patients, and the continued oncologic treatment in cancer patients as well as preservation and improvement of quality of life. Non-surgical approaches (e.g. topical antiseptics, systemic antibiotics, painkillers) are applicable at all stages and can contribute to controlling pain and infection and encouraging spontaneous bone sequestrum exfoliation. Resective bone surgical approaches can have high success rates at all stages.

Recommended Reading

Reid IR, Billington EO. Drug therapy for osteoporosis in older adults. Lancet. 2022;399(10329):1080–92.

Ruggiero SL, Dodson TB, Aghaloo T, Carlson ER, Ward BB, Kademani D. American Association of Oral and Maxillofacial Surgeons' Position Paper on Medication-Related Osteonecrosis of the Jaws-2022 Update. J Oral Maxillofac Surg. 2022;80(5):920–43.

Immunology and the Oral Cavity

74

Guangzhao Guan and Ajith Polonowita

74.1 Introduction

The oral cavity is part of the mucosal immune system and plays an essential role in innate and adaptive immune responses. Immune dysfunction may be classified into hypersensitivity or overactivity, autoimmunity or self-reactivity, and immunodeficiency or underactivity.

74.2 Hypersensitivity Responses

74.2.1 Angioedema

Angioedema (angioneurotic oedema) (Fig. 74.1) is a rare hive-like, localised subcutaneous (or submucosal) swelling, which is caused by increased vascular permeability in interstitial tissues of the respiratory and gastrointestinal tract.

Angioedema (inherited or acquired) can be classified into at least eight types and include:

1. Hereditary angioedema
2. Acquired C1esterase inhibitor deficiency
3. Inherited angioedema with normal C1esterase inhibitor levels
4. Angiotensin converting enzyme inhibitor (ACEI)-associated angioedema

G. Guan
Faculty of Dentistry, The University of Otago, Dunedin, New Zealand

A. Polonowita (✉)
Faculty of Dentistry, The University of Otago, Dunedin, New Zealand

Christchurch Hospital, Canterbury, New Zealand
e-mail: ajith.polonowita@otago.ac.nz

R. Balasubramaniam et al. (eds.), *Oral Medicine - A Clinical Guide*,
https://doi.org/10.1007/978-3-031-36797-7_74

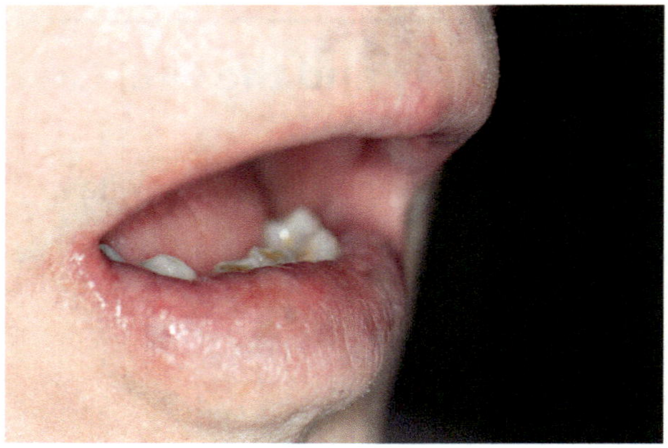

Fig. 74.1 Angioedema

5. Idiopathic angioedema
6. Allergic angioedema
7. NSAID-associated angioedema
8. Angioedema with urticarial vasculitis

The most commonly affected sites are the lips, palms, soles, limbs, truck, and genitalia. Allergy-associated angioedema is usually transient. However, it can be life-threatening (anaphylaxis) when severe angioedema affects the tongue, soft palate, pharynx, larynx, respiratory tract or results in airway obstruction. If signs of airway compromise are detected, endotracheal intubation is likely to be necessary, and the airway should be managed accordingly.

74.3 Immune-Mediated Dermatologic Diseases and Autoimmunity

A number of autoimmune diseases, such as pemphigus vulgaris and mucous membrane pemphigoid, occur in which antibodies against tissue antigens cause immunopathological damage by activating type II hypersensitivity. The diagnosis is based on the clinical manifestations, histological examination, direct/indirect immunofluorescence, demonstration of tissue-bound and circulating autoantibodies, immunoblotting, or quantitative immunoassay. The management of autoimmune diseases can be challenging and may require a multidisciplinary approach. Erythema multiforme (considered a type IV hypersensitivity reaction) is an acute, relatively common immune-mediated condition characterised by serosanguinous exudates on the lips and the appearance of distinctive target-like lesions on the skin. Oral lichenoid contact lesion could be considered a type IV hypersensitivity reaction. It may present as a unilateral lesion rather than bilateral lesions, which can usually be seen in oral lichen planus. Although the aetiology of plasma cell gingivitis (see Chap. 24) is unknown, it has been postulated to be a hypersensitive reaction to allergens.

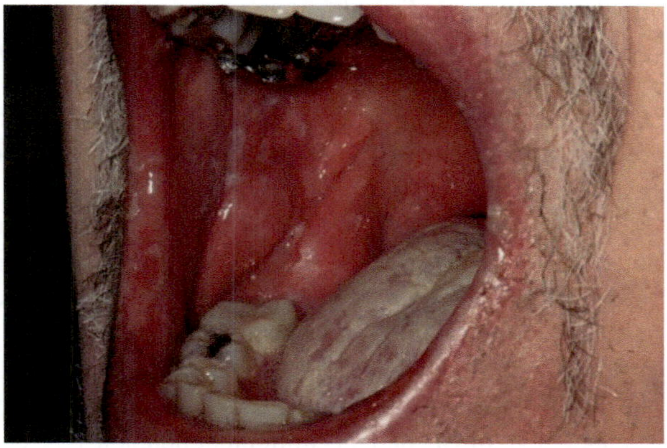

Fig. 74.2 Pemphigus vulgaris presenting as painful blisters and erosions of the oral mucosa

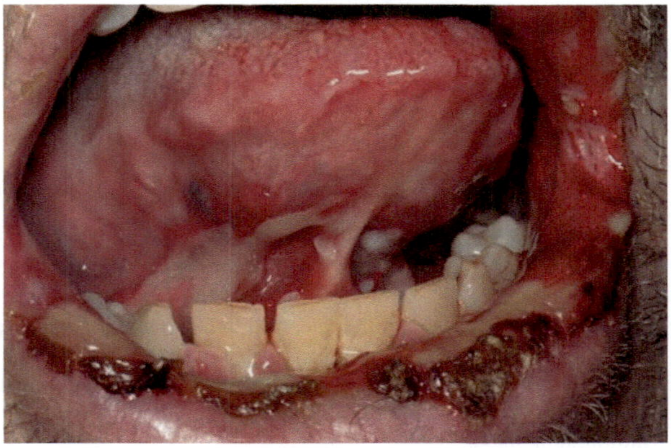

Fig. 74.3 Erythema multiforme presenting as diffuse ulceration of the oral mucosa and haemorrhagic crusts on the lips

A number of autoimmune diseases that manifest in the oral cavity have been attributed to hypersensitivity reactions. Type II hypersensitivity has been implicated in the aetiopathogenesis of pemphigus vulgaris (Fig. 74.2) and mucous membrane pemphigoid. In these conditions, autoantibodies against tissue antigens cause immunopathological damage.

Type IV hypersensitivity has been implicated in erythema multiforme (Fig. 74.3) and oral lichenoid contact lesions. Additionally, although the aetiology of plasma cell gingivitis has not been fully elucidated, it has been postulated to be associated with hypersensitivity to certain allergens.

74.4 Immunodeficiency Disorders

Immunodeficiency disorders can be grouped into primary (inherited) and secondary (acquired disease- or therapy-associated).

Primary immunodeficiencies (PIDs) are a group of more than 400 disorders that can lead to mild to life-threatening infections, swellings, and autoimmune problems. PID symptoms often appear in infancy, but some may not present until adulthood. General clinical features may include recurrent infections, infection with unusual micro-organisms (e.g. Aspergillus, Pneumocystis), infection of unusual sites, chronic diarrhoea, failure to thrive, chronic osteomyelitis/deep-seated abscesses, and structural damage (e.g. bronchiectasis). Specific oral manifestations are not common in PID. The management of PID involves treating the underlying cause of the PID and preventing and treating infections.

Secondary immunodeficiencies, such as HIV, are far more common than primary immunodeficiencies. These deficiencies manifest clinically by an increased frequency or unusual complications of infections. The secondary immunodeficiencies have a wide spectrum of presentation, depending on the magnitude of the offending external condition and on the host susceptibility. Malnutrition, both under and over-nutrition, is the most common secondary immunodeficiency worldwide. Management of secondary immunodeficiencies is dependent on the associated cause. Common oral manifestations of immunodeficiencies are often related to opportunistic infection or exaggerated presentation of commensal pathogens and are summarised in Table 74.1.

Table 74.1 Oral manifestations of hypersensitivity, autoimmunity and immunodeficiencies

Aetiology	Clinical presentation
Hypersensitivity	Allergic angioedema, allergy contact cheilitis
Immune-mediated diseases	Pemphigus vulgaris, mucous membrane pemphigoid, erythema multiforme, oral lichenoid lesion, and plasma cell gingivitis
Autoimmunity	Pemphigus vulgaris, mucous membrane pemphigoid, systemic lupus erythematosus, chronic ulcerative stomatitis, rheumatoid arthritis, and juvenile idiopathic arthritis
Immunodeficiency disorders	
Primary	Dysmorphic face with hypertelorism, mid-face anomalies, and arched palate
	Anodontia/hypodontia, delayed formation and eruption of teeth, aberrant tooth shape, and enamel hypoplasia along with enamel hypocalcification
	Early-onset aggressive periodontitis
	Recurrent aphthous-like ulceration
	Necrotizing stomatitis
	Chronic multifocal oral candidosis, angular cheilitis, hyperplastic candidiosis
Secondary	Oral hairy leukoplakia (EBV)
	Oral candidosis such as linear gingival erythema, pseudomembranous candidosis, and erythematous candidosis
	Necrotizing ulcerative stomatitis/gingivitis
	Kaposi sarcoma
	Non-Hodgkin lymphoma
	Linear gingival erythema
Malignancy	Lymphomas (non-Hodgkin lymphoma is commonly associated with HIV infection and immunosuppressive therapy)
	Kaposi sarcoma
Others	Aphthous-like ulcers, HIV-associated salivary gland disease

Recommended Reading

Black RE, Allen LH, Bhutta ZA, Caulfield LE, de Onis M, Ezzati M, et al. Maternal and child undernutrition: global and regional exposures and health consequences. Lancet. 2008;371(9608):243–60.

Cheng YS, Gould A, Kurago Z, Fantasia J, Muller S. Diagnosis of oral lichen planus: a position paper of the American Academy of Oral and Maxillofacial Pathology. Oral Surg Oral Med Oral Pathol Oral Radiol. 2016;122(3):332–54.

Higgins GT, Allan RB, Hall R, Field EA, Kaye SB. Development of ocular disease in patients with mucous membrane pemphigoid involving the oral mucosa. Br J Ophthalmol. 2006;90(8):964–7.

Neville BW, Damm DD, Allen CM, Chi AC. Oral and Maxillofacial Pathology. 4th ed. Missouri: WB Saunders, Elsevier; 2016. p. 439–40.

Sánchez-Borges M, Asero R, Ansotegui IJ, Baiardini I, Bernstein JA, Canonica GW, et al. Diagnosis and treatment of urticaria and angioedema: a worldwide perspective. World Allergy Organ J. 2012;5(11):125–47.

Spickett G. Oxford Handbook of Clinical Immunology and Allergy. 4th ed. Oxford, England: Oxford University Press; 2020.

Szczawinska-Poplonyk A, Gerreth K, Breborowicz A, Borysewicz-Lewicka M. Oral manifestations of primary immune deficiencies in children. Oral Surg Oral Med Oral Pathol Oral Radiol Endod. 2009;108(3):e9-20.

Tan HP, Smaldone MC, Shapiro R. Immunosuppressive preconditioning or induction regimens : evidence to date. Drugs. 2006;66(12):1535–45.

Tangye SG, Al-Herz W, Bousfiha A, Chatila T, Cunningham-Rundles C, Etzioni A, et al. Human Inborn Errors of Immunity: 2019 Update on the Classification from the International Union of Immunological Societies Expert Committee. J Clin Immunol. 2020;40(1):24–64.

Zuraw BL. Clinical practice. Hereditary angioedema. N Engl J Med. 2008;359(10):1027–36.

Connective Tissue Disorders

75

Joseph Ryan, Kristy Yap, and Tami Yap

75.1 Introduction

Connective tissues diseases (CTDs) have an abnormality in structure or function of one or more elements of connective tissue particularly collagen and elastin. In oral tissue, collagen is present in the alveolar bone, periodontal ligament, pulp, cementum, temporomandibular joint, dentine, gingiva, and basal bone. Systemic lupus erythematosus, dermatomyositis, Sjogren's syndrome, and mixed connective tissue disorders are discussed here. Rheumatoid arthritis is discussed in Chap. 76, and scleroderma is discussed in Chap. 77.

75.2 Systemic Lupus Erythematosus (SLE)

SLE is a chronic inflammatory multi-organ autoimmune disease with a wide range of clinical presentations. A loss of self-tolerance results from abnormal immunological function and the production of autoantibodies, leading to the formation of immune complexes. The reported prevalence of SLE is approximately 20–150 cases per 100,000 persons. Oral lesions in SLE and cutaneous forms of LE often resemble

J. Ryan
School of Dentistry and Medical Sciences, Charles Sturt University, Orange, NSW, Australia

Capital Oral & Facial Surgery, Deakin, ACT, Australia

K. Yap
Monash Health, Clayton, VIC, Australia

T. Yap (✉)
Melbourne Dental School, The University of Melbourne, Carlton, VIC, Australia
e-mail: tspyap@unimelb.edu.au

R. Balasubramaniam et al. (eds.), *Oral Medicine - A Clinical Guide*,
https://doi.org/10.1007/978-3-031-36797-7_75

that seen in oral lichen planus (OLP) both clinically and histologically, presenting as ulcerations, erythema, white plaques, spots, or striation. However dissimilarly to OLP, lesions in SLE are often unilateral and found on the palate. Further oral manifestations of SLE may include cheilitis, nonspecific erythema, salivary gland hypofunction, purpure, discoid lesions, and petechiae. Cases of trigeminal and facial sensory neuropathy have been reported with SLE. For further details, see Chap. 18 on Lupus Erythematosus.

75.3 Dermatomyositis

Dermatomyositis is characterised is characterized by the combination of a cutaneous rash and progressive muscle weakness. The reported incidence of dermatomyositis is 9.64 cases per million persons at an average age of 40 years with a slight female predominance. Dermatomyositis is postulated to develop in genetically susceptible individuals combined with environmental factors. The signs and symptoms of dermatomyositis can appear suddenly or develop gradually over time and include a chrematistic violet-coloured or red painful or pruritic rash developing most commonly on the face, eyelids, knuckles, elbows, and knees. This is accompanied by muscle weakness generally involving the muscles closest to the trunk. The orofacial muscles can be affected with higher frequency of dysphagia and dysgeusia and temporomandibular disorders featuring muscle contracture and myalgia.

75.4 Sjogren's Syndrome (SS)

SS is categorised as either primary or secondary SS, if occurring with another CTD. The disease process involves autoantibody production and chronic lymphocytic infiltration of exocrine glands, as well as multiple body organ systems manifesting as xerostomia, dry eyes, fatigue, vasculitis, neuropathy, arthralgias, kidney disorders, and lung disease. Primary SS is among the most prevalent rheumatic disorders affecting women to men in a 9:1 ratio. Presentation may include major salivary gland enlargement. Lack of saliva and thickened saliva may result in caries, oral infections, taste alterations, inability to eat certain foods or poor appetite, difficulty chewing and swallowing, oral burning, mucosal irritation, and effects on speech. Prostheses may be poorly tolerated. Patients have a lifetime risk estimated to be 5–15% of lymphoma. For further details, see Chap. 30 on Sjogren's syndrome.

75.5 Mixed Connective Tissue Disease (MCTD)

MCTD is a multisystem disorder with overlapping features of systemic lupus erythematosus, scleroderma, and polymyositis and is differentiated from them by a high titre of antibody to ribonucleoprotein. Orofacial manifestations of MCTD include trigeminal neuralgia-like pain, neuropathy, features suggestive of Sjogren's syndrome, and lymphadenopathy.

Connective tissues disorders are usually managed by an overseeing rheumatologist with involvement of multiple disciplines. Any disease which affects collagen will directly or indirectly affect oral tissues and its functions.

Recommended Reading

Crincoli V, Cannavale M, Cazzolla AP, Dioguardi M, Piancino MG, Di Comite M. Temporomandibular Disorders and Oral Features in Idiopathic Inflammatory Myopathies (IIMs) Patients: An Observational Study. Int J Med Sci. 2021;18(14):3158–70.

Kudsi M, Nahas LD, Alsawah R, Hamsho A, Omar A. The prevalence of oral mucosal lesions and related factors in systemic lupus erythematosus patients. Arthritis Res Ther. 2021;23(1):229.

Menzies S, O'Shea F, Galvin S, Wynne B. Oral manifestations of lupus. Ir J Med Sci. 2018;187(1):91–3.

Pandey A, Pandey M, Pandey V, Ravindran V. Oral manifestations of autoimmune connective tissue diseases. Indian J Rheumatol. 2018;13(4),264.

Vivino FB, Bunya VY, Massaro-Giordano G, Johr CR, Giattino SL, Schorpion A, et al. Sjogren's syndrome: An update on disease pathogenesis, clinical manifestations and treatment. Clin Immunol. 2019;203:81–121.

Arthropathies of the Temporomandibular Joint

76

Joseph Ryan, Kristy Yap, and Tami Yap

76.1 Introduction

Arthropathy refers to any disease of the joints, and arthritis is one type of arthropathy which is defined as "inflammation of the articular surfaces of the joint". Several types of arthritis can affect the temporomandibular joint (TMJ). By far the most common in adults is degenerative arthritis referred to as osteoarthritis; however, inflammatory arthropathies remain an important cause of morbidity. Juvenile idiopathic arthritis refers to all forms of arthritis presenting before age 16 with TMJ involvement in 90% of cases. Signs and symptoms of TMJ arthritis include pain, loss of joint function, joint instability, mandibular dysfunction, anterior open bite and loss of posterior mandibular height.

76.2 Osteoarthritis

Osteoarthritis is a condition that occurs due to age-related degenerative changes in a joint. It is not classically an inflammatory entity and features joint arthroses which may be secondarily inflamed. The precise cause of osteoarthritis is unknown, but it is generally thought to be due to mechanical overloading of the joint. The

J. Ryan
School of Dentistry and Medical Sciences, Charles Sturt University, Orange, NSW, Australia

Capital Oral and Facial Surgery, Deakin, ACT, Australia

K. Yap
Monash Health, Clayton, VIC, Australia

T. Yap (✉)
Melbourne Dental School, The University of Melbourne, Carlton, VIC, Australia
e-mail: tspyap@unimelb.edu.au

R. Balasubramaniam et al. (eds.), *Oral Medicine - A Clinical Guide*,
https://doi.org/10.1007/978-3-031-36797-7_76

prevalence of osteoarthritis of the TMJ on clinical and MRI in patients with TMJ pain is up to 25%. The patient with osteoarthritis usually reports unilateral joint pain that is aggravated by and may limit mandibular movement, peaking in the late afternoon or evening. Secondary central mechanisms of pain facilitation may be present. Crepitation can typically be felt over the joint. Palpation of the condyle and manual loading increases the pain. The diagnosis is supported radiographically and may include structural changes in the subarticular bone of the condyle or fossa (flattening, osteophytes, erosions). Treatment aims to alleviate pain and decrease joint loading.

76.3 Inflammatory Arthropathies and Spondyloarthropathies

Rheumatoid arthritis (RA) features persistent inflammatory synovitis leading to the destruction of the articular surfaces and subarticular bone. Roughly 50% of patients with RA will present with TMJ complaints. The diagnosis of RA is based upon joint involvement, serology (ACPA and rheumatoid factors), acute phase reactants (C-reactive protein and erythrocyte sedimentation rate) and duration of symptoms. TMJ condylar changes on MRI scans include erosion, osteophyte formation, abnormal condylar shape, disc alterations, bone marrow oedema, synovitis and decreased movement.

The prevalence of **ankylosing spondylitis (AS)**, the most common spondyloarthropathy, is 0.1–0.2% in the general population and is related to the prevalence of HLA-B27 with a peak incidence of 15–40 years of age. This condition affects the axial skeleton often resulting in chronic inflammatory back and neck pain. The TMJ is involved in approximately 10% of cases.

Psoriatic arthritis (PsA) is a seronegative arthropathy that affects approximately 6–30% of patients with psoriasis, which affects 1.2% of the general population. Psoriatic arthritis has a multifactorial aetiology including a genetic link with HLA-B27. In PsA, the joints are generally affected in an asymmetrical distribution with interphalangeal joints most commonly affected.

Management of inflammatory arthritis is aimed at reducing the autoimmune inflammatory response pharmacologically with an overseeing rheumatologist. Supportive therapy for RA is directed towards pain reduction. Oral appliances may decrease loading and parafunction-related forces on the articular surfaces. Gentle physical therapy to maintain joint mobility is important, since hypomobility is often a consequence of this disorder. Arthrocentesis, intra-articular corticosteroids and arthroscopic procedures may play a role with reduction of severe acute symptoms. Ankylosis is a possible long-term complication of arthritis and requires surgical management. Gross loss of condylar support can cause major occlusal changes which can be challenging to manage.

Recommended Reading

González-Chávez SA, Pacheco-Tena C, Campos Torres RM, Quiñonez-Flores CM, Reyes-Cordero G, Caraveo Frescas TJ. Temporomandibular and Odontological Abnormalities in Patients with Rheumatoid Arthritis. Reumatol Clin (Engl Ed). 2020;16(4):262–71.

Gualtierotti R, Marzano AV, Spadari F, Cugno M. Main Oral Manifestations in Immune-Mediated and Inflammatory Rheumatic Diseases. J Clin Med. 2018;8(1):21.

Hamburger J. Orofacial manifestations in patients with inflammatory rheumatic diseases. Best Pract Res Clin Rheumatol. 2016;30(5):826–50.

Kalladka M, Quek S, Heir G, Eliav E, Mupparapu M, Viswanath A. Temporomandibular joint osteoarthritis: diagnosis and longterm conservative management: a topic review. J Indian Prosthodont Soc. 2014;14(1):6–15.

Sodhi A, Naik S, Pai A, Anuradha A. Rheumatoid arthritis affecting temporomandibular joint. Contemp Clin Dent. 2015;6(1):124–7.

Scleroderma

77

Joseph Ryan, Kristy Yap, Mat Lim, and Tami Yap

77.1 Introduction

Scleroderma (systemic sclerosis) is a connective tissue disease of unknown origin characterised by small vessel vasculopathy, autoantibody production and excessive deposition of collagen in the skin and internal organs. Its aetiology may be secondary to immunologic mechanisms, vascular endothelial cell injury and activation of fibroblasts.

There are two major recognised patterns that the illness can take—diffuse or limited disease. In diffuse scleroderma, skin thickening occurs more rapidly and involves more skin areas than in limited disease. In addition, individuals with diffuse scleroderma have a higher risk of developing lung fibrosis while those with limited disease are at risk of pulmonary arterial hypertension. Limited scleroderma has previously been referred to as CREST syndrome (Calcinosis, Raynaud phenomenon, oesophageal dysfunction, sclerodactyly and telangiectasia).

J. Ryan
School of Dentistry and Medical Sciences, Charles Sturt University, Orange, NSW, Australia

Capital Oral and Facial Surgery, Orange, NSW, Australia

K. Yap
Monash Health, Clayton, VIC, Australia

M. Lim
Melbourne Dental School, The University of Melbourne, Carlton, VIC, Australia

Dental Unit, Alfred Health, Prahran, VIC, Australia

T. Yap (✉)
Melbourne Dental School, The University of Melbourne, Carlton, VIC, Australia
e-mail: tspyap@unimelb.edu.au

R. Balasubramaniam et al. (eds.), *Oral Medicine - A Clinical Guide*,
https://doi.org/10.1007/978-3-031-36797-7_77

Scleroderma is rare, affecting only 1 in 100,000 but has a very high morbidity rate and represents the highest case-specific mortality of any rheumatic disorder.

77.2 Clinical Presentation and Investigations

The most common oral manifestation of scleroderma is microstomia as a result of sclerosis of the perioral tissues. Many patients also experience xerostomia and loss of mobility of the tongue, with difficulty eating, accelerated dental decay, and increased predisposition to low-grade erythematous oral candidosis. Loss of expression lines of the face leading to a mask-like an appearance, thinned lips and a "purse-string" appearance to the mouth is yet another complication. Other findings include mucous membrane blanching, and buccal mucosa fibrosis with loss of normal elasticity. There may also be telangiectasia and ulceration. Thickening of the periodontal ligaments (PDL) (more exaggerated in the posterior region) may be noted radiographically as widening of the PDL space occurring at the expense of alveolar bone. This may be the result of the masticatory muscles becoming bulkier secondary to fibrosis leading to increased occlusal forces.

Scleroderma is associated with impaired periodontal health. Periodontal disease and scleroderma share many common features. They are characterised by a comparable disease course (prominent acute inflammation during early stages and fibrosis, atrophy and tissue destruction during more advanced stages) and by similar aetiopathogenic pathways (microvascular alterations, fibrosis, and inflammation with increased serum levels of proinflammatory cytokines). Further to this, various disease-related factors, such as xerostomia that promote the development of dental plaque, impaired oral hygiene due to reduced manual dexterity and mouth opening, or microvascular alterations that can lead to tissue ischemia could account for the higher prevalence of periodontal disease recorded in scleroderma patients.

Diagnosis of scleroderma is predominantly based on a detailed medical history and clinical examination. Patients can exhibit a wide range of clinical features including skin thickening, calcinosis, lung fibrosis, arthralgia, myocardial lesions and renal insufficiency. Serological tests which support the diagnosis include a positive antinuclear antibody in an anticentromere pattern, anti-topoisomerase I (anti-Scl-70) antibody and anti-RNA polymerase III antibody.

77.3 Management and Prognosis

Management of scleroderma is targeted to systems involved. Scleroderma has a major impact on the orofacial region in approximately 80% of patients. Orofacial involvement often goes underdiagnosed and is poorly managed.

Patients should visit a dentist frequently for the maintenance of oral health. Priorities include preventive care against caries and progressive periodontitis. Regular mouth stretching exercises are helpful in maintaining remaining mouth movement. Treatment planning requires consideration of potential loss of access of posterior teeth over time for both home and professional care.

Recommended Reading

Jung S, Martin T, Schmittbuhl M, Huck O. The spectrum of orofacial manifestations in systemic sclerosis: a challenging management. Oral Dis. 2017;23(4):424–39.

Veale BJ, Jablonski RY, Frech TM, Pauling JD. Orofacial manifestations of systemic sclerosis. Br Dent J. 2016;221(6):305–10.

Vincent C, Agard C, Barbarot S, N'guyen JM, Planchon B, Durant C, et al. [Orofacial manifestations of systemic sclerosis: A study of 30 consecutive patients]. Rev Stomatol Chir Maxillofac. 2010;111(3):128–34.

Zhang S, Zhu J, Zhu Y, Zhang X, Wu R, Li S, et al. Oral manifestations of patients with systemic sclerosis: a meta-analysis for case-controlled studies. BMC Oral Health. 2021;21(1):250.

78 Inflammatory Bowel Disease

Sue-Ching Yeoh, Michelle Kang, and Kenelm Kwong

78.1 Introduction

Inflammatory bowel disease (IBD) includes Crohn's disease (CrD) and ulcerative colitis (UC). The mechanism behind IBD involves a dysregulated inflammatory response, in genetically predisposed individuals, to an unknown environmental trigger that interacts with the intestinal flora. Any part of the gastrointestinal tract can be affected, including the oral cavity. Although UC and CrD are generally considered as distinct conditions with separate clinical, anatomical, and histological findings, they likely represent a continuous spectrum of disease and severity.

S.-C. Yeoh (✉)
Sydney Oral Medicine, Wahroonga, NSW, Australia

Royal Prince Alfred Hospital, Camperdown, NSW, Australia

Chris O'Brien Lifehouse, Camperdown, NSW, Australia

Sydney Dental School, University of Sydney, Sydney, Australia
e-mail: Sue-Ching.Yeoh@sydney.edu.au

M. Kang
Sydney Oral Medicine, Wahroonga, NSW, Australia

Sydney Dental Hospital, Surry Hills, NSW, Australia

Hunter New England Local Health District, Newcastle, NSW, Australia

School of Dentistry and Medical Sciences, Charles Sturt University, Orange, NSW, Australia

K. Kwong
Sydney Dental Hospital, Surry Hills, NSW, Australia

R. Balasubramaniam et al. (eds.), *Oral Medicine - A Clinical Guide*,
https://doi.org/10.1007/978-3-031-36797-7_78

The annual incidence of both CrD and UC is 1 and 20 per 100,000 population with a prevalence of 25–300 per 100,000. Both conditions may occur at any age, however there is a bimodal peak age of onset of 20–40 and 60–70 years of age. There is no gender predilection. Up to 15% of persons with IBD have a first-degree relative who also has IBD. IBD is more prevalent among Ashkenazi Jews. The estimated prevalence of oral lesions in patients with IBD is between 20 and 50%.

78.2 Oral Manifestations

In most cases, intestinal involvement precedes the oral lesions. However, oral lesions may be the primary presenting signs in 5–10% of IBD patients. Oral lesions are more common in CrD than UC. Children are more often affected, and there appears to be a male dominance. Oral disease is more common in CrD patients with proximal gastrointestinal tract and/or perianal involvement. Oral lesions may become more severe during active gastrointestinal disease, and up to 30% of patients may continue to manifest oral lesions despite disease control at other sites. Oral lesions seen in relation to inflammatory bowel conditions can be broadly classified as "specific" or "nonspecific" (Table 78.1).

Table 78.1 Oral lesions associated with inflammatory bowel conditions

CrD specific lesions	Nonspecific oral lesions
Mucosal tags (Fig. 78.1) • Usually labial and buccal vestibules, retromolar region • <75% show noncaseating granulomas on histopathology • No direct association with intestinal CrD activity	Aphthous-like ulceration • In <10% of patients with UC • 20–30% of those with CrD
Cobblestoning of the buccal mucosa (Fig. 78.2) • Also fissured swelling and corrugation • Hyperplastic appearance • Mainly posteriorly	Pyostomatitis vegetans • Multiple miliary white or yellow pustules, erythematous, and oedematous base • Pustules may rupture and coalesce, forming "snailtrack" ulcers • Commonly on labial gingiva, labial and buccal mucosa • May occur in CrD but more frequently in UC
Mucogingivitis • Gingiva appears oedematous, granular, and hyperplastic, with or without ulceration • May involve entire gingival height	Angular cheilitis (Fig. 78.1)
Facial swelling • Lip swelling, vertical fissures, deep linear ulcerations usually in the buccal sulci with hyperplastic folds • Midline lip fissuring may occur in CrD • No clear association with intestinal CrD activity	Persistent submandibular lymphadenopathy

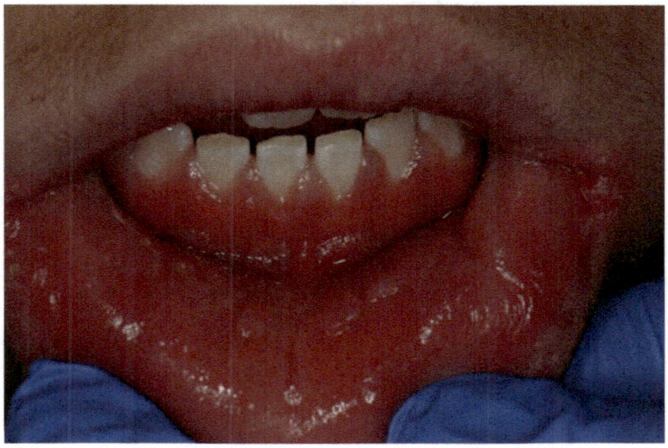

Fig. 78.1 Mucosal tags and angular cheilitis

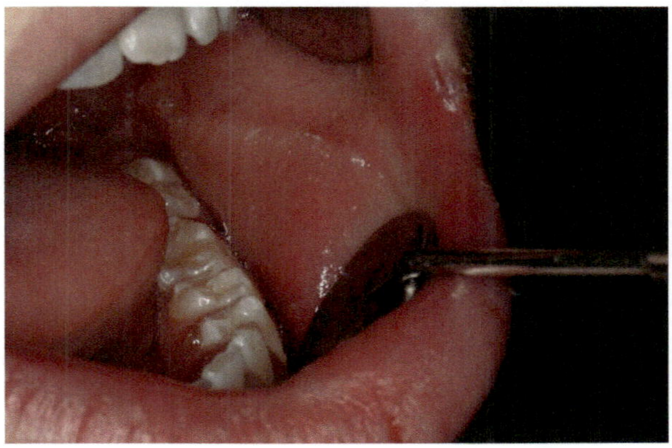

Fig. 78.2 Cobblestoning of the posterior buccal mucosa

The specific lesions of CrD contain granulomatous changes observed on histological examination. They are less common than nonspecific lesions and can occur either concomitantly with intestinal symptoms or before gut presentation by several years.

It is important to note that oral and facial lesions may on occasions precede or be indicative of subclinical inflammatory bowel disease.

There can be overlap in oral manifestations of CrD and UC with nonspecific oral lesions dominating UC presentations.

78.3 Management of Oral Manifestations and Prognosis

Management of the oral manifestations of IBD usually involves collaboration with the treating IBD physician who is managing the underlying intestinal disease. Oral symptoms may require topical and/or systemic corticosteroids, immunosuppressive drugs (such as azathioprine and methotrexate), and biologic agents.

Increased mortality is associated with CrD with surgical complications and malnourishment as the major causes of direct CrD-related mortality. No definitive evidence has been found to suggest a higher mortality among patients with UC.

Recommended Reading

Lourenço SV, Hussein TP, Bologna SB, Sipahi AM, Nico MM. Oral manifestations of inflammatory bowel disease: a review based on the observation of six cases. J Eur Acad Dermatol Venereol. 2010;24(2):204–7.

Molodecky NA, Soon IS, Rabi DM, Ghali WA, Ferris M, Chernoff G, et al. Increasing incidence and prevalence of the inflammatory bowel diseases with time, based on systematic review. Gastroenterology. 2012;142(1):46-54.e42; quiz e30.

Pittock S, Drumm B, Fleming P, McDermott M, Imrie C, Flint S, et al. The oral cavity in Crohn's disease. J Pediatr. 2001;138(5):767–71.

Coeliac Disease

79

Sue-Ching Yeoh, Michelle Kang, and Kenelm Kwong

79.1 Introduction

Coeliac disease (CD) is a chronic autoimmune disorder of the small intestine triggered by the ingestion of gluten and related prolamines in genetically predisposed patients. The clinical spectrum of CD is varied and includes the classic presentation of malabsorption with diarrhoea, nonclassical extraintestinal features, subclinical or asymptomatic forms, and potential disease characterised by positive serology with a normal intestinal mucosa on biopsy.

S.-C. Yeoh (✉)
Sydney Oral Medicine, Wahroonga, NSW, Australia

Royal Prince Alfred Hospital, Camperdown, NSW, Australia

Chris O'Brien Lifehouse, Camperdown, NSW, Australia

Sydney Dental School, University of Sydney, Sydney, Australia
e-mail: Sue-Ching.Yeoh@sydney.edu.au

M. Kang
Sydney Oral Medicine, Wahroonga, NSW, Australia

Sydney Dental Hospital, Surry Hills, NSW, Australia

Hunter New England Local Health District, Newcastle, NSW, Australia

School of Dentistry and Medical Sciences, Charles Sturt University, Orange, NSW, Australia

K. Kwong
Sydney Dental Hospital, Surry Hills, NSW, Australia

R. Balasubramaniam et al. (eds.), *Oral Medicine - A Clinical Guide*,
https://doi.org/10.1007/978-3-031-36797-7_79

The prevalence of CD in western populations has been reported to range between 1 and 1.5%. The prevalence increases to 3–6% in patients with type I diabetes mellitus and up to 20% in first-degree relatives of coeliac patients, making the overall prevalence approximately one in 100 of the population.

79.2 Oral Manifestations

The prevalence of recurrent aphthous-like ulceration in patients with CD has been reported to be significantly higher than in the general population. These typically present as ovoid or round shallow tender ulcers affecting nonkeratinised mucosa. A yellow fibrin pseudomembrane and characteristic erythematous halo are usually present (Fig. 79.1). Symptoms may be exacerbated with spicy and acidic food. Oral ulcers are generally effectively managed with adherence to a strict gluten-free diet, with significant improvement, if not complete remission. Reappearance of the ulcers coincides with reintroduction of dietary gluten.

Enamel hypoplasia and/or hypomineralisation defects are noted in many patients with CD. The precise cause of these enamel defects remains unclear. It is hypothesised that a gluten induced immunological process occurs between 6 months and 7 years in the enamel-producing organ resulting in defective enamel formation. The coeliac-type enamel lesions, like the CD itself, were strongly associated with the HLA allele DR3.

The distribution of defects closely correlates with age of diagnosis and are linked to stages of odontogenesis occurring at the time of active/undiagnosed CD. Permanent incisors are the most commonly affected, followed by molars, canines then premolars.

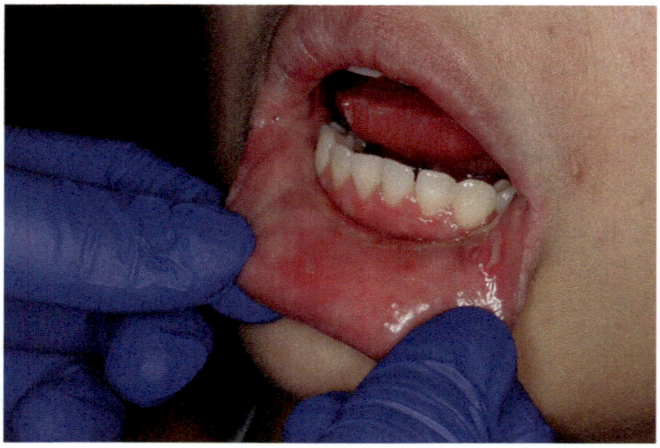

Fig. 79.1 Aphthous-like ulceration in a patient with poorly controlled coeliac disease

79.3 Management of Oral Manifestations and Prognosis

At present, a strict life-long gluten-free diet is the only available effective treatment of CD. In reality, complete avoidance of gluten is difficult due to hidden gluten in food contaminants.

Various management strategies have been described in an attempt to ameliorate or induce remission of aphthous-like ulceration in patients with CD. These include the use of topical corticosteroids and systemic medications such as pentoxifylline, doxycycline, zinc sulphate, colchicine, clofazimine, prednisone, montelukast sodium, sulfones, thalidomide, methotrexate, cyclosporin, chlorambucil, infliximab, etanercept, interferon-α, azathioprine, and levamisole.

Early diagnosis and preventive care are essential for the successful management of developmental enamel defects. Children with a family history of CD should be assessed for enamel defects as soon as the teeth erupt. The main problems associated with enamel hypoplasia are compromised aesthetics, tooth sensitivity, and increased risk for caries and tooth wear. Children with extensive enamel defects usually require interdisciplinary management possibly involving extraction, complex restorations, or orthodontics.

Recommended Reading

Aine L. Dental enamel defects and dental maturity in children and adolescents with coeliac disease. Proc Finn Dent Soc. 1986;82 Suppl 3:1–71.

Aydemir S, Tekin NS, Aktunç E, Numanoğlu G, Ustündağ Y. Celiac disease in patients having recurrent aphthous stomatitis. Turk J Gastroenterol. 2004;15(3):192–5.

Bucci P, Carile F, Sangianantoni A, D'Angiò F, Santarelli A, Lo Muzio L. Oral aphthous ulcers and dental enamel defects in children with coeliac disease. Acta Paediatr. 2006;95(2):203–7.

Other Gastrointestinal Conditions 80

Kenelm Kwong, Michelle Kang, and Sue-Ching Yeoh

80.1 Introduction

Aside from inflammatory bowel disease and coeliac diseases, other gastrointestinal conditions can have orofacial manifestations, and oral mucosal disease may involve other gastrointestinal sites. A selection of conditions is discussed.

80.2 Gastro-Oesophageal Reflux Disease

Gastro-oesophageal reflux disease (GORD) is a common condition affecting 15% of the adult population, characterized by troublesome symptoms or complications that arise from the reflux of acidic stomach contents into the oesophagus. Gastrointestinal symptoms include heart burn, chest pain, bloating, and belching. Patients may complain of hoarseness or sore throat. Asthma may also be

K. Kwong (✉)
Sydney Dental Hospital, Surry Hills, NSW, Australia
e-mail: Kenelm.kwong@health.nsw.gov.au

M. Kang
Sydney Dental Hospital, Surry Hills, NSW, Australia

Sydney Oral Medicine, Wahroonga, NSW, Australia

Hunter New England Local Health District, Newcastle, NSW, Australia

School of Dentistry and Medical Sciences, Charles Sturt University, Orange, NSW, Australia

S.-C. Yeoh
Sydney Oral Medicine, Wahroonga, NSW, Australia

Royal Prince Alfred Hospital, Camperdown, NSW, Australia

Chris O'Brien Lifehouse, Camperdown, NSW, Australia

Sydney Dental School, University of Sydney, Camperdown, Australia

R. Balasubramaniam et al. (eds.), *Oral Medicine - A Clinical Guide*,
https://doi.org/10.1007/978-3-031-36797-7_80

exacerbated, and chest infections may arise, from the aspiration of the refluxate. GORD can be complicated by oesophageal erosions and scarring which may lead to dysphagia, and progression to oesophageal adenocarcinoma.

Orally, GORD manifests chiefly as dental erosion, particularly of the posterior teeth as well as the palatal surfaces of maxillary incisor teeth. This appears initially as a smooth glazed enamel surface, followed by increased translucency of the thinned enamel, ultimately leading to cupping erosion lesions of the posterior occlusal tooth surfaces. If unaddressed, GORD can result in dentinal sensitivity and extensive tooth surface loss, necessitating extensive dental reconstruction. Patients may also complain of bad breath, a recurrent sour taste, or a burning sensation in their mouth. Additionally, GORD has been associated with obstructive sleep apnoea and sleep bruxism.

80.3 Autoimmune Gastritis

Autoimmune gastritis is an autoimmune disorder affecting the gastric mucosa, leading to impaired absorption of dietary cobalamin and subsequent vitamin B12 deficiency. The reduced production of intrinsic factor associated with autoimmune gastritis can lead to pernicious anaemia, causing pallor, tiredness, and shortness of breath. Vitamin B12 deficiency can also cause demyelinating neuropathy which can manifest as paresthesia, weakness, altered reflexes, balance problems, gait changes, memory loss, confusion, and psychosis. Long-term complications of autoimmune gastritis include gastric adenocarcinoma and carcinoid tumours.

The main oral manifestation of autoimmune gastritis and vitamin B12 deficiency is atrophic glossitis or stomatitis. Other associations include symptoms of oral burning or dysesthesia and recurrent aphthous ulceration.

Serological tests can assist in the diagnosis of autoimmune gastritis and/or pernicious anaemia which include a full blood count, peripheral blood smear, vitamin B12 assay, and detection of auto-antibodies against intrinsic factors and gastric parietal cells. Conditions associated with vitamin B12 deficiency are managed with life-long supplementation of vitamin B12.

80.4 Liver Diseases

Hepatitis C may lead to a Sjögren-like lymphocytic sialadenitis, although hyposalivation has not been consistently reported. Hepatitis C is also associated with an increased prevalence of oral lichen planus in certain geographical regions.

Liver cirrhosis can manifest as sialadenosis with bilateral parotid swelling, halitosis (fetor hepaticus), as well as haemorrhagic tendencies.

80.5 Mucous Membrane Pemphigoid

Mucous membrane pemphigoid (MMP) is associated with healing with scarring. The oesophagus is affected in 15% of MMP patients. Patients may be asymptomatic or may experience odynophagia, dysphagia, or haematemesis. Endoscopic investigations may reveal longitudinal folding of the oesophageal mucosa and diffuse non-specific inflammatory changes, ulcerations, oesophageal webbing, and stricture formation. Presence of scarring may guide required immunosuppression. Surgery or oesophageal dilatation may also be indicated.

80.6 Lichen Planus

Patients with oral lichen planus may also have oesophageal involvement which may be asymptomatic or cause odynophagia and dysphagia. Endoscopic findings are often limited to the proximal and mid-oesophagus; these may include lacy white papules, oesophageal webs, erosions, and stricture formation. Oesophageal lichen planus may require management with systemic therapy and endoscopic dilatation.

80.7 Conclusion

Gastrointestinal conditions may have orofacial manifestations, and immune-mediated oral mucosal disease can manifest in the oesophagus. Dental practitioners should familiarise themselves with relevant signs and symptoms of these conditions and consider on-referral where appropriate.

Recommended Reading

Carrozzo M. Oral diseases associated with hepatitis C virus infection. Part 1. sialadenitis and salivary glands lymphoma. Oral Dis. 2008;14(2):123–30.

Carrozzo M. Oral diseases associated with hepatitis C virus infection. Part 2: lichen planus and other diseases. Oral Dis. 2008;14(3):217–28.

Chandan VS, Murray JA, Abraham SC. Esophageal lichen planus. Arch Pathol Lab Med. 2008;132(6):1026–9.

Field EA, Speechley JA, Rugman FR, Varga E, Tyldesley WR. Oral signs and symptoms in patients with undiagnosed vitamin B12 deficiency. J Oral Pathol Med. 1995;24(10):468–70.

Ranjitkar S, Smales RJ, Kaidonis JA. Oral manifestations of gastroesophageal reflux disease. J Gastroenterol Hepatol. 2012;27(1):21–7.

Neurological, Neuromuscular, and Neuro-Sensory Disorders

81

Mathew Lim and Tami Yap

81.1 Introduction

Conditions that impact on the function of the nervous system can have both orofacial manifestations, including orofacial pain, and oral mucosal presentations.

81.2 Epilepsy

Epilepsy is a collective term used for a group of common neurological conditions whereby affected individuals have a tendency to have recurrent seizures. The aim of treatment is ideally to suppress seizures with patient education, identification and avoidance of trigger factors, and treatment of identifiable predisposing pathology or disease. Several anticonvulsants are associated with drug-induced gingival hyperplasia including carbamazepine, levetiracetam, phenytoin, topiramate, and valproic acid. Seizures are also considered a risk factor for sleep and awake bruxism.

M. Lim (✉)
Dental Services, Alfred Health, Melbourne, VIC, Australia

Alfred Hospital Clinical School, Monash University, Melbourne, VIC, Australia

Melbourne Dental School, The University of Melbourne, Carlton, VIC, Australia

Royal Melbourne Hospital, Parkville, VIC, Australia
e-mail: mathew.lim@alfred.org.au

T. Yap
Dental Services, Alfred Health, Melbourne, VIC, Australia

Melbourne Dental School, The University of Melbourne, Carlton, VIC, Australia

Royal Melbourne Hospital, Parkville, VIC, Australia

Royal Dental Hospital Melbourne, Carlton, VIC, Australia

R. Balasubramaniam et al. (eds.), *Oral Medicine - A Clinical Guide*,
https://doi.org/10.1007/978-3-031-36797-7_81

81.3 Cerebrovascular Disease

A cerebrovascular accident (CVA), or stroke, is the sudden onset of loss of focal neurological function in the brain, retina, or spinal cord. Strokes can be broadly categorised as either thromboembolic or haemorrhagic. A key component of acute management for cerebrovascular events is antiplatelet or anticoagulant therapy. A complication of stroke that may include orofacial involvement is central post-stroke pain, a central neuropathic pain resulting from lesion or dysfunction in the central nervous system.

81.4 Progressive Neuromuscular Conditions

Neuromuscular conditions are characterised by progressive motor symptoms but are not linked by aetiology, however present similar challenges due to increasing levels of disability which may also extend to cognitive changes. These conditions are incurable, and management is focused on improving symptoms and supporting quality of life.

81.4.1 Parkinson's Disease

Parkinson's disease is a progressive neurodegenerative condition, mostly of older adults, characterised by the classical clinical presentation of extrapyramidal motor symptoms, including tremor, bradykinesia, and rigidity, and behavioural changes. Patients may present with awake bruxism.

81.4.2 Huntington's Disease

Huntington's disease is a relatively rare progressive neurodegenerative disorder characterised by chorea and dystonia, lack of coordination, emotional disturbances, and progressive dementia caused by a mutation of chromosome 4 (Huntingtin gene) inherited in an autosomal dominant pattern. Patients may present with awake teeth clenching and grinding, and jaw bracing.

81.4.3 Multiple Sclerosis

Multiple sclerosis (MS) is a chronic, progressive, neurological disease resulting in impairments to motor coordination, visual sensation, and other nerve signals that are caused by autoimmune damage to the myelin sheath around nerve axons. Up to 1 in 20 MS patients suffer neuropathic pain including orofacial pain. Bilateral trigeminal neuralgia (TN), characterised by paroxysmal, electric shock-like pain within the distribution of the trigeminal nerve and triggered by innocuous stimuli, can be an initial presentation of MS. MS is the most common cause of secondary TN.

81.5 Motor Neurone Disease

Motor neurone disease (MND) is a group of uncommon but fatal neurodegenerative conditions whereby progressive impairments result from damage to upper or lower motor neurones in the central nervous system. The cause and pathogenesis remain largely unknown. Bruxism and neuromuscular conditions, both movement disorders, have been associated but causal relationships have yet to be clarified. Neuromuscular conditions can also be associated with musculoskeletal pain, including temporomandibular disorders.

81.6 Myasthenia Gravis

Myasthenia gravis is an uncommon chronic autoimmune disorder characterised by muscular weakness and fatigue that is exacerbated by exertion and stress, which improves with rest. Muscle weakness can affect the muscles of facial expression as well as the muscles of mastication.

81.7 Taste Disturbances

Prevalence of true taste disturbance is significantly less than olfactory impairment, which is considered to be approximately 5% of the population, increasing with age. Reduced taste, or hypogeusia, has been found to be lower than 5% with complete loss of taste very rare. Dysgeusia is defined as a change in taste which is unpleasant, often described as bitterness or a metallic taste. Taste disturbances may associate with vitamin deficiency, increasing age, medications as well as xerostomia and reduced salivary flow, with potential interplay between these factors. Taste changes may also be a presentation on the spectrum of oral dysaesthesia.

Recommended Reading

Carr AS, Cardwell CR, McCarron PO, McConville J. A systematic review of population based epidemiological studies in Myasthenia Gravis. BMC Neurol. 2010;10:46.

Di Stefano G, Maarbjerg S, Truini A. Trigeminal neuralgia secondary to multiple sclerosis: from the clinical picture to the treatment options. J Headache Pain. 2019;20(1):20.

Fiest KM, Sauro KM, Wiebe S, Patten SB, Kwon CS, Dykeman J, et al. Prevalence and incidence of epilepsy: A systematic review and meta-analysis of international studies. Neurology. 2017;88(3):296–303.

Hatahira H, Abe J, Hane Y, Matsui T, Sasaoka S, Motooka Y, et al. Drug-induced gingival hyperplasia: a retrospective study using spontaneous reporting system databases. J Pharm Health Care Sci. 2017;3:19.

Henry JL, Lalloo C, Yashpal K. Central poststroke pain: an abstruse outcome. Pain Res Manag. 2008;13(1):41–9.

Kiernan MC, Talman P, Henderson RD, Harris R. Establishment of an Australian motor neurone disease registry. Med J Aust. 2006;184(7):367–8.

Mackay J, Mensah GA. The Atlas of Heart Disease and Stroke. Geneva, Switzerland: World Health Organization; 2004.

Walker FO. Huntington's disease. Lancet. 2007;369(9557):218–28.

Welge-Lüssen A, Dörig P, Wolfensberger M, Krone F, Hummel T. A study about the frequency of taste disorders. J Neurol. 2011;258(3):386–92.

82 Bloodborne Infectious Diseases

Antonio Celentano and Li-Qiao Rachel Ling

82.1 Introduction

Bloodborne infectious diseases are caused by pathogens that are carried in the blood, of which most notably are hepatitis B (HBV), hepatitis C (HCV), and human immunodeficiency virus (HIV). Some of these diseases, mainly HCV and HIV, can lead to a wide cohort of oral manifestations, such as salivary gland hypofunction, xerostomia, candidosis, and neuropathic orofacial pain, among others.

Transmission of bloodborne diseases is through percutaneous or permucosal contact with infected blood. This can be from sexual contact, blood transfusions, shared needles, and from mother to child during pregnancy, childbirth, or breast feeding. While HBV and HCV are not directly cytopathic, both cause liver damage from immune-mediated attempts to eradicate infected liver cells. HIV directly targets the immune system by infecting CD4+ T lymphocytes, increasing the risk of opportunistic infection.

In 2019, an estimated 230,154 people (0.90% of Australia's population) were living with chronic hepatitis B (CHB). Nevertheless, there is a sustained decline of HBV infection among younger people (<25 years) due to infant and catch-up vaccination programs. HCV prevalence in Australia is instead estimated at around 1%,

A. Celentano (✉) · L.-Q. R. Ling
Melbourne Dental School, The University of Melbourne, Carlton, VIC, Australia
e-mail: antonio.celentano@unimelb.edu.au

R. Balasubramaniam et al. (eds.), *Oral Medicine - A Clinical Guide*,
https://doi.org/10.1007/978-3-031-36797-7_82

with approximately 230,000 HCV cases. Overall, Australia's HIV prevalence among the general population remains one of the lowest among developed countries, with approximately 26,000 people currently living with HIV, and about 1000 new cases diagnosed each year. HIV-positive patients are 6 times more likely to have concurrent HCV infection, though treatment does not differ from HCV-monoinfected patients.

82.2 Clinical Presentation and Investigations

Hepatitis B and C infections are typically asymptomatic in the acute phase. However, 30% of HBV cases will develop jaundice and 0.1–0.5% will develop acute liver failure; other symptoms include lethargy, dark urine, and abdominal pain. Chronic hepatitis can lead to liver cirrhosis, failure, and hepatic cancer. Oral lichen planus (OLP) and sialadenitis have been reported as predominant and exclusive extrahepatic manifestations of HCV infection. These could therefore serve as a clue of HCV infection in asymptomatic patients, thus enabling early diagnosis and better prognosis. Due to the weakened immune system seen in HIV-positive patients, there are also commonly associated oral presentations. The ten encountered consistently, in descending prevalence, are as follows: oral candidiasis, oral hairy leucoplakia, herpes simplex virus infection, Kaposi's sarcoma, nonspecific ulceration, aphthous ulcers, periodontal disease, salivary gland disease, oral melanotic hyperpigmentation, and oral warts. Recognising oral lesions is essential since it could be the first signs of HIV infection, predictors of disease progression, indicator for decreased of the cluster-differentiated 4 (CD4+) T-cell count as well as indicator for increasing viral load.

Acute hepatitis must be differentiated from other nonspecific viral illnesses, and from other disorders causing jaundice. Patients suspected of having hepatitis are tested initially with liver tests, including aminotransferases, bilirubin, and alkaline phosphatase. Screening for HBV and HCV will then include serological tests, allowing determination of susceptibility, active infection, or immunity through vaccination or past infection. These include hepatitis B surface antigen (HBsAg), IgM antibody to hepatitis B core (IgM anti-HBc), antibody to HCV (anti-HCV), and Hepatitis C RNA (HCV-RNA) polymerase chain reaction (PCR). Different antibody markers can signal the change between acute and chronic HBV infection, while reverse transcription PCR is required to differentiate hepatitis C stage. Prognosis and response to treatment can be directed by additional investigations such as serum fibrosis panels, genotype testing, and liver biopsy. ELISA test is instead used to detect HIV infection. If an ELISA test is positive, the Western blot test is usually administered immediately after to confirm the diagnosis. Monitoring of the CD4 cell count is also crucial soon after diagnosis with HIV, before beginning HIV treatment and for as long as the CD4 count is low. It provides important information about disease progression and the immune system.

82.3 Management and Prognosis

A vaccine course is available and scheduled for HBV but not for hepatitis C and HIV. Most adults will spontaneously clear acute HBV infection—about 2–6% of adults will go on to develop chronic infection. In these patients, antiviral medications such as tenofovir and interferon can be used to reduce the severity of liver cirrhosis but is not curative. Age at time of infection also affects prognosis—90% of infants with hepatitis B go on to become chronically infected. Around 30% of HCV infected persons spontaneously clear the virus within 6 months of infection without any treatment, while the remaining 70% will develop chronic HCV infection. Of those with chronic HCV infection, the risk of cirrhosis ranges from 15 to 30% within 20 years. Direct-acting antivirals are highly effective at treating hepatitis C, though late diagnosis and liver failure can require liver transplantation. The prognosis in patients with untreated HIV infection is poor, with an overall mortality rate of more than 90%. The average time from infection to death is 8–10 years, although individual variability ranges from less than 1 year to long-term nonprogression. ART remains the most effective treatment for HIV; while not curative, viral load can be managed with the goal of preventing progression to acquired immune deficiency syndrome (AIDS).

Recommended Reading

Lai CL, Ratziu V, Yuen MF, Poynard T. Viral hepatitis B. Lancet. 2003;362(9401):2089–94.

Lok AS, McMahon BJ, Brown RS, Wong JB, Ahmed AT, Farah W, et al. Antiviral therapy for chronic hepatitis B viral infection in adults: A systematic review and meta-analysis. Hepatology. 2016;63(1):284–306.

Patton LL. Current strategies for prevention of oral manifestations of human immunodeficiency virus. Oral Surg Oral Med Oral Pathol Oral Radiol. 2016;121(1):29–38.

Platt L, Easterbrook P, Gower E, McDonald B, Sabin K, McGowan C, et al. Prevalence and burden of HCV co-infection in people living with HIV: a global systematic review and meta-analysis. Lancet Infect Dis. 2016;16(7):797–808.

Reichart PA, Gelderblom HR, Khongkhunthian P, Schmidt-Westhausen A. Ebola virus disease: any risk for oral and maxillofacial surgery? An overview. Oral Maxillofac Surg. 2016;20(2):111–4.

Akintoye S.O, Greenberg M.S. Recurrent Aphthous Stomatitis. Dent Clin North Am. 2014;58:281–297. https://doi.org/10.1016/j.cden.2013.12.002

Sexually Transmitted Infections

83

Antonio Celentano and Tamara Matthyssen

83.1 Introduction

Sexually transmitted infections (STIs) are caused by bacteria, viruses and parasites and can occur at both urogenital and extragenital sites. While spread primarily through sexual activity, STIs can also be transmitted non-sexually (e.g. mothers to infants during childbirth, shared needles or contact with infectious lesions). Due to increasing rates of infection and cases of multidrug resistance, STIs remains a major public health issue.

STIs are caused by a variety of microorganisms that are primarily transmitted through direct contact with mucosal membranes (Table 83.1). The infection is transmitted to the oral cavity mostly through orogenital contact during sexual activity. These organisms can survive in the host for many years and have developed multiple mechanisms for evading the immune system, such as the suppression of antigen presentation, paralysis of phagocytic neutrophils and antigenic and phase variation.

Since 2015, the USA has observed an increase in rates of gonorrhoea (56%), chlamydia (19%) and syphilis (75%), while rates of HSV-2 were observed to have declined. In 2020, there was an estimated 374 million new cases of infection due to chlamydia, gonorrhoea, syphilis, and trichomoniasis alone. An estimated 491.5 million people were living instead with HSV-2 infection in 2016, equivalent to 13.2% of the world's population aged 15–49 years.

A. Celentano (✉) · T. Matthyssen
Melbourne Dental School, The University of Melbourne, Carlton, VIC, Australia
e-mail: antonio.celentano@unimelb.edu.au

R. Balasubramaniam et al. (eds.), *Oral Medicine - A Clinical Guide*,
https://doi.org/10.1007/978-3-031-36797-7_83

Table 83.1 Microrganisms resulting sexually transmitted infections

STI	Causative agent	Possible oral manifestation
Gonorrhoea	*Neisseria gonorrhoeae* (Gram-negative bacteria)	Atypical fiery erythema and sometimes superficial ulcer formation covered with a greyish or yellowish-exudate
Chlamydia	*Chlamydia trachomatis* (Gram-negative bacteria)	Mostly orally asymptomatic. When symptomatic erosive lesions of the oral mucosa and labial mucosa possibly followed by haemorrhagic crusts; tonsillitis, sore throat worsened with swallowing
Syphilis (Chap. 13)	*Treponema pallidum* (spirochaete bacteria)	Multiple painless irregularly shaped or aphthous-like ulcers with whitish edges on the tongue, lips, and lingual mucosa and oropharynx; mucous patches (mostly noted in secondary syphilis
Herpes	Herpes simplex virus	Both HSV-1 and HSV-2 can have manifestations in either the oral or genital regions and both lead to an initial primary infection followed by less severe secondary infection. Of the two, HSV-1 is by far the most common *Primary HSV infection*: Prodromal symptoms accompanied by generalised acute marginal gingivitis, sore throat, tonsillitis, and onset of numerous small, thin-walled vesicles. HSV-2 infection is characterised by a similar clinical picture to that of HSV-1, *h*owever, the illness is usually less severe and not as prolonged as that caused by HSV-1 *Recurrent intraoral herpes*: Small ulcers preceded by vesicles and commonly appear on the hard palate *Recurrent herpetic labialis (RHL)*: (also known as cold sores) cluster of small vesicles affecting the lips which break in 1–2 days, forming small ulcers of 1–3 mm that spontaneously heal without scarring within 6–10 days
HPV	Human papillomavirus	*Condyloma acuminatum* might develop in orofacial area and present as lesions are multiple painless, whitish and cauliflower shaped
Hepatitis B	Hepatitis B virus	Orally asymptomatic. Systemic symptoms in adults include jaundice and fulminant liver failure (30% and 0.3% of cases, respectively)
HIV/AIDS	Human immunodeficiency viruses	Candidosis: Usually pseudomembranous (Fig. 83.1a) Hairy leukoplakia (Fig. 83.1b), Kaposi's sarcoma *Oral mucosal hyperpigmentation*: Multiple light to dark brown macules, or as patchy and diffuse which can affect any oral site, but mainly the gingiva (probably caused by different medications used to treat HIV/AIDS); *salivary gland hypofunction and xerostomia; neuropathic orofacial pain*

83.2 Clinical Presentation and Investigations

Infected patients often show very few or no symptoms, for example, an international study showed that 86–92% of women and 55–86% of men with urogenital gonorrhoea infection were asymptomatic. When symptoms do occur, these can commonly include urethritis, cervicitis and dysuria among others (predominantly observed in gonorrhoea, chlamydia and trichomoniasis). Oral practitioners may play a role in first diagnosis if oral symptoms and signs are present, especially if the presentation mimics other oral mucosal diseases. Oral presentations of selected STIs are summarised in Table 83.1. Infection with an STI associate will increase the risk of acquiring HIV, adverse birth outcomes, certain cancers (e.g. cervical, liver, oropharyngeal), pelvic inflammatory disease and infertility in women.

Diagnosis of most STIs relies on nucleic acid amplification tests from swabs, first-catch urines, cervical fluid specimen, or serological specimen. Other methods include Gram-stained smears, microscopy, and antibody tests. Culture is also used to determine antibiotic susceptibility. Investigation of oral tissue specimens may take place when other differentials are considered. Presence of culprit microbes may require specific stains or methods, for example, a direct smear for Gram staining is usually sufficient to identify Gonorrhoea as soon as the swab specimen is collected from the oral mucosa. The swab can be rolled gently onto the slide to preserve cellular morphology and over an area less than 1 cm^2. *Treponema pallidum* instead can be visualised through Warthin Starry stain, while *Chlamydia trachomatis* may require Giemsa stain or iodine.

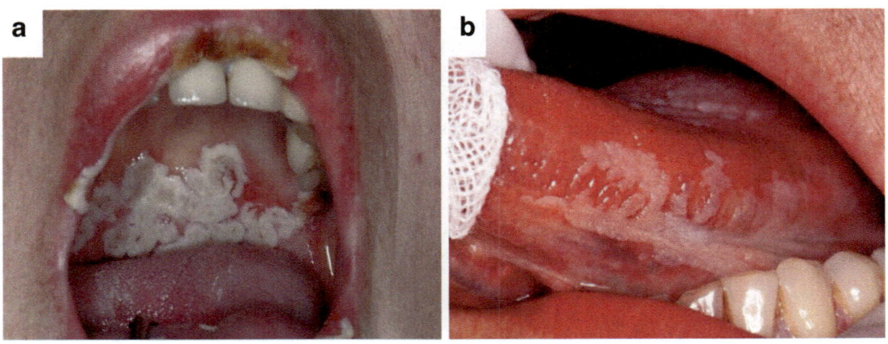

Fig. 83.1 (**a**) Diffuse pseudomembranous candidosis in HIV infection. (**b**) Hairy leukoplakia in a patient with HIV infection: white adherent plaque with linear folds affecting the left lateral border of the tongue and left ventral mucosa

83.3 Management and Prognosis

Some viral STIs are associated with latency, such as HPV, hepatitis B, and HSV-2 cannot be cured. HPV and hepatitis B can be prevented through vaccination, while antivirals such as acyclovir, valacyclovir, and famciclovir can be used to treat HSV-2 symptoms. STI caused by bacteria such as gonorrhoea, chlamydia, syphilis, and trichomoniasis can be treated with antibiotics. For gonorrhoea, treatment with ceftriaxone is the first line of defence (>95% of patients successfully treated) or alternatively gentamicin plus azithromycin. For pharyngeal gonococcal infection, there is no alternative to ceftriaxone. Chlamydia can be treated with doxycycline (except in the case of pregnancy in which amoxicillin can be used). Benzathine penicillin G is effective against syphilis, with doxycycline as an alternative (except during pregnancy). Trichomonas can be treated with metronidazole but at different dosages for men and women. Treatment for STIs usually occurs under the care of sexual health or infectious disease experts. Contact tracing forms part of the holistic management. While various treatments currently exist for most STIs, the emergence of multidrug resistant organisms and the secondary implications of infection can result in poorer prognosis.

Recommended Reading

Gondivkar S, Gadbail A, Sarode GS, Sarode SC, Patil S, Awan KH. Infectious diseases of oral cavity. Dis Mon. 2019;65(6):164–84.

Scurtu LG, Jinga V, Simionescu O. Fascinating Molecular and Immune Escape Mechanisms in the Treatment of STIs (Syphilis, Gonorrhea, Chlamydia, and Herpes Simplex). Int J Mol Sci. 2022;23(7):3550.

Tuddenham S, Hamill MM, Ghanem KG. Diagnosis and Treatment of Sexually Transmitted Infections: A Review. JAMA. 2022;327(2):161–72.

Unemo M, Jensen JS. Antimicrobial-resistant sexually transmitted infections: gonorrhoea and Mycoplasma genitalium. Nat Rev Urol. 2017;14(3):139–52.

Unemo M, Seifert HS, Hook EW, Hawkes S, Ndowa F, Dillon JR. Gonorrhoea. Nat Rev Dis Primers. 2019;5(1):79.

Wong WF, Chambers JP, Gupta R, Arulanandam BP. Chlamydia and Its Many Ways of Escaping the Host Immune System. J Pathog. 2019;2019:8604958.

Oral Complications of Cancer Therapy 84

Agnieszka Frydrych

84.1 Introduction

Cancer therapies (CTs) including surgery, radiation therapy (RT), chemotherapy and the novel targeted cancer therapies (TCT) often result in debilitating and, at times, life-threatening adverse effects. Oral complications depend on cancer type, location and treatment and have been summarised in Table 84.1. Mucositis and radiation related injuries are covered in Chaps. 85 and 86.

84.2 Oral Mucosal Complications

84.2.1 Stomatitis

Stomatitis has been associated with the use of TCT including monoclonal antibodies, signal transduction and angiogenesis inhibitors, apoptosis inducing drugs, gene expression modulators, hormone therapies and immunotherapy. Stomatitis may resemble aphthous ulceration, herpetic stomatitis or erythema multiforme-like eruption. Management includes basic oral care and the use of corticosteroids and analgesics.

84.2.2 Infections

Opportunistic infections are a common complication of CTs and contribute to mortality and morbidity. These are summarised in Table 84.2.

A. Frydrych (✉)
UWA Dental School, The University of Western Australia, Nedlands, WA, Australia
e-mail: agnieszka.frydtrych@uwa.edu.au

R. Balasubramaniam et al. (eds.), *Oral Medicine - A Clinical Guide*,
https://doi.org/10.1007/978-3-031-36797-7_84

Table 84.1 Early and late oral complications of cancer therapy

	Oral complications E—Early presentation L—Late presentation				
Treatment modality	Mucosal	Salivary gland	Musculoskeletal	Neurological	Other
Surgery	• Scarring (L)	• Resection of single glands not usually associated with significant salivary gland hypofunction	• Trismus (E, L) • Disfigurement (E)	• Neuropathy (E, L)	• Dysphagia (E, L) • Speech disorder (E, L) • Chronic fistulas (E)
Radiation therapy	• Radiation mucositis (E) • Fibrosis (L)	• Salivary gland hypofunction (E, L) • Xerostomia (E, L)	• Osteoradionecrosis (L) • Myofibrotic contracture (L)	• Dysgeusia (E, L) • Neuropathy (L)	• Infections (E, L) • Dysphagia (E, L) • Speech disorder (E, L) • Dental caries and periodontal disease (L) • Malignancy (L)
Chemotherapy	• Mucositis (E, L) • Pigmentation (L)	• Salivary gland hypofunction (E, L) • Xerostomia (E, L)	• Osteonecrosis (L)	• Dysgeusia (E, L) • Neuropathy (L)	• Infections (E, L) • Dysphagia (E, L) • Dental caries and periodontal disease (L) • Malignancy (L)
Targeted cancer therapies	• Stomatitis (E, L) • Hyperkeratotic lesions (L) • Pigmentation (L)	• Salivary gland hypofunction (E, L) • Xerostomia (E, L)	• Osteonecrosis (L)	• Dysgeusia (E, L) • Neuropathy (L)	• Infections (E, L) • Dysphagia (E, L) • Dental caries and periodontal disease (L) • Malignancy (L)

Table 84.2 Opportunistic infections in cancer patients

Infection	Presentation	Common infective agents	Management	Potential complications
Viral	Local mucosal reactivation with or without systemic manifestations; may be atypical and mimic mucositis or aphthous ulceration	Herpes simplex virus-1	• Therapeutic or preventive acyclovir and valacyclovir	Life-threatening encephalitis, disseminated infection
Fungal	Pseudomembranous, chronic hyperplastic, erythematous candidosis, angular cheilitis	*Candida* spp.	• Therapeutic or preventive topical and systemic antifungals	Life-threatening systemic dissemination
Bacterial	Painful, well circumscribed ulceration	Gram-negative bacilli (*Pseudomonas aeruginosa*, *Klebsiella pneumoniae*) and Gram-positive cocci (*S. aureus*, *S. epidermidis* and *S. pyogenes*)	• Culture and sensitivity testing based systemic antibiotic selection • Preventive pre-therapy dental evaluation and treatment	Sepsis

84.2.3 Other

Oral hyperkeratotic lesions, gingival hyperplasia and mucosal pigmentation have been associated with the use of TCT and chemotherapy. Complications such as graft versus host disease (GVHD) and those related to haematological abnormalities including thrombocytopenia and neutropenia are detailed in Chaps. 23 and 66, respectively.

84.3 Xerostomia and Salivary Gland Hypofunction

Xerostomia and salivary gland hypofunction (SGH) may be transient or permanent, depending on the cancer type and treatment regimen. Xerostomia and SGH are associated with increased risk of oral infections, including dental caries, impaired oral functioning and reduced life quality.

Management focuses on palliation of symptoms and preventive measures including good oral hygiene, elimination of sources of mucosal trauma, adequate hydration, use of oral moisturising agents, salivary stimulation and implementation of caries preventive protocols.

84.4 Musculoskeletal Complications

84.4.1 Medication-Related Osteonecrosis of the Jaw (MRONJ)

Bone-modifying agents (BMA) (e.g. bisphosphonates and denosumab) and angiogenic inhibitors are used in the management of patients with skeletal metastases with the aim of reducing pathologic fractures, spinal cord compression and the need for RT or surgery to bone. Prevention, clinical presentation and management of MRONJ are discussed in Chap. 73.

84.4.2 Dysphagia

Problems with swallowing may result in food sticking in the throat and/or aspiration which may lead to pneumonia. Although a common complication of head and neck cancer (HNC) treatment, it also occurs in patients with other solid tumours. Dysphagia is most often the result of tissue loss or damage as a result of the tumour and/or its treatment.

84.5 Sensory Disturbances

84.5.1 Dysgeusia

Abnormal sense of taste may result in malnutrition and significant morbidity. Damage to sensory receptor cells and abnormal neuronal activities are responsible. Nutritional counselling may be helpful to some patients in minimising the symptoms of dysgeusia.

84.5.2 Orofacial Pain

Over 80% of HNC patients report pain during treatment, which is largely the result of mucositis. Sensory nerve damage may also contribute to orofacial pain, and persistent pain is reported by about one-third of cancer patients beyond the 6-month period post treatment completion. Pain management should distinguish between and address nociceptive and neuropathic pain to achieve optimal pain control.

84.5.3 Other Neuropathies

CTs such as surgery, chemotherapy and TCT can damage sensory nerves and/or alter function resulting in dysesthesia or paraesthesia.

84.6 Additional Considerations

Persistent SGH can lead to dental complications. HNC patients treated with RT and those suffering from chronic GVHD are the most vulnerable and at risk of developing dental caries and progressive periodontal disease. Preventive strategies focus on managing the SGH; good oral hygiene; fluoride and remineralising agent use; dietary modification and preventive dental care.

Impaired mastication and speech can be associated with surgical and nonsurgical CTs. Anatomical alterations due to tumour resections, neuropathies, tissue fibrosis, SGH and tooth loss are responsible. Disfigurement and resultant psychological implications are important consequences of surgical treatment of oral cancer.

Orofacial asymmetry and dental abnormalities such as failure of teeth to develop, delayed eruption and altered tooth form may be seen in children who undergo cancer treatment.

Finally, second primary cancers, including of the oral cavity, may complicate CTs. All patients should therefore undergo regular, at least annual, comprehensive oral examination. Pre-cancer therapy dental evaluation and treatment as well as the regular and lifelong dental care has the potential to prevent or reduce the impact of oral complications of cancer treatments.

Recommended Reading

Carrozzo M, Eriksen JG, Bensadoun RJ, Boers-Doets CB, Lalla RV, Peterson DE. Oral Mucosal Injury Caused by Targeted Cancer Therapies. J Natl Cancer Inst Monogr. 2019 ;2019(53):lgz012.

Elad S, Zadik Y, Hewson I, Hovan A, Correa ME, Logan R, et al. A systematic review of viral infections associated with oral involvement in cancer patients: a spotlight on Herpesviridea. Support Care Cancer. 2010;18(8):993–1006.

Elad S, Zadik Y, Yarom N. Oral Complications of Nonsurgical Cancer Therapies. Atlas Oral Maxillofac Surg Clin North Am. 2017;25(2):133–47.

Epstein JB, Hong C, Logan RM, Barasch A, Gordon SM, Oberle-Edwards L, et al. A systematic review of orofacial pain in patients receiving cancer therapy. Support Care Cancer. 2010;18(8):1023–31.

Frowen J, Hughes R, Skeat J. The prevalence of patient-reported dysphagia and oral complications in cancer patients. Support Care Cancer. 2020;28(3):1141–50.

Hovan AJ, Williams PM, Stevenson-Moore P, Wahlin YB, Ohrn KE, Elting LS, et al. A systematic review of dysgeusia induced by cancer therapies. Support Care Cancer. 2010;18(8):1081–7.

Jensen SB, Pedersen AM, Vissink A, Andersen E, Brown CG, Davies AN, et al. A systematic review of salivary gland hypofunction and xerostomia induced by cancer therapies: prevalence, severity and impact on quality of life. Support Care Cancer. 2010;18(8):1039–60.

Lalla RV, Latortue MC, Hong CH, Ariyawardana A, D'Amato-Palumbo S, Fischer DJ, et al. A systematic review of oral fungal infections in patients receiving cancer therapy. Support Care Cancer. 2010;18(8):985–92.

Mosel DD, Bauer RL, Lynch DP, Hwang ST. Oral complications in the treatment of cancer patients. Oral Dis. 2011;;17(6):550–9.

Yarom N, Shapiro CL, Peterson DE, Van Poznak CH, Bohlke K, Ruggiero SL, et al. Medication-Related Osteonecrosis of the Jaw: MASCC/ISOO/ASCO Clinical Practice Guideline. J Clin Oncol. 2019;37(25):2270–90.

Oral Mucositis

85

Agnieszka Frydrych

85.1 Introduction

Oral mucositis (OM) is a common complication of radiation therapy (RT) to the head and neck (H&N); combination of radiation and chemotherapy (RT-CT); as well as chemotherapy (CT), particularly with conditioning regimens in haemopoietic stem cell transplantation (HSCT). It is associated with considerable pain, increased opioid consumption, local and systemic infection risk, and hospitalisation in addition to reduced tolerability of cancer treatment, an impaired nutritional status, and quality of life. In some cases, OM may compromise treatment outcomes and prognosis.

The pathogenesis of OM is multifactorial. Evidence suggests that mucositis involves a series of biologically complex and interactive events that occur in epithelial and connective tissues of the mucosa. Chemotherapeutic agents or RT generate oxidative stress and reactive oxygen species (ROS) (also called oxygen free radicals,) and this process appears to be the primary event in most pathways leading to mucositis. Chemotherapy or RT may initiate mucositis directly by causing DNA strand breaks, through the generation of ROS, or through enzymatic or transcription factor (nuclear factor kappa B (NFkB)) activation in multiple cellular elements within the mucosa. Activation of NFkB can result in the upregulation of many genes including those that result in the production of proinflammatory cytokines tumor necrosis factor α, interleukin (IL)-1β and IL-6. These events result in tissue damage.

Oral mucositis affects almost all patients undergoing RT to the H&N region, with the incidence being highest among those undergoing RT-CT and 75–100% of HSCT patients.

A. Frydrych (✉)
UWA Dental School, The University of Western Australia, Nedlands, WA, Australia
e-mail: agnieszka.frydrych@uwa.edu.au

R. Balasubramaniam et al. (eds.), *Oral Medicine - A Clinical Guide*,
https://doi.org/10.1007/978-3-031-36797-7_85

85.2 Clinical Presentation and Investigations

The severity of OM varies between individuals and is dependent on treatment regimens and patient factors such as age, gender, underlying disease and genetic predisposition. The World Health Organization (WHO) has developed a simple scale for the purpose of recording the extent and severity and OM (Table 85.1).

Mucositis due to CT is characterised by inflammation and painful mucosal ulceration usually limited to non-keratinised surfaces and typically arises within 2 weeks after initiation of CT. In RT-induced OM, first clinical signs become apparent by the end of the first week of treatment. Lesions are confined to the field of RT, manifesting initially as mucosal erythema and eventually progressing to painful ulceration (Fig. 85.1). Lesions of OM typically heal within 2–4 weeks after the last dose of CT or RT.

Diagnosis of OM is clinical and based on the known use of CT/RT as well as the appearance, timing, and location of oral lesions. Most common differential diagnoses include viral infections and, in the case of patients who have received allogeneic HSCT, acute graft-vs-host disease.

Table 85.1 The World Health Organization scale for oral mucositis

Grade	Description
0	No oral mucositis
1	Erythema and soreness
2	Ulcers, able to eat solids
3	Ulcers, requires liquid diet (due to mucositis)
4	Ulcers, alimentation not possible (due to mucositis)

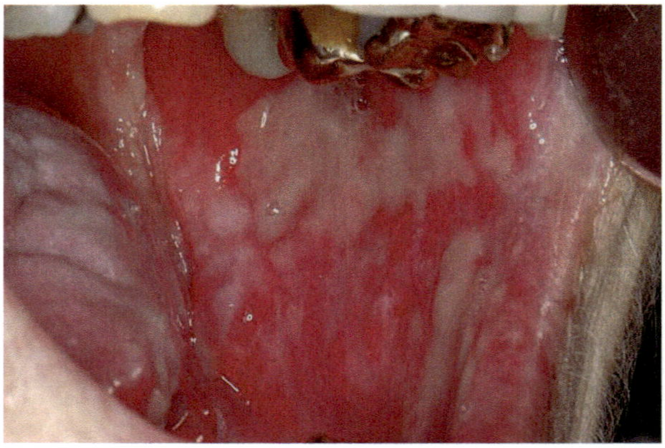

Fig. 85.1 Oral mucositis in a patient undergoing radiotherapy for the treatment of tongue squamous cell carcinoma

85.3 Management and Prognosis

Oral mucositis can be complicated by the presence of neutropenia and predispose to bacteremia, septicemia, and fungemia, which may be life threatening.

Management of OM may be approached from a preventative or a treatment perspective. Table 85.2 illustrates the current evidence-based guidelines and expert opinions (where a guideline is not possible due to limited/inconsistent data) developed by the Multinational Association of Supportive Care in Cancer/International Society of Oral Oncology partnership (MASCC/ISOO).

Table 85.2 Prevention and treatment of oral mucositis: evidence-based guidelines (G) and expert opinion (EO) developed by MASCC/ISOO

Analgesics
Topical 0.2% morphine mouthwash is suggested for the treatment of OM-associated pain in H&N cancer patients treated with RT-CT (G)
Basic oral care (BOC)
• Dental assessment and treatment is desirable to reduce the risk for local and systemic infections from odontogenic sources (EO) • Oral hygiene protocols including recommendations with regard to bland mouth rinses, toothbrushes, and flossing procedures are beneficial for the prevention of OM during CT, H&N RT, and HSCT (G) • Educating patients about BOC is appropriate and may improve self-management and adherence to the recommended protocols during cancer treatment (EO) • Saline and sodium bicarbonate rinses increase oral clearance and may be helpful in maintaining oral hygiene and improving comfort (EO)
Photobiomodulation
The use of low-level laser therapy is recommended for the prevention of OM in adult patients receiving HSCT conditioned with high dose CT, with or without total body irradiation and in adult patients receiving RT to the H&N or RT and CT for H&N cancer (G)
Anti-inflammatory agents
Benzydamine mouthwash is recommended for the prevention of OM in patients with H&N cancer receiving moderate dose of RT (<50 Gy) and is suggested for the prevention of OM in patients with H&N cancer receiving RT and CT (G)
Growth factors
The use of keratinocyte growth factor-1 intravenously is recommended for the prevention of OM in patients undergoing HSCT with a conditioning regimen that includes high dose CT and total body irradiation (G)
Natural and miscellaneous agents
Oral glutamine is suggested (with caution) for the prevention of OM due to RT-CT in H&N cancer (G)
Honey is suggested for the prevention of OM in H&N cancer patients treated with either RT or RT-CT (G) *Note: Data regarding dental caries in studies concerning honey-based protocols for the prevention of OM is lacking*
Cryotherapy
Oral cryotherapy is recommended for the prevention of OM in patients undergoing HSCT when the conditioning includes high dose melphalan and in patients receiving bolus 5-fluorouracil for solid tumors (G)

Recommended Reading

Al-Dasooqi N, Sonis ST, Bowen JM, Bateman E, Blijlevens N, Gibson RJ, et al. Emerging evidence on the pathobiology of mucositis. Support Care Cancer. 2013;21(11):3233–41..

Elad S, Cheng KKF, Lalla RV, Yarom N, Hong C, Logan RM, et al. MASCC/ISOO clinical practice guidelines for the management of mucositis secondary to cancer therapy. Cancer. 2020;126(19):4423-31.

Hong CHL, Gueiros LA, Fulton JS, Cheng KKF, Kandwal A, Galiti D, et al. Systematic review of basic oral care for the management of oral mucositis in cancer patients and clinical practice guidelines. Support Care Cancer. 2019;27(10):3949–67.

Lalla RV, Sonis ST, Peterson DE. Management of oral mucositis in patients who have cancer. Dent Clin North Am. 2008;52(1):61–77, viii.

Pulito C, Cristaudo A, Porta C, Zapperi S, Blandino G, Morrone A, et al. Oral mucositis: the hidden side of cancer therapy. J Exp Clin Cancer Res. 2020;39(1):210.

Scully C, Sonis S, Diz PD. Oral mucositis. Oral Dis. 2006;12(3):229–41

Sonis ST, Elting LS, Keefe D, Peterson DE, Schubert M, Hauer-Jensen M, et al. Perspectives on cancer therapy-induced mucosal injury: pathogenesis, measurement, epidemiology, and consequences for patients. Cancer. 2004;100(9):1995–2025.

Stringer AM, Logan RM. The role of oral flora in the development of chemotherapy-induced oral mucositis. J Oral Pathol Med. 2015;44(2):81–7

Radiotherapy and the Oral Cavity

86

Agnieszka Frydrych

86.1 Introduction

Radiotherapy (RT) plays a vital role in the management of head and neck cancer (HNC), with most patients receiving a total dose of 50–70 Gy. All tissues in the irradiated field are affected. Tissue-sparing techniques such as intensity-modulated RT have enhanced cytocidal efficiency while reducing damage to healthy tissues.

Ionising radiation produces its therapeutic effect and associated adverse effects by damaging the genetic material of cells, resulting in cell death. Early (acute) effects occur in tissues with rapid turnover rates, while in tissues with slower turnover, damage may not be apparent for months or years after treatment. Complications depend on the size of the irradiated field and dose of RT.

About 75% of all HNC patients are offered RT with either curative or palliative intent, and over 80% of those patients will endure acute oral complications.

86.2 Clinical Presentation and Investigations

86.2.1 Acute Effects

1. **Oral mucosa.** Mucositis manifests as mucosal erythema, eventually progressing to ulceration and pseudomembrane as the dose of radiation increases. Candidosis is the most common infection during and shortly after RT.
2. **Taste.** Partial or complete loss of taste is experienced by most patients undergoing RT, with a gradual return to normal or near normal levels within 1 year post treatment for the majority of individuals.

A. Frydrych (✉)
UWA Dental School, The University of Western Australia, Nedlands, WA, Australia
e-mail: agnieszka.frydrych@uwa.edu.au

R. Balasubramaniam et al. (eds.), *Oral Medicine - A Clinical Guide*,
https://doi.org/10.1007/978-3-031-36797-7_86

3. **Salivary glands.** Doses above 50 Gy are associated with irreversible injury. Aside from the reduced salivary flow, saliva may become viscous and the mucosae dry. Consequently, speech, mastication, deglutition, oral comfort and even sleep can be compromised and denture tolerance may be reduced. Oral malodour may develop.

86.2.2 Late Effects

1. **Dentition.** Radiation-related dental caries (RRDC) is an aggressive form of caries, predominantly related to the RT-induced salivary gland hypofunction. Dental caries may become evident within 3 months of RT and lead to complete loss of dentition within 1 year.
2. **Periodontium.** RT-induced changes include radiographic widening of the periodontal ligament space and destruction of bony trabeculae. There is increased risk of periodontal attachment loss, tooth loss and osteoradionecrosis (ORN).
3. **Bone.** Irradiated bone responds poorly to trauma and infection, which may lead to ORN, the risk of which is 5–15% and higher in the mandible than the maxilla (Fig. 86.1). It presents as necrotic, exposed bone of at least 3 months duration, with no evidence of healing, in the absence of local neoplastic recurrence or metastatic disease. The risk is lifelong and increases with time. Surgery, including dental extraction, periodontal disease, other infection and trauma associated with poorly fitting prosthesis are risk factors.
4. **Muscles and joints.** Irradiation of muscles and ligaments may lead to myofibrotic contractures, which generally develop 3–6 months post completion of RT and are frequently permanent.
5. **Other tissues.** RT-induced damage to vessels and nerves may give rise to lymphoedema and neuropathy.

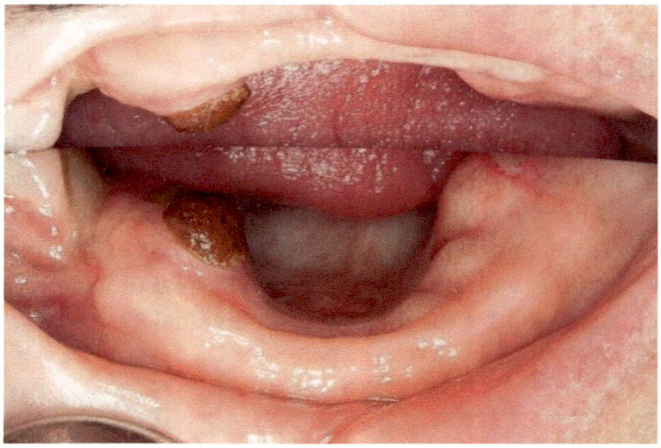

Fig. 86.1 Osteoradionecrosis affecting the right mandibular lingual torus

Diagnosis of RT-related injuries is clinical and based on the known use of this treatment modality as well as the appearance, timing, and location of the injuries. Residual/recurrent malignancy should be ruled out.

86.3 Management and Prognosis

In view of complexities of patient care, all HNC patients should be managed in a multidisciplinary team setting. Multidisciplinary teams include surgeons, oncologists, dieticians, speech pathologist, and dentists/dental specialists. Dental management includes three phases and is summarized in Table 86.1.

Table 86.1 Dental management of head and neck patients treated with radiotherapy

Pre-RT assessment
• Comprehensive clinical and radiographic examination • Identification of factors likely to increase risk of oral complications • Risk modification, e.g. hygiene prophylaxis, restorations, denture adjustments and/or prophylactic dental extractions • Discussion of oral hygiene, denture care and caries preventive protocols • Consideration of stent use to manage radiation back scatter in patients with metallic restorations
Management during RT
• ***Oral mucositis***—Maintenance of good oral hygiene, use of bland mouth washes, anti-inflammatory agents, analgesics and low-level laser therapy • ***Oral candidosis***—Treated with topical antifungal agents. When systemic therapy is indicated, fluconazole is the first drug of choice
Post-RT care
• ***Xerostomia and salivary gland hypofunction***—Oral lubricants, antimicrobial salivary substitutes and salivary stimulation using chewing gum and cholinergic agonists can provide symptomatic relief and offer some dental protection • ***Dental caries***—Prevention involves meticulous oral hygiene, fluoride and other remineralizing agent and chlorhexidine use, dietary modifications and lifelong dental care. Resin composite is the material of choice for established RRDC • ***Periodontal disease***—Regular preventative dental care minimizes the risk of disease progression • ***Myofibrotic contractures***—Stretching/strengthening exercises during and after RT are necessary to maintain adequate mouth opening • ***Osteoradionecrosis***—Removable prosthesis should be avoided unless essential, and denture provision delayed by 6–12 months post-RT. Caries prevention and regular dental care decrease the risk of extractions and ORN. Treatment of ORN ranges from the use of antiseptic rinses through to debridement, sequestrectomy, resection, and free flaps, with or without the use of adjuncts such as hyperbaric oxygen or pentoxifylline, tocopherol and clodronate

Recommended Reading

Bhandari S, Soni BW, Bahl A, Ghoshal S. Radiotherapy-induced oral morbidities in head and neck cancer patients. Spec Care Dentist. 2020;40(3):238–50.

Kufta K, Forman M, Swisher-McClure S, Sollecito TP, Panchal N. Pre-Radiation dental considerations and management for head and neck cancer patients. Oral Oncol. 2018;76:42–51.

Levi LE, Lalla RV. Dental Treatment Planning for the Patient with Oral Cancer. Dent Clin North Am. 2018;62(1):121–30.

Lin A. Radiation Therapy for Oral Cavity and Oropharyngeal Cancers. Dent Clin North Am. 2018;62(1):99–109.

Palmier NR, Migliorati CA, Prado-Ribeiro AC, de Oliveira MCQ, Vechiato Filho AJ, de Goes MF, et al. Radiation-related caries: current diagnostic, prognostic, and management paradigms. Oral Surg Oral Med Oral Pathol Oral Radiol. 2020;130(1):52–62.

Ratko TA, Douglas GW, de Souza JA, Belinson SE, Aronson N. Radiotherapy Treatments for Head and Neck Cancer Update. Radiotherapy Treatments for Head and Neck Cancer Update. AHRQ Comparative Effectiveness Reviews. Rockville (MD)2014.

Specht L. Oral complications in the head and neck radiation patient. Introduction and scope of the problem. Support Care Cancer. 2002;10(1):36–9.

Sroussi HY, Epstein JB, Bensadoun RJ, Saunders DP, Lalla RV, Migliorati CA, et al. Common oral complications of head and neck cancer radiation therapy: mucositis, infections, saliva change, fibrosis, sensory dysfunctions, dental caries, periodontal disease, and osteoradionecrosis. Cancer Med. 2017;6(12):2918–31.

White JM, Panchal NH, Wehler CJ, Bestgen SC, Colón JE, Desai HG, et al. Department of Veterans Affairs Consensus: Preradiation dental treatment guidelines for patients with head and neck cancer. Head Neck. 2019;41(5):1153–60.

Part IV

Orofacial Pain and Related Disorders

Odontogenic Pain

87

Jeremy Lau

87.1 Introduction

The most common reason for emergency dental attendance is due to acute odontogenic pain. Acute odontogenic pain can be extremely severe and debilitating. Chronic odontogenic pain tends to be associated with milder symptoms.

Odontogenic pain serves a protective biological function. Common sources of pain include the dental pulp and periapical tissues. In early signs of pulp reaction to insults including dentinal hypersensitivity and reversible pulpitis, the most commonly accepted theory to date regarding transmission of pain from the dentino-pulpal complex is the hydrodynamic theory. Pain provoked by thermal, chemical, or mechanical stimuli is a consequence of fluid displacement within the dental tubules and odontoblasts, which subsequently activates nociceptors. Nociceptors transduce a signal via myelinated Aδ or unmyelinated C fibres, and the signal propagates to the trigeminal spinal complex before synapsing in the thalamus and the cerebral cortex. In the event where there is overwhelming inflammation, the structural integrity of the pulp is lost. This leads to necrosis followed by infection. Pain caused by inflammatory events in the pulp involves a range of non-specific mediators of inflammation including histamine, bradykinin, serotonin, interleukin, and arachidonic acid.

The prevalence of orofacial pain is reported to be between 13 and 19% of the general population. The overwhelming majority (95%) of orofacial pain is odontogenic pain, resulting from caries, periodontal disease, or tooth fracture. Even though odontogenic pain can affect all ages, the age group with the most frequent presentation is between 17 and 29 years.

J. Lau (✉)
UWA Dental School, The University of Western Australia, Nedlands, WA, Australia
e-mail: jeremy.lau@uwa.edu.au

R. Balasubramaniam et al. (eds.), *Oral Medicine - A Clinical Guide*,
https://doi.org/10.1007/978-3-031-36797-7_87

87.2 Clinical Presentation and Investigations

History, including onset and timing of odontogenic pain can lead to diagnosis. Some common pain descriptors or symptoms include lingering or constant pain, poorly localised pain, sharp pain, temperature sensitive, dull and aching sensation, pain on biting, referred pain, and limited mouth opening. Common associated signs include the breakdown of tooth structure, the presence of dental biofilm, periodontal pocketing, or the presence of cracks. It is important to differentiate between odontogenic causes of pain and non-odontogenic causes of pain (Table 87.1). Specific presentation of each condition is beyond the scope of this chapter.

There are a range of clinical investigations that can assist in establishing a diagnosis. Clinical tests and investigations would typically include:

- History of the presenting concern
- Medical history
- Imaging
- Examination
- Clinical tests:
  - Pulp sensitivity tests
  - Percussion tests
  - Triplex spray
  - Palpation
  - Probing
  - Mobility test
  - FracFinder or transilluminator

Table 87.1 Common differential diagnosis of odontogenic pain and non-odontogenic pain

Categories of odontogenic conditions	Differential diagnosis of common painful odontogenic conditions
Endodontic conditions (typically caused by caries, cracked tooth, trauma)	Reversible pulpitis Irreversible pulpitis Necrobiosis Pulp necrosis Pulpless, infected root canal system Degenerative changes Previous root canal treatment—with infection
Periodontal disease	Gingivitis Periodontitis Pericoronitis Necrotising periodontal disease Endo-periodontal lesions Periodontal abscesses Peri-implantitis
Physical and chemical injuries of teeth	Attrition Abrasion Erosion Abfraction Cracked tooth syndrome Facial trauma Postsurgical pain

Table 87.1 Continued

Categories of non-odontogenic pains	Differential diagnosis of painful non-odontogenic conditions
	Masticatory or cervical myofascial pain Sinus disorders related pain Orofacial pain attributed to lesions or disease of the cranial nerves • Pain attributed to lesion or disease of the trigeminal nerve such as trigeminal neuralgia or post-traumatic trigeminal neuropathic pain • Pain attribute to lesions or disease of the glossopharyngeal nerve such as glossopharyngeal neuralgia Orofacial pain resembling presentations of primary headaches • Orofacial migraine • Tension type orofacial pain • Trigeminal autonomic orofacial pain • Neurovascular orofacial pain Idiopathic orofacial pain • Burning mouth syndrome • Persistent idiopathic facial pain • Persistent idiopathic dentoalveolar pain • Constant unilateral facial pain with additional attacks Orofacial pain from visceral chest infections • Myocardial infarction • Angina pectoris Orofacial pain from visceral throat disease • Subacute thyroiditis • Eagle syndrome Nociplastic pain

87.3 Management and Prognosis

Management of acute odontogenic pain involves addressing associated odontogenic pathology. Often this will involve restorative treatment with direct or indirect restorations, periodontal debridement, or removal of the tooth or pulpal complex. Risk factors for further odontogenic pain should be addressed. Odontogenic pain often resolves once the pathological source is addressed. At times, analgesics can be considered for pain relief. Adjunctive antimicrobial therapy may be appropriate in certain situations such as severe superficial infections with swelling and systemic signs and symptoms, or unresponsive or deep infections. Where there is spreading odontogenic infection with severe or systematic features, it is imperative to provide appropriate support for airway, breathing, and circulation. Patients may require hospital admission for management.

Recommended Reading

Abbott PV, Yu C. A clinical classification of the status of the pulp and the root canal system. Aust Dent J. 2007;52(1 Suppl):S17–31.

Balasubramaniam R, Turner LN, Fischer D, Klasser GD, Okeson JP. Non-odontogenic toothache revisited. Open J Stomatol. 2011;1(3):92–102.

DeAngelis AF, Barrowman RA, Harrod R, Nastri AL. Review article: Maxillofacial emergencies: oral pain and odontogenic infections. Emerg Med Australas. 2014;26(4):336–42.

Henry MA, Hargreaves KM. Peripheral mechanisms of odontogenic pain. Dent Clin North Am. 2007;51(1):19–44, v.

Horst OV, Cunha-Cruz J, Zhou L, Manning W, Mancl L, DeRouen TA. Prevalence of pain in the orofacial regions in patients visiting general dentists in the Northwest Practice-based REsearch Collaborative in Evidence-based DENTistry research network. J Am Dent Assoc. 2015;146(10):721–8.e3.

International Classification of Orofacial Pain, 1st edition (ICOP). Cephalalgia. 2020;40(2):129–221.

Gbenga Omitola O, Olabisi Arigbede A. Prevalence and Pattern of Pain Presentation among Patients Attending a Tertiary Dental Center in a Southern Region of Nigeria. J Dent Res Dent Clin Dent Prospects. 2010;4(2):42–6.

Papapanou PN, Sanz M, Buduneli N, Dietrich T, Feres M, Fine DH, et al. Periodontitis: Consensus report of workgroup 2 of the 2017 World Workshop on the Classification of Periodontal and Peri-Implant Diseases and Conditions. J Periodontol. 2018;89 Suppl 1:S173–S182.

Renton T. Dental (Odontogenic) Pain. Rev Pain. 2011;5(1):2–7.

Rosenberg PA. Endodontic pain. Endod Topics. 2014;30(1):75–98.

Sinus Disorder-Related Orofacial Pain

88

Supeetha Suntharamoorthy

88.1 Introduction

Sinus disorder-related orofacial pain refers to pain originating from an environmental or host-related disorder in the paranasal sinuses and the nasal cavity and presenting in the craniofacial and dentoalveolar structures. The condition may be acute if present for less than 3 months or chronic if present for longer.

Sinusitis is the third most commonly occurring respiratory condition in Australia, accounting for 1.4 of every 100 visits to the medical practitioner. In 2017–2018, around two million or 8.4% of Australians experienced chronic sinusitis with a higher prevalence in people aged over 45 years and in females (9.7%) than males (7.1%). The incidence of non-odontogenic sinusitis is estimated to be 60–90% with approximately 11% of patients experiencing referred dental pain.

Orofacial pain of sinus origin commonly arises from a viral upper respiratory tract infection and may be complicated by bacterial and fungal infection, allergy, anatomical defects, immunodeficiency, hormones, medications, recreational drugs, growths (polyps, cysts, neoplasms) and conditions that affect mucociliary function such as cystic fibrosis. The ensuing inflammation, termed sinusitis or rhinosinusitis, and occlusion of the sinus ostia activate the surrounding nociceptors initiating facial pain through the trigeminothalamic tract (Fig. 88.1). Branches of the maxillary division of the trigeminal nerve that innervate the maxillary teeth in close anatomical proximity to the maxillary sinuses may also be activated triggering pain mimicking dental pain.

S. Suntharamoorthy (✉)
UWA Dental School, University of Western Australia, Nedlands, WA, Australia
e-mail: supeetha.suntharamoorthy@uwa.edu.au

R. Balasubramaniam et al. (eds.), *Oral Medicine - A Clinical Guide*,
https://doi.org/10.1007/978-3-031-36797-7_88

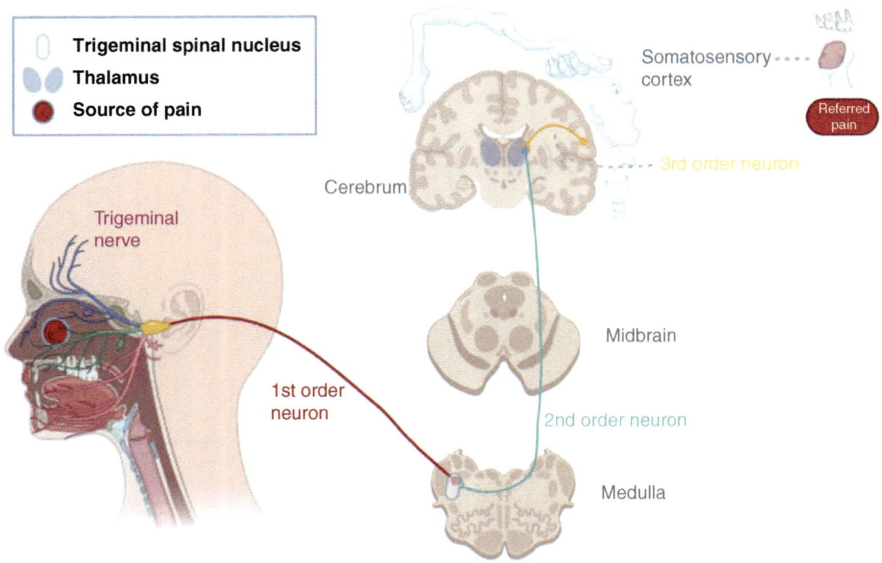

Fig. 88.1 Trigeminothalamic pain pathway. Branches of the trigeminal nerve: *V1* Ophthalmic, *V2* Maxillary, *V3* Mandibular

88.2 Clinical Presentation and Investigations

The clinical presentation includes unilateral or bilateral facial pain with a feeling of continuous pressure and fullness along the infraorbital and maxillary regions that may be accompanied by:

- *Major symptoms:* nasal congestion, purulent or bloody nasal discharge, postnasal drip, hyposmia, fever.
- *Minor symptoms:* headache, oedema, lymphadenopathy, halitosis, sore throat, dysphonia, cough, fatigue, earache, toothache of one or more maxillary teeth that is exacerbated on cold stimuli, chewing, clenching, coughing, sneezing and bending forward.
- Radiographic opacity in the sinus cavity with thickening of the mucosal lining and/or marginal destruction.

Diagnosis relies on the following investigations:

- A detailed history.
- Clinical examination:
  - Extra-oral exam: percussion of sinuses, palpation of lymph nodes, visual assessment for oedema and erythema of sinuses.
  - Intra-oral exam of oropharynx, oral cavity and dentition.

  - Diagnostic tests: pulp sensibility, local anaesthetic administration to exclude dental pathology, bending forward to assess quality of pain from a change in pressure, response to nasal decongestants or antihistamines.
- Radiographic exam: dental imaging, CT scan of sinuses.
- Laboratory tests: culture of nasal secretion.
- ENT evaluation: ultrasonography, transillumination, endoscopy.

Differential diagnosis of sinus disorder-related orofacial pain includes odontogenic pain, neoplasia such as squamous cell carcinoma, nasopharyngeal carcinoma, adenocarcinoma, undifferentiated carcinoma, neuroblastoma, melanoma, lymphoma and sarcoma; myofascial pain, neurovascular pain, neuropathic pain, cardiogenic pain and nociplastic pain.

88.3 Management and Prognosis

Sinusitis of viral origin is commonly managed with an analgesic, nasal decongestant and saline irrigation. Antibiotics or antifungals may be prescribed in cases of bacterial or fungal infection, whilst chronic cases may be supplemented with corticosteroids and antihistamines. Referral to an otolaryngologist is recommended for recurrent sinusitis. Prognosis is conditional to the severity of the sino-nasal disorder and its management.

Recommended Reading

Australian Bureau of Statistics. Australian health survey: first results, 2017–18 [Internet]. Canberra (ACT): Commonwealth of Australia; 2018. Report No.: 4364055001DO003_20172018. [cited 2023 November 1]. https://www.abs.gov.au/statistics/health/health-conditions-and-risks/national-health-survey-first-results/2017-18#about-thenational-health-survey.

Battisti AS, Modi P, Pangia J. Sinusitis. In: StatPearls [Internet]. Treasure Island (FL): StatPearls Publishing; 2023. Available at: https://www.ncbi.nlm.nih.gov/books/NBK470383/. Accessed 10 Nov 2023.

De Laat A. Differential diagnosis of toothache to prevent erroneous and unnecessary dental treatment. J Oral Rehabil. 2020;47(6):775–81.

Ferguson M. Rhinosinusitis in oral medicine and dentistry. Aust Dent J. 2014;59(3):289–95.

Kim KM, Byun JS, Jung JK, Choi JK. Non-odontogenic toothache caused by acute maxillary sinusitis: a case report. J Oral Med Pain. 2016;41(2):80–4.

Little RE, Long CM, Loehrl TA, Poetker DM. Odontogenic sinusitis: A review of the current literature. Laryngoscope Investig Otolaryngol. 2018;3(2):110–4.

Morcom S, Phillips N, Pastuszek A, Timperley D. Sinusitis. Aust Fam Physician. 2016;45(6):374–7.

Shirazi N, Bist SS, Selvi TN, Harsh M. Spectrum of Sinonasal Tumors: A 10-year Experience at a Tertiary Care Hospital in North India. Oman Med J. 2015;30(6):435–40.

Cardiogenic Orofacial Pain

89

Supeetha Suntharamoorthy

89.1 Introduction

Cardiogenic orofacial pain refers to pain originating from the heart as a result of ischaemia, infarction or inflammation and presenting in the chest, shoulder, back, abdomen, arm and craniofacial structures. It is a form of *referred* pain of cardiac origin in contrast to *radiating* pain, the latter of which refers to pain spreading from the cardiac region to other areas of the body. The pain can occur spontaneously or with physical activity.

One in every 15 patients experience orofacial pain as the sole symptom during a cardiac ischaemic episode (angina) and 1 in every 20 patients an acute myocardial infarction. In 32–40% of patients, orofacial pain occurs concomitantly with other common pain sites such as the chest, shoulder, back and arms, however, in the absence of chest pain, the craniofacial region is the most dominant site of pain. Prevalence is higher in females than males and in those with predisposing cardiovascular risk factors.

Orofacial pain of cardiac origin is induced by cardiovascular conditions, commonly ischaemic heart disease and more rarely aortic dissection, pericarditis and congenital heart diseases such as atrial septal defect. The referral pattern of cardiogenic pain correlates to the distribution of the visceral afferent fibres of the vagus and sympathetic nerves from within the heart and the cardiac plexus. During a pathophysiological event, cardiac pain transmitted through these nerves converges with the trigeminal nerve in the upper cervical spine ensuing the relaying of pain in the head, neck and jaw regions via the spinothalamic tract (Fig. 89.1).

S. Suntharamoorthy (✉)
UWA Dental School, The University of Western Australia, Nedlands, WA, Australia
e-mail: supeetha.suntharamoorthy@uwa.edu.au

R. Balasubramaniam et al. (eds.), *Oral Medicine - A Clinical Guide*,
https://doi.org/10.1007/978-3-031-36797-7_89

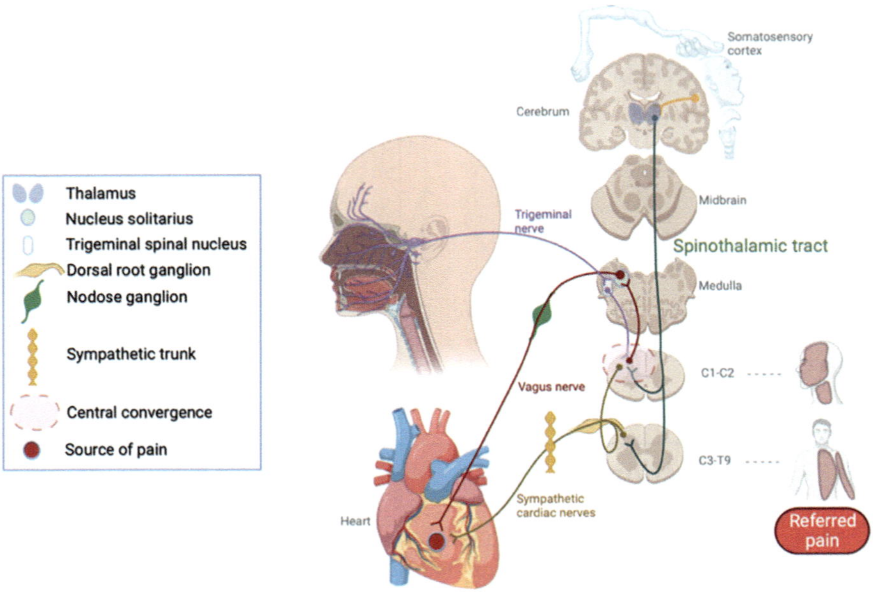

Fig. 89.1 Referral pathway of cardiac pain

89.2 Clinical Presentation and Investigations

The characteristic clinical features of cardiogenic orofacial pain that differentiate it from odontogenic pain include:

- **Location (in order of most common to least):** throat, mandible (left > right), temporomandibular joints/ears, neck, teeth (mandibular > maxillary), temporal, face
- **Symmetry:** bilateral > unilateral (6:1)
- **Character:** pressure, burning, dull, deep, diffuse, sharp
- **Intensity:** slight to severe
- **Frequency:** intermittent, constant
- **Duration:** short (minutes), lingering
- **Mode of onset:** spontaneously or on physical exertion
- **Other symptoms:** pain in the chest, arms, shoulder, back and abdomen, sweating, nausea, vomiting, vertigo, breathlessness, palpitations, weakness, numbness

Pain lasting for more than 20 min or exacerbated symptoms of angina, arrhythmia and hypotension are critical features requiring urgent medical attention.

Diagnosis of cardiogenic orofacial pain relies on identifying the source of pain through the following investigations:

- A detailed history
- Clinical examination:
  - Vitals: blood pressure, heart rate, respiratory rate, temperature, weight, height
  - Extra-oral and intra-oral exam
  - Diagnostic tests: pulp sensibility, dental imaging, local anaesthetic administration to exclude local pathology
- Cardiological evaluation: exercise test, stress test, electrocardiography, laboratory tests and coronary angiography

Differential diagnosis of cardiogenic orofacial pain includes odontogenic pain, neoplasia such as oral squamous cell carcinomas, neoplasms of the heart, lungs, mediastinum and cranium as well as metastatic neoplasms, myelomas, lymphomas and leukaemia; myofascial pain, sinus pain, neurovascular pain, neuropathic pain and nociplastic pain.

89.3 Management and Prognosis

Treatment of cardiogenic orofacial pain relies on prompt referral to an emergency department during an acute event or a cardiologist for evaluation and management. Prognosis is conditional to the severity of the cardiovascular condition and its management.

Recommended Reading

Danesh-Sani SH, Danesh-Sani SA, Zia R, Faghihi S. Incidence of craniofacial pain of cardiac origin: results from a prospective multicentre study. Aust Dent J. 2012;57(3):355–8.

López-López J, Adserias-Garriga MJ, Garcia-Vicente L, Jané-Salas E, Chimenos-Küstner E, Pereferrer-Kleiner D. Orofacial pain of cardiac origin, serial of clinical cases. Med Oral Patol Oral Cir Bucal. 2012;17(4):e633–7.

Kreiner M, Okeson J, Tanco V, Waldenström A, Isberg A. Orofacial Pain and Toothache as the Sole Symptom of an Acute Myocardial Infarction Entails a Major Risk of Misdiagnosis and Death. J Oral Facial Pain Headache. 2020;34(1):53–60.

Ramachandran S, Arulneyam J, Arokiaraj MC. Tooth pain can be indicative of undiagnosed atrial septal defect. Res Rev: J Dent Sci 2014;2(2):104–105.

Romero-Reyes M, Salvemini D. Cancer and orofacial pain. Med Oral Patol Oral Cir Bucal. 2016;21(6):e665–e671.

Sharav Y, Benoliel R. Orofacial pain and headache. 2nd Illinois: Quintessence Publishing; 2015. 654

Non-odontogenic Pain Related to Pathology

90

Paul Mueller and Ramesh Balasubramaniam

90.1 Introduction

Pain associated with peripheral sensitisation related to acute tissue injury is a protective and appropriate function of nociceptors to CNS pathways. Associated phenomena of referred pain, allodynia and hyperalgesia resulting from sensitisation and recruitment of previously silent nociceptive receptors are purposed to protect against further tissue damage. This chapter considers pain relating to pathology of the oral mucosa, bones of the orofacial complex and salivary glands.

90.2 Oral Mucosal Pain and Pain Related to Cancer

Pain associated with oral mucosal pathology is more commonly related to epithelial atrophy, erosion or ulceration. Common description of symptoms includes rawness, burning, stinging, exacerbation with hot, acidic, salty or spicy foods, irritation with dentifrices, and tenderness to touch. These signs and symptoms may be observed in conditions such as oral dermatoses.

Presence of pathology affects the threshold of nociceptors. Secretion of nociceptive mediators by presence of pathology, namely: (1) endothelin-1 results in sensitisation of silent nociceptor, (2) proteases results in TRPV1-dependent thermal hyperalgesia and TRPV4-dependent mechanical allodynia, and (3) nerve growth

P. Mueller · R. Balasubramaniam (✉)
UWA Dental School, The University of Western Australia, Nedlands, WA, Australia
e-mail: ramesh.balasubramaniam@uwa.edu.au

R. Balasubramaniam et al. (eds.), *Oral Medicine - A Clinical Guide*,
https://doi.org/10.1007/978-3-031-36797-7_90

factor results in thermal hyperalgesia and mechanical allodynia with acute and chronic exposure respectively.

Nociceptive afferent fibres (A-delta and C-fibres) carry noxious stimuli of the oral mucosa through the trigeminothalamic pathway to the primary somatosensory cortex. Dense innervation within the oral cavity affects two-point discrimination resulting in increased precision when localising the origin of pain. Referral of pain may be experienced with the convergence of multiple primary afferent nociceptors in the dorsal horn of the trigeminal nucleus caudalis. Primary afferent fibres often co-synapse with converging afferent fibres of other anatomical regions to second-order neurons, which may be responsible for the experience of referred pain and subsequent misdiagnosis of the true pathological source, particularly with visceral pathology.

90.3 Pain from Intraosseous Pathology

Pain originating from intraosseous pathology of the orofacial complex also follows the trigeminothalamic pathway to the primary somatosensory cortex and may be associated with osteomyelitis, osteonecrosis, osseous cystic involvement, intraosseous neoplasms such as osteosarcoma, and locoregional spread of oral cancers with osseous involvement. The periosteum is the most densely innervated component of bone. Bone marrow, however, contains the greatest number of nociceptive afferent fibres, followed by mineralised bone. Periosteal and intraosseous pathology can involve many afferent neurons; pain is generally experienced in pathology with an inflammatory response, possibly relating to secondary infection, with resultant neuropeptide release affecting nociceptive pathways.

Neoplasms and metastasis of oral cancer involving bones of the orofacial complex may result in pain, with inflammatory processes initiated by cancer-induced tissue damage. Tumours can locally secrete inflammatory cytokines and growth factors, resulting in a nociceptive neurophysiological mechanism as described with pain associated with oral mucosal pathology. Mechanical pressure associated with bone cancers from tumour growth, tumour acidosis, and tissue hypoxia facilitate nociception which present as pain and hyperalgesia.

90.4 Salivary Gland Pain

Pain may be associated with non-neoplastic salivary gland pathology such as sialolithiasis, mucous retention cyst, sialodenosis, sialadenitis, and necrotising sialometaplasia. Sialolithiasis results in tissue distension, and the subsequent inflammation leads to the release of cytokines involved in hyperalgesia and allodynia. Neoplastic pathology is rarely associated with pain, with the exception of mucoepidermoid carcinoma and adenoid cystic carcinoma. Adenoid cystic carcinoma is associated with perineural invasion of nerve growth factor, which is similar to oral mucosal pain, thus results in thermal hyperalgesia and mechanical allodynia.

90.5 Conclusion

Pathology of the oral mucosa, bones of the orofacial complex, and salivary gland are often associated with pain. As such, pain associated with oral pathology as a source of non-odontogenic pain must be part of the diagnostic sieve when assessing patients with orofacial pain complaints.

Recommended Reading

Chung MK, Wang S, Oh SL, Kim YS. Acute and Chronic Pain from Facial Skin and Oral Mucosa: Unique Neurobiology and Challenging Treatment. Int J Mol Sci. 2021;22(11):5810.

Dohi T, Morita K, Abdin MJ, Kitayama T, Motoyama N, Zohn R. Mechanisms of pain sensitization and the treatments. In: Hiroshima Conference on Education and Science in Dentistry; 2006 Jan 8; Hiroshima, Japan. 2006. p. 87–94.

Neurovascular Orofacial Pain

91

Guru O

91.1 Introduction

The International Classification of Orofacial Pain (ICOP) describes neurovascular orofacial pain (NVOP) as 'attacks of variable duration of moderate or severe intraoral pain, without head pain, often accompanied by toothache like symptoms, with mild autonomic and/or migraine-like symptoms'.

NVOP is a poorly understood disease entity. It is thought to share many pathophysiological underpinnings as migraine. An important concept in the pathogenesis of migraine is neurogenic inflammation and its effects on the vasculature. The antidromic release of nitric oxide, substance P, calcitonin gene-related peptide and vasoactive intestinal polypeptide by afferent C fibres and parasympathetic fibres results in vasodilation and plasma extravasation. This in turn sensitises and activates primary afferents associated with the vessels, which play a role in the pain experienced in migraine. Similar mechanisms are thought to occur in the orofacial sensory supply for NVOP.

Guru O (✉)
UWA Dental School, University of Western Australia, Nedlands, WA, Australia
e-mail: guru.o@uwa.edu.au

R. Balasubramaniam et al. (eds.), *Oral Medicine - A Clinical Guide*,
https://doi.org/10.1007/978-3-031-36797-7_91

91.2 Clinical Presentation and Investigations

NVOP is an extremely rare condition and with diagnostic criteria not previously clearly defined and as such only a hand full of articles have provided data on this condition. From the limited reports that are available, NVOP appears to affect patients aged 40–50 years of age with females being more affected (3.2:1 ratio).

In addition to the diagnostic criteria for NVOP given by ICOP, clinical reports suggest sensitivity to cold in adjacent teeth and mucosa during and in-between attacks. Moreover, referral to adjacent sites may occur if pain is severe. NVOP can also be subclassified into a short-lasting variant with relatively short attacks (1–4 hours) and a long-lasting variant (>4 hours).

Differentials for NVOP include other neurovascular pain diagnoses restricted to the maxillofacial region or those referred that refer into it, such as posttraumatic trigeminal neuropathic pain, persistent idiopathic facial pain, persistent idiopathic dentoalveolar pain, or odontogenic pain. As such, an accurate pain history is paramount, and a pain diary is invaluable. Appropriate imaging such as a CT maxilla or mandible, and MRI brain and skull base must be carried out to rule out secondary causes of facial pain (e.g. space occupying lesions, multiple sclerosis). Furthermore, the use of therapeutics effective in migraine treatment can be diagnostic.

91.3 Management and Prognosis

Treatment for NVOP parallels migraine therapy and as such is either abortive or preventative. Abortive treatment should be utilized when there are less than 4 attacks per month or to supplement preventative treatment when there is breakthrough pain. Paracetamol 1000 mg or ibuprofen 400–800 mg is first line, failing which triptans (e.g. rizatriptan wafers 10 mg) may be trialed. Abortive medications should be taken as early as possible when symptoms are noted. Preventative treatment is indicated when:

- Attacks are frequent (>4 attacks per month)
- Significant interference in quality of life
- Contraindication, failure, adverse effects, or overuse of abortive treatments (e.g. >15 days of NSAIDs or paracetamol per month, >10 days for triptans, opioids, combination analgesics, and ergot derivatives).
- Patient preference for fewer attacks

Preventatives that have a high level of efficacy include anti-epileptics (topiramate, sodium valproate) and beta blockers (metoprolol, propranolol, timolol). Preventative medication selection should be based on the highest level of evidence for effectiveness and is guided by other comorbid conditions and side effects. As such, consultation with the patient's general medical practitioner or referral to a neurologist may be necessary.

Recommended Reading

The American Headache Society Position Statement On Integrating New Migraine Treatments Into Clinical Practice. Headache. 2019;59(1):1–18.

Benoliel R, Sharav Y, Eliav E. Neurovascular orofacial pain. J Am Dent Assoc. 2010;141(9):1094–6.

Dodick DW. A Phase-by-Phase Review of Migraine Pathophysiology. Headache. 2018;58(Suppl 1):4–16.

Ha H, Gonzalez A. Migraine Headache Prophylaxis. Am Fam Physician. 2019;99(1):17–24.

Haviv Y, Zini A, Keshet N, Almoznino G, Benoliel R, Sharav Y. Features of Neurovascular Orofacial Pain Compared to Painful Posttraumatic Trigeminal Neuropathy. J Oral Facial Pain Headache. 2020;34(2):121–128.

International Classification of Orofacial Pain, 1st edition (ICOP). Cephalalgia. 2020;40(2):129–221.

Nociplastic Orofacial Pain

92

Bridget McGowan and Ramesh Balasubramaniam

92.1 Introduction

Nociplastic pain is "pain that arises from altered nociception despite no clear evidence of actual or threatened tissue damage causing the activation of peripheral nociceptors, or evidence for disease or lesion of the somatosensory system causing the pain". It can occurs in isolation, or in a mixed-pain state in chronic orofacial pain that is nociceptive or neuropathic in origin. Nociplastic pain is an evolving categorisation system that gives validity for pain complaints previously described by stigmatised terms such as psychogenic or functional.

Nociplastic pain affects 5–15% of the population. Chronic pain affects between 7% and 40% of the population of which the craniofacial region represents the second most afflicted region, with a prevalence of 16.1% in the last 12 months (56% females, mean age 45–64 years).

The aetiology and pathogenesis of nociplastic pain are unknown. It is postulated to be attributed to dynamic interplay between altered nociception, and peripheral and/or central nervous system sensitisation, that results in amplification and/or decreased inhibition of pain stimuli at multiple levels in the nervous system, causing hyper-responsiveness to painful and non-painful stimuli.

These mechanisms are key features of chronic overlapping pain conditions (COPCs) or functional somatic syndromes (FSS). COPCs are characterised by overlapping signs and symptoms of chronic widespread pain, along with somatoform symptoms, such as sleep disorders, digestive disorders and cognitive dysfunction. COPCs represent conditions such as fibromyalgia, irritable bowel syndrome,

B. McGowan
Casuarina Square Dental Care, Casuarina, NT, Australia

R. Balasubramaniam (✉)
UWA Dental School, The University of Western Australia, Nedlands, WA, Australia
e-mail: ramesh.balasubramaniam@uwa.edu.au

R. Balasubramaniam et al. (eds.), *Oral Medicine - A Clinical Guide*,
https://doi.org/10.1007/978-3-031-36797-7_92

temporomandibular disorders, vulvodynia, myalgic encephalomyelitis, painful bladder syndrome, endometriosis, chronic tension-type headache and chronic lower back pain.

In the orofacial region, augmented trigeminal pain processing and central sensitisation are mechanisms in nociplastic pains, which share several common characteristics with COPCs. Patients with orofacial pains report more somatic symptoms, and 63.8% of these patients have chronic functional conditions. Fibromyalgia and xerostomia, in addition to psychiatric, gastrointestinal, otorhinolaryngologic, sleep and digestive disturbances have been associated with orofacial pains.

92.2 Clinical Presentation and Investigations

In the orofacial region, nociplastic pains include persistent pain related to temporomandibular disorders, burning mouth syndrome, persistent idiopathic facial pain and persistent idiopathic dentoalveolar pain. These conditions are covered in Chaps. 93, 103, 104 and 105, respectively.

The nociplastic pain clinical criteria was developed by International Association for the Study of Pain in 2021 (Table 92.1). The criteria are supported by expert opinion, however are yet to be clinically validated.

Nociplastic pain is often misdiagnosed as having an organic cause. Distinguishing nociplastic pain from nociceptive and neuropathic pains is essential. Diagnosis can be achieved by clinical history, focussing on pain characteristics, presence of other somatic and psychological symptoms, and physical examination to identify underlying differential diagnostic and comorbid conditions. For nociplastic orofacial pain, exclusion of all other causes of odontogenic and non-odontogenic orofacial pains is necessary. Of significance, the diagnosis of somatoform pain disorders has been removed from the Diagnostic and Statistical Manual of Mental Disorders fifth edition (DSM-V), as pain can no longer be diagnosed as a psychiatric condition. A clinical decision-making tree to categorise nociplastic pain by The International Association for the Study of Pain is applicable to the orofacial region (Fig. 92.1).

Table 92.1 International association for the study of pain clinical criteria for nociplastic pain

Mandatory criteria: *Possible nociplastic pain* • Patients have to report pain of at least 3 months duration • Patients have to report a regional rather than discrete pain distribution • Patients have to report pain that cannot entirely be explained by nociceptive mechanisms • Patients have to report pain that cannot entirely be explained by neuropathic mechanisms • Patients have to show clinical signs of pain hypersensitivity (i.e. evoked pain hypersensitivity phenomena such as static or dynamic mechanical allodynia, heat or cold allodynia, and/or painful after-sensations after any of the mentioned evoked pain hypersensitivity assessments) that are present at least in the region of pain

Table 92.1 (continued)

Optional criteria: *Probable nociplastic pain* – Patients present with a history of pain hypersensitivity in the region of pain (i.e. sensitivity to touch, movement, pressure or heat/cold) – Patients present at least one of the defined comorbidities: – Increased sensitivity to sound, light and/or odours – Sleep disturbances with frequent nocturnal awakenings – Fatigue or cognitive problems

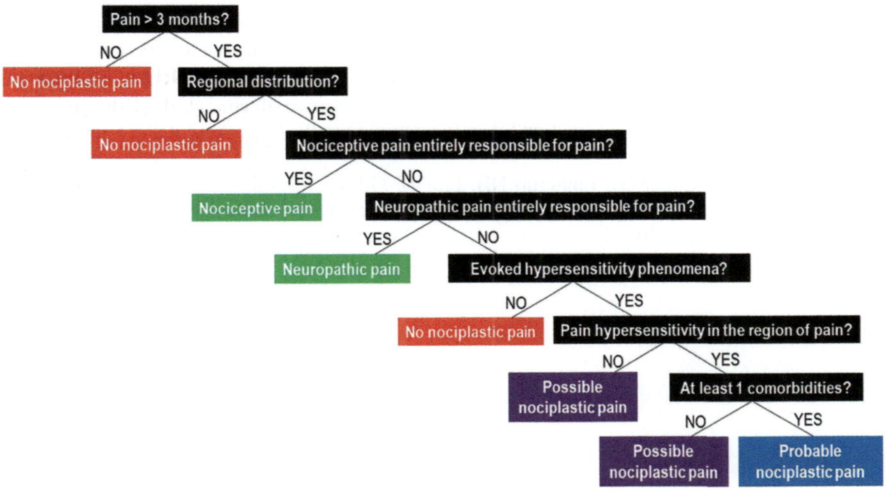

Fig. 92.1 Clinical decision-making tree of the International Association for the Study of Pain clinical criteria for nociplastic pain. (Reproduced with permission from original publication: Nijs J, Lahousse A, Kapreli E, Bilika P, Saracoglu I, Malfliet A, et al. Nociplastic Pain Criteria or Recognition of Central Sensitization? Pain Phenotyping in the Past, Present and Future. J Clin Med. 2021;10(15):3203.)

92.3 Management and Prognosis

Management of nociplastic pain focuses on personalising treatment modalities to improve symptoms and function, rather than eliminate pain. A comprehensive biopsychosocial approach is essential and may include:

- Non-pharmacological therapies including patient education, stress and sleep management, exercise and psychotherapy.
- Pharmacological therapies including antidepressants, antiepileptics, anaesthetics, naltrexone and botulinum toxin, have been used with some success.
- Addressing relevant underlying comorbidities, such as fibromyalgia and irritable bowel syndrome and comorbid symptoms, such as fatigue, cognitive problems, sleep disturbances and sensitivity to light, sound and/or odours.

Prognostically, 50–75% of patients show improvement, whereas 10–30% are recalcitrant to treatment.

Recommended Reading

de Siqueira SRDT, de Siqueira JTT, Teixeira MJ. Association between craniofacial pain complaints, somatoform symptoms and chronic diseases. Arch Oral Biol. 2021;122:104892.

Fitzcharles MA, Cohen SP, Clauw DJ, Littlejohn G, Usui C, Häuser W. Nociplastic pain: towards an understanding of prevalent pain conditions. Lancet. 2021;397(10289):2098–110.

Herrero Babiloni A, Nixdorf DR, Moana-Filho EJ. Persistent dentoalveolar pain disorder: A putative intraoral chronic overlapping pain condition. Oral Dis. 2020;26(8):1601–9.

Kosek E, Clauw D, Nijs J, Baron R, Gilron I, Harris RE, et al. Chronic nociplastic pain affecting the musculoskeletal system: clinical criteria and grading system. Pain. 2021;162(11):2629–34.

Nijs J, Lahousse A, Kapreli E, Bilika P, Saraçoğlu İ, Malfliet A, et al. Nociplastic Pain Criteria or Recognition of Central Sensitization? Pain Phenotyping in the Past, Present and Future. J Clin Med. 2021;10(15):3203.

Raja SN, Carr DB, Cohen M, Finnerup NB, Flor H, Gibson S, et al. The revised International Association for the Study of Pain definition of pain: concepts, challenges, and compromises. Pain. 2020;161(9):1976–82.

Pain-Related Temporomandibular Disorders

93

Alissa Jacobs, Lalima Tiwari, and Jacinta Vu

93.1 Introduction

Pain-related temporomandibular disorders (TMD-P) are a spectrum of disorders affecting the temporomandibular joint (TMJ) and its supporting structures; broadly, myalgia, arthralgia and headache attributed to TMD.

Masticatory myalgia is pain of a muscular origin that is affected by jaw function or parafunctional activities. The pain occurs in the jaw, temple, ear or pre-auricular area.

Masticatory myalgia can be divided into 3 subcategories, namely local myalgia, myofascial pain and myofascial pain with referral. TMJ arthralgia refers to pain from the TMJ that is affected by jaw function or parafunctional activities. Headache attributed to temporomandibular disorders is a headache of any type in the temporal area secondary to another TMD-P that is affected by jaw function or parafunctional activities.

There is no universal cause for TMD-P; however, there are known risk factors such as direct trauma (blow to jaw), indirect trauma (acceleration-deceleration injury), prolonged microtrauma (bruxism) and systemic factors (joint laxity). Psychosocial factors are likely to aggravate the pain in at risk patients, impacting the capacity to adapt and function.

TMD-P are the second most common painful musculoskeletal condition (behind chronic back pain), affecting 5–12% of the population.

A. Jacobs
Fiona Stanley Hospital, Murdoch, WA, Australia
e-mail: alissa.jacobs@health.wa.gov.au

L. Tiwari (✉)
UWA Dental School, The University of Western Australia, Nedlands, WA, Australia
e-mail: lalima.tiwari@uwa.edu.au

J. Vu
Perth Children's Hospital, Nedlands, WA, Australia

R. Balasubramaniam et al. (eds.), *Oral Medicine - A Clinical Guide*,
https://doi.org/10.1007/978-3-031-36797-7_93

Table 93.1 Clinical presentation pain related to temporomandibular disorders (adapted from diagnostic criteria for temporomandibular disorders)

Masticatory myalgia
• Confirmation of location of pain in a masticatory muscle • Familiar pain with muscle palpation (temporalis or masseter) or maximum opening – *Local myalgia*—pain localised to the site of palpation – *Myofascial pain*—pain extends beyond the site of palpation whilst remaining within the boundary of the muscle – *Myofascial pain with referral*—pain extends beyond the boundary of the muscle being palpated
Arthralgia
• Confirmation of pain located in the area of the TMJ • Reporting of familiar pain in the TMJ during palpation of or around the lateral pole of the TMJ or maximum opening, right and left lateral or protrusive movements
Headache attributed to TMD
• Confirmation of pain in the area of the temporalis muscle on examination • Reporting of familiar headache pain in the temporal region during palpation of the temporalis or maximum opening, right and left lateral or protrusive movements

93.2 Clinical Presentation and Investigations

All TMD-P will present with a history of pain in the masticatory system aggravated by jaw function. Clinical examination can help elicit the nature of the pain (Table 93.1).

Diagnosis of TMD-P is based on history and clinical examination including provocation and palpation of the joint and muscles as outlined above.

The palpation pressure for myalgia is 1 kilogram for 2 seconds, however, to differentiate between the three subclasses (local, myofascial and myofascial with referral) pressure should be applied for 5 seconds to assess spreading or referred pain if present.

93.3 Management and Prognosis

Treatment of TMD-P aims to manage a patient's symptoms rather than cure. Treatment should be non-surgical and tailored to an individual (Table 93.2).

Acute painful TMD is often self-limiting and responsive to treatment. Chronic TMD-P is often associated with central sensitisation, psychosocial distress and has a greater impact on activities of daily living and quality of life. It is often recalcitrant to treatment (Table 93.3).

Table 93.2 Treatment of pain related to temporomandibular disorders

Education and reassurance regarding the diagnosis and prognosis
Soft diet, habit awareness and reversal, pain free jaw movement, sleep hygiene
Home exercise program
Application of heat (muscular pain) or ice (joint pain)
Physiotherapy with an expert in TMD
Oral appliance therapy
Pharmacological:
Short-term anti-inflammatories and muscle relaxants for acute pain
Pain modulating medication (antidepressants, anticonvulsants) for persistent pain
Cognitive behavioural therapy and psychotherapy to manage contributing psychosocial factors especially for chronic TMD

Table 93.3 Risk factors for progression to chronic painful temporomandibular disorders

Time taken to access treatment
High pain intensity and catastrophising behaviour
Significant active psychosocial burden
Acute traumatic childhood event
Sleep dysfunction and disorders
Hormonal factors (female sex hormones)
Medical comorbidities (e.g. depression, chronic back pain, fibromyalgia, irritable bowel syndrome)

Recommended Reading

De Leeuw R, Klasser GD, editors. Orofacial pain: guidelines for assessment, diagnosis, and management. Hanover Park, IL, USA: Quintessence Publishing Company, Incorporated; 2018.

Ohrbach R, Michelotti A. The Role of Stress in the Etiology of Oral Parafunction and Myofascial Pain. Oral Maxillofac Surg Clin North Am. 2018;30(3):369–79.

Schiffman E, Ohrbach R. Executive summary of the Diagnostic Criteria for Temporomandibular Disorders for clinical and research applications. J Am Dent Assoc. 2016;147(6):438–45.

Schiffman E, Ohrbach R, Truelove E, Look J, Anderson G, Goulet JP, et al. Diagnostic Criteria for Temporomandibular Disorders (DC/TMD) for Clinical and Research Applications: recommendations of the International RDC/TMD Consortium Network* and Orofacial Pain Special Interest Group†. J Oral Facial Pain Headache. 2014;28(1):6–27.

Therapeutic Guidelines Limited. Oral and dental expert group. Management of Temporomandibular Disorders. In: Therapeutic Guidelines: Oral and Dental. Version 3. Melbourne: Therapeutic Guidelines Limited; 2019. p. 145.

Wieckiewicz M, Boening K, Wiland P, Shiau YY, Paradowska-Stolarz A. Reported concepts for the treatment modalities and pain management of temporomandibular disorders. J Headache Pain. 2015;16:106.

Intra-articular Temporomandibular Joint Disorders

94

Alissa Jacobs, Lalima Tiwari, and Jacinta Vu

94.1 Introduction

Intra-articular disorders of the temporomandibular joint (TMJ) include disc displacement (DD), osteoarthritis (OA) and subluxation of the TMJ.

A DD is an intracapsular disorder involving the disc-condyle complex. The articular disc can be displaced in an anterior, medial or lateral to the condylar head on opening. It may or may not be reduced on closing. A posterior disc displacement can occur; however, it is rare.

Degenerative joint disease (DJD) includes osteoarthritis (symptomatic) and osteoarthrosis (asymptomatic) of the TMJ. DJD involving the joint is characterised by deterioration of the articular tissue with concomitant osseous changes within the condyle and/or articular surface. Subluxation is a hypermobility disorder involving the disc-condyle complex and the articular eminence whereby in the open mouth position, the disc-condyle complex is positioned anterior the articular eminence and can return only with re-posturing the mandible or self-manipulation.

Intra-articular disorders of the TMJ are thought to be caused by trauma, however exact mechanism remains unknown (Table 94.1):

The precise numbers of individuals affected by intra-articular TMJ disorders are difficult to ascertain with prevalence reported as low as 17.5% and as high as 74.7%.

A. Jacobs
Fiona Stanley Hospital, Murdoch, WA, Australia
e-mail: alissa.jacobs@health.wa.gov.au

L. Tiwari (✉)
UWA Dental School, The University of Western Australia, Nedlands, WA, Australia
e-mail: lalima.tiwari@uwa.edu.au

J. Vu
Perth Children's Hospital, Nedlands, WA, Australia

R. Balasubramaniam et al. (eds.), *Oral Medicine - A Clinical Guide*,
https://doi.org/10.1007/978-3-031-36797-7_94

Table 94.1 Risk factors for intra-articular temporomandibular joint disorders

Microtrauma
Small repetitive injuries to the joint as seen in bruxism
Direct trauma
A blow to the jaw, wide and prolonged mouth opening
Indirect trauma
Acceleration-deceleration injury such as a motor vehicle accident
Systemic
Tissue laxity such as joint hypermobility and Ehlers–Danlos syndrome

Table 94.2 Clinical presentation intra-articular temporomandibular disorders (adapted from diagnostic criteria for temporomandibular disorders)

Disc displacement
Disc displacement with reduction
• Clicking, popping and snapping noises reported or observed
• A corrected deviation on mouth opening is often seen
Disc displacement with reduction with intermittent locking
• As above for disc displacement with reduction plus jaw locking with limited mouth opening then unlocking reported or observed
Disc displacement without reduction with limited opening
• Persistent limited opening (<40 mm) that doesn't change with manipulation
Disc displacement without reduction without limited opening
• Previous persistent limited opening severe enough to interfere with ability to eat
• Current opening >40 mm
• Uncorrected deviation on opening
Degenerative joint disease
• Noises reported or observed
• Crepitus detected with palpation during TMJ movement
Subluxation
• Locking or catching in a wide mouth position requiring manipulation to close

94.2 Clinical Presentation and Investigations

To diagnose intra-articular TMJ disorders, a supporting history is required, with joint clicking sounds and locking within the previous 30 days considered significant. Clinical examination assists in diagnosis of all disorders. Observed opening pattern has poor sensitivity and specificity in diagnosing subclass of intra-articular TMD but remain part of the evaluation (Table 94.2).

Diagnosis of intra-articular TMJ disorders are based on history taking and clinical presentation and remains the cornerstone of diagnosis. Supplementary imaging (MRI, CT) has been recommended to characterise limited mandibular opening to confirm intra-articular disorders and rule out sinister pathology.

94.3 Management and Prognosis

Treatment aims to manage symptoms of concern rather than cure. Treatment should be tailored to the specific diagnosis and commence conservatively. Asymptomatic and functional TMJ disc displacement does not require any intervention. Evidence does not support progressive worsening of a disc displacement in most patients. Treatment involves education and reassurance regarding the diagnosis and prognosis, self-care such as habit reversal, soft diet and pain-free jaw function, jaw exercises program and oral appliance therapy. In cases of persistent pain and dysfunction despite non-surgical treatments, intra-articular corticosteroid and surgical intervention such as TMJ arthrocentesis and arthroscopy, and rarely arthroplasty and joint replacement may be necessary.

Recommended Reading

Deodato F, Trusendi R, Giorgetti R, Scalese MU. Predisposition for temporomandibular joint disorders: loose ligaments. Cranio. 2006;24(3):179–83.

Dimitroulis G. Management of temporomandibular joint disorders: A surgeon's perspective. Aust Dent J. 2018;63 Suppl 1:S79–S90.

Schiffman E, Ohrbach R, Truelove E, Look J, Anderson G, Goulet JP, et al. Diagnostic Criteria for Temporomandibular Disorders (DC/TMD) for Clinical and Research Applications: recommendations of the International RDC/TMD Consortium Network* and Orofacial Pain Special Interest Group†. J Oral Facial Pain Headache. 2014;28(1):6–27.

Tanaka E, Detamore MS, Mercuri LG. Degenerative disorders of the temporomandibular joint: etiology, diagnosis, and treatment. J Dent Res. 2008;87(4):296–307.

Therapeutic Guidelines Limited. Management of temporomandibular disorders. In: Oral and Dental Expert Group. Therapeutic Guidelines: oral and dental. Version 3. Melbourne: Therapeutic Guidelines Limited; 2019. p. 145.

Tension-Type Headache

95

Srividya Iyer and Ramesh Balasubramaniam

95.1 Introduction

Tension-type headache (TTH), the most common headache, is described as a mild to moderate pain that feels like a tight band around the head.

Global and societal burden of TTH is higher than that of migraine with prevalence ranging from 46% to 78%. TTH has a female to male predilection of 5:4, with the average age of onset of 25–30 years. Peak prevalence occurs between the ages 30 and 39 years and decreasing slightly with age.

The underlying aetiology of TTH is unclear and is believed to be a result of interactions between genetic, hormonal, environmental and neuromuscular factors leading to the activation of the ascending trigeminothalamic pathway. Pericranial myofascial mechanisms are postulated to be of significance in episodic TTH. Sensitisation of pain pathways in the central nervous system resulting from prolonged nociceptive stimuli from pericranial myofascial tissues appear to be responsible for chronic TTH.

95.2 Clinical Presentation and Investigations

TTH varies significantly in frequency, intensity and duration, from rare, short-lasting episodes of discomfort to frequent, long-lasting or even continuous disabling headaches. TTH associated pain is usually described as dull, pressure-like and

S. Iyer
Melbourne Dental School, The University of Melbourne, Carlton, VIC, Australia

R. Balasubramaniam (✉)
UWA Dental School, The University of Western Australia, Nedlands, WA, Australia
e-mail: ramesh.balasubramaniam@uwa.edu.au

R. Balasubramaniam et al. (eds.), *Oral Medicine - A Clinical Guide*,
https://doi.org/10.1007/978-3-031-36797-7_95

Table 95.1 Diagnostic criteria for frequent episodic tension-type headache (International Classification of Headache Disorders-3)

1. At least 10 episodes of headache occurring on 1–14 days/month on average for >3 months (≥12 and <180 days/year) and fulfilling criteria B-D
2. Lasting from 30 min to 7 days
3. At least two of the following four characteristics:
   - Bilateral location
   - Pressing or tightening (non-pulsating) quality
   - Mild or moderate intensity
   - Not aggravated by routine physical activity such as walking or climbing stairs
4. Both of the following:
   - No nausea or vomiting
   - No more than one of photophobia or phonophobia
5. Not better accounted for by another ICHD-3 diagnosis

constricting, giving a sense of fullness in the head similar to wearing a tight band, lasting between 30 min and 7 days. It may or may not be associated with pericranial tenderness. Infrequent episodic TTH is characterised by at least 10 episodes of headache occurring on less than 1 day per month on average (<12 days/year); frequent episodic TTH is characterised by headaches between 1 and 14 days every month; and chronic TTH occurs at least 180 days per year. Stress, lack of sleep and hunger are among the most common precipitants reported by individuals with TTH. In 90% of patients, the pain presents bilaterally.

It is pertinent to exclude the other causative disorders and hence a thorough clinical history along with physical and neurological examination is necessary for diagnosis (Table 95.1). In addition to its characteristic distribution and intermittent nature, no nausea or vomiting (anorexia may occur) and no more than one episode of photophobia or phonophobia should be present as per the diagnostic criteria.

Medication overuse headache, chronic post-traumatic headache, temporomandibular disorders, idiopathic intracranial hypertension, brain tumour and cervical spondylosis are some of the differential diagnoses of TTH.

95.3 Management and Prognosis

Treatment includes acute treatment for individual attacks and preventive treatment to minimise the frequency of attacks (Table 95.2).

Table 95.2 Common tension-type headache treatment strategies

Pharmacological treatments	Non-Pharmacological treatments
Acute/abortive treatments *Simple analgesics* • Aspirin (500–1000 mg single dose) • Acetaminophen (1000 mg single dose) • Ibuprofen (400 mg single dose) • Naproxen sodium (375–550 mg single dose) • Ketoprofen(25–50 mg single dose) • Diclofenac potassium (50–100 mg single dose) *Combination medications* • Analgesics with caffeine 130 mg, codeine 30 mg or tranquilisers (clonazepam 0.5–1 mg/day) **Prophylactic treatments for chronic TTH** *Tricyclic antidepressants* • Amitriptyline (10–75 mg /day) *Other antidepressants* • Venlafaxine (150 mg/day) • Mirtazapine (15–30 mg/day) *Anticonvulsants* • Gabapentin (300–900 mg/day) • Topiramate (100 mg/day) *Botulinum toxin A (50–100 U/treatment)*	**Preventive strategies** *Lifestyle changes* • Avoid alcohol • Nutrient dense diet • Sleep hygiene *Physical exercise* *Physical therapy* • Quantum molecular resonance • Electromyographic biofeedback *Psycho-behavioural treatment* • Cognitive behavioural therapy • Hypnotherapy • Psychotherapy *Complementary medicine* • Acupuncture • Massage

Recommended Reading

Bentivegna E, Luciani M, Paragliola V, Baldari F, Lamberti PA, Conforti G, et al. Recent advancements in tension-type headache: a narrative review. Expert Rev Neurother. 2021;21(7):793–803.

Headache Classification Committee of the International Headache Society (IHS) The International Classification of Headache Disorders, 3rd edition. Cephalalgia. 2018;38(1):1–211.

Spierings EL, Ranke AH, Honkoop PC. Precipitating and aggravating factors of migraine versus tension-type headache. Headache. 2001;41(6):554-8.

Zwart JA, Dyb G, Stovner LJ, Sand T, Holmen TL. The validity of 'recognition-based' headache diagnoses in adolescents. Data from the Nord-Trøndelag Health Study 1995–97, Head-HUNT-Youth. Cephalalgia. 2003;23(3):223–9.

Migraine

96

Srividya Iyer and Ramesh Balasubramaniam

96.1 Introduction

Migraine is a complex disabling headache disorder with varying neurological presentations. Symptoms usually manifest as a throbbing, unilateral headache. The headache can be accompanied by nausea, disturbed vision, dizziness, numbness or tingling in extremities or face and sensitivity to light (photophobia), sound (phonophobia), and smells (hyperosmia).

Migraine is a major global health problem that affects over 10% of the population. Prevalence increases from the first to fourth decades and thereafter declines. In the adolescents and adults group, women (18%) are three times more affected then men (6%). Severity and frequency in women may be influenced by the hormonal changes. Majority experience migraine without aura (70–90%) compared to those with aura (10–30%). Chronic migraine is rarer at 1–2% of migraine sufferers and may be bilateral.

One of the most important aspects of the pathophysiology of migraine is familial clustering of the disorder. Current opinion favours a primarily neural cause, involving feedback loops through innervation of cranial arteries in the trigeminovascular system. The release of vasoactive substances around brain vessels causing

S. Iyer
Melbourne Dental School, The University of Melbourne, Melbourne, VIC, Australia

R. Balasubramaniam (✉)
UWA Dental School, The University of Western Australia, Nedlands, WA, Australia
e-mail: ramesh.balasubramaniam@uwa.edu.au

R. Balasubramaniam et al. (eds.), *Oral Medicine - A Clinical Guide*,
https://doi.org/10.1007/978-3-031-36797-7_96

sensitisation of trigeminal afferents, vasodilatation and pain has been thought to initiate migraine for many decades. There has been ongoing debate whether vascular factors are of primary importance, or if the blood vessels are just bystanders.

96.2 Clinical Presentation and Investigations

Almost 90% of migraine patients report triggers or precipitating factors that may initiate an attack. However, exposure to a trigger or precipitant may not always result in the onset of a headache attack. Common triggers are stress, anxiety, foods (such as chocolate, cheese), alcoholic beverages, menstruation, fatigue, alteration in sleeping pattern, missed meals, changes in weather, head trauma, and intense lights or smells. Serious complications are rare and include status migrainous, persistent aura without infarction, migrainous infarction, and migraine aura-triggered seizure.

Migraine is diagnosed based on criteria of the 3rd Edition of the International Classification of Headache Disorders (ICHD-3). Migraine with aura is primarily characterized by the transient focal neurological symptoms that usually precede or sometimes accompany the headache. The diagnosis of migraine without aura may be confirmed when criteria are met and after organic disease is excluded. The patient will have experienced a minimum of five attacks, each lasting between 4 and 72 h (untreated or unsuccessfully treated); with at least two of the following four characteristics: unilateral location, pulsating quality, moderate or severe pain. Other headache disorders can have similar presentations as migraine (Table 96.1). Diagnosis is based on clinical history and presentation. Imaging may be necessary to rule out symptomatic intracranial pathology.

Table 96.1 Clinical features of other headache disorders with similar presentations as migraine

Parameter	Tension type headache (TTH)	Cluster headache	Medication overuse headache	Paroxysmal hemicrania	Hemicrania continua	SUNCT[a]
Gender ratio (F:M)	1.5:1	1:4	1:3	1.6–2.4:1	1.8:1	1:1.5
Pain location	Fronto—Temporal, holocephalic	Temporal, periorbital, maxillary	Nonspecific	Orbit/temple	Orbit/temple/ hemicrania	Periorbital
Pain characteristics	Dull/aching Mild–moderate	Boring/sharp/ stabbing/throbbing Severe	Varies depending on the primary headache disorder and drug overused	Throbbing/sharp Severe	Throbbing/ sharp/pressure Mild, moderate, or severe	Burning/ stabbing/sharp Severe
Attack frequency	1–15/month	1/alternate day—8/day	15 or more days per month	1–40/day (0.5/day for more than half the time)	Daily and continuous	3–200/day

[a] *S*hort-lasting, *u*nilateral, *n*euralgiform, headache attacks with *c*onjunctival injection and *t*earing (SUNCT)

Table 96.2 Migraine treatment strategies

Pharmacological

NSAIDs - ibuprofen, aspirin, diclofenac
Acetaminophen (APAP) -paracetamol
Opioids - codeine with APAP, oxycodone, butorphanol nasal spray
Antiemetics- metoclopramide, domperidone
Beta - blockers - propranolol, metoprolol, atenolol
Calcium channel antagonists- verapamil, flunarizine
Tricyclic antidepressants- amitriptyline, nortriptyline
Anticonvulsant- topiramate, sodium valproate, gabapentin
Triptans- sumatriptan, rizatriptan, zolmitriptan
Serotonin -norepinephrine reuptake inhibitors (SNRIs)- venlafaxine, duloxetine
Ergot alkaloids- ergotamine
Botulinum toxin type A- Onabotulinum toxin A
CGRP monoclonal antibodies - fremanezumab, erenumab, galcanezumab, eptinezumab

Non-pharmacological

Psychological interventions
•Relaxation
•Biofeedback
•Cognitive behavioral therapy

Complimentary and alternative interventions
Nutraceuticals e.g. caffeine
Manual therapies
Acupuncture

Self management and lifestyle interventions
•Identification of individual triggering factors
•Healthy lifestyle
•Exercise
•Hydration
•Good sleep hygiene
•Diet optimisation

96.3 Management and Prognosis

The treatment of migraine consists of preventive and abortive therapies which can be pharmacological and non-pharmacological in nature (Table 96.2).

Recommended Reading

Ashina M, Katsarava Z, Do TP, Buse DC, Pozo-Rosich P, Özge A, et al. Migraine: epidemiology and systems of care. Lancet. 2021;397(10283):1485–95.

Goadsby PJ, Holland PR, Martins-Oliveira M, Hoffmann J, Schankin C, Akerman S. Pathophysiology of Migraine: A Disorder of Sensory Processing. Physiol Rev. 2017;97(2):553–622.

Joutel A, Ducros A, Vahedi K, Labauge P, Delrieu O, Pinsard N, et al. Genetic heterogeneity of familial hemiplegic migraine. Am J Hum Genet. 1994;55(6):1166–72.

Headache Classification Committee of the International Headache Society (IHS) The International Classification of Headache Disorders, 3rd edition. Cephalalgia. 2018;38(1):1–211.

Russell MB. Genetic epidemiology of migraine and cluster headache. Cephalalgia. 1997;17(6):683–701.

Stewart WF, Wood C, Reed ML, Roy J, Lipton RB, Lipton RB, et al. Cumulative lifetime migraine incidence in women and men. Cephalalgia. 2008;28(11):1170–8.

Trigeminal Autonomic Cephalalgias

97

Anita Lim and Ramesh Balasubramaniam

97.1 Introduction

Trigeminal autonomic cephalalgias (TACs) are a group of primary headaches that share the core features of unilateral headache with ipsilateral cranial autonomic symptoms. The TACs are cluster headache; paroxysmal hemicrania; hemicrania continua and short-lasting unilateral neuralgiform headache attacks (SUNHA), an umbrella term for short-lasting unilateral neuralgiform headache attacks with conjunctival injection and tearing (SUNCT) and short-lasting unilateral neuralgiform headache attacks with cranial autonomic symptoms (SUNA). The prevalence of TACs is less than 0.1% of the population, of which cluster headache is the most common. TACs cause significant disability and may persist for many years to the detriment of the patient's quality of life.

Cranial autonomic symptoms result from the activation of the trigeminal autonomic reflex and the subsequent increase in parasympathetic tone. The role of higher brain centres, such as the hypothalamus, is not fully understood. Sleep disturbances, endocrinological and systemic autonomic homeostasis have also been implicated.

A. Lim
Melbourne Dental School, The University of Melbourne, Carlton, VIC, Australia

R. Balasubramaniam (✉)
UWA Dental School, The University of Western Australia, Nedlands, WA, Australia
e-mail: ramesh.balasubramaniam@uwa.edu.au

R. Balasubramaniam et al. (eds.), *Oral Medicine - A Clinical Guide*,
https://doi.org/10.1007/978-3-031-36797-7_97

97.2 Clinical Presentation and Investigations

TACs generally involve the orbito-temporal region and are typically characterised by severe pain of a throbbing, stabbing and/or sharp quality of varying frequencies and duration (Table 97.1). Hemicrania continua is a less severe headache with intense pain exacerbations, however never completely resolves. Cluster headache, paroxysmal hemicrania and hemicrania continua patients are often restless or agitated during an attack. During cluster headache, the pain intensity can be so excruciating that patients are unable to lie down, and characteristically pace or rock side-to-side.

TAC attacks can be spontaneous or triggered. The seasonal propensity and circadian rhythmicity of cluster headache differentiate it from other TACs, with attacks occurring mostly in spring and autumn, and at night, waking patients from sleep. Paroxysmal hemicrania and hemicrania continua share common triggers of alcohol, exercise, neck movement, warm environment, irregular sleep, tiredness, stress or relaxation after stress. SUNHA attacks can be triggered by cutaneous stimulation such as touching the face, eating or tooth brushing.

Cranial autonomic features present ipsilateral to the headache and may include conjunctival injection, lacrimation, nasal congestion, rhinorrhoea, forehead and facial sweating, miosis, ptosis and/or eyelid oedema. In SUNHA, there may also be forehead and facial flushing and/or a sensation of fullness in the ear. In hemicrania continua, cranial autonomic features are present during the periods of pain exacerbation. Photophobia and phonophobia, and less commonly nausea, vomiting and osmophobia may be present.

Cluster headache may present with a migrainous aura. However, TACs tend to resolve within 2–3 hours compared to migraine, which typically lasts 4–72 hours if

Table 97.1 Comparison of clinical characteristics of trigeminal autonomic cephalalgias

Parameter	Cluster headache	Paroxysmal hemicrania	SUNHA	Hemicrania continua
Gender ratio (F:M)	1:4.3	1:1	1:1.5	1.6:1
Age of onset (years)	20–30 s	30–40 s	40–50 s	30–40 s
Pain location	Temporal, (retro-) orbital, maxillary teeth	Temporal, frontal, occipital	(retro-)orbital, temporal, frontal, nose	Temporal, (retro-) orbital, parietal, occipital, frontal
Pain intensity	Severe	High	Moderate to high	Mild to moderate with attacks of greater intensity
Pain quality	Piercing, throbbing, burning	Sharp, stabbing, throbbing	Stabbing, electric shock, sharp	Throbbing, sharp, dull, pressure
Pain duration	15–180 minutes	2–30 minutes	1 second–10 minutes	Hours to days
Pain frequency	Every other day to 8/day	1–40/day	3–200/day	Daily

untreated. There is significant symptom overlap between SUNCT and trigeminal neuralgia, however SUNCT lacks a refractory period.

TAC pain referral to orofacial structures is infrequent but not uncommon, and may mimic toothache, sinusitis, temporomandibular disorders and trigeminal neuralgia. Neurovascular orofacial pain, a unilateral pain of the middle to lower face with mild cranial autonomic symptoms, closely resembles TACs but has a toothache-like and/or pulsating quality.

97.3 Management and Prognosis

Acute cluster headache attacks are treated by inhalation of pure oxygen, with verapamil used for long-term prophylaxis. Indomethacin 150–225 mg daily is both diagnostic and treatment for paroxysmal hemicrania and hemicrania continua, distinguishing them from the other TACs. SUNCT and SUNA attacks respond well to intravenous lignocaine if tolerated. Lamotrigine is the preferred prophylaxis for SUNCT, followed by topiramate. SUNA patients are less responsive to lamotrigine, and topiramate has no effect. Gabapentin and carbamazepine are effective prophylactics in 20–40% of SUNCT and SUNA patients.

Recommended Reading

Bahra A, May A, Goadsby PJ. Cluster headache: a prospective clinical study with diagnostic implications. Neurology. 2002;58(3):354–61.

Barloese MCJ. The pathophysiology of the trigeminal autonomic cephalalgias, with clinical implications. Clin Auton Res. 2018;28(3):315–24.

Cittadini E, Goadsby PJ. Hemicrania continua: a clinical study of 39 patients with diagnostic implications. Brain. 2010;133(Pt 7):1973–86.

Cittadini E, Matharu MS, Goadsby PJ. Paroxysmal hemicrania: a prospective clinical study of 31 cases. Brain. 2008;131(Pt 4):1142–55.

Cohen AS, Matharu MS, Goadsby PJ. Short-lasting unilateral neuralgiform headache attacks with conjunctival injection and tearing (SUNCT) or cranial autonomic features (SUNA)–a prospective clinical study of SUNCT and SUNA. Brain. 2006;129(Pt 10):2746–60.

Fischera M, Marziniak M, Gralow I, Evers S. The incidence and prevalence of cluster headache: a meta-analysis of populationbased studies. Cephalalgia. 2008;28(6):614–8.

International Classification of Orofacial Pain, 1st edition (ICOP). Cephalalgia. 2020;40(2):129–221.

May A, Leone M, Afra J, Linde M, Sándor PS, Evers S, et al. EFNS guidelines on the treatment of cluster headache and other trigeminal-autonomic cephalalgias. Eur J Neurol. 2006;13(10):1066–77.

Weng HY, Cohen AS, Schankin C, Goadsby PJ. Phenotypic and treatment outcome data on SUNCT and SUNA, including a randomised placebo-controlled trial. Cephalalgia. 2018;38(9):1554–63.

Headache Disorders Related to Orofacial Pain

98

Susan Wong and Ramesh Balasubramaniam

98.1 Introduction

Headaches affect almost half of the world's population. Diagnosis is often challenging due to complex and heterogenous presentation. The International Classification of Headache Disorders (ICHD-3) classifies an exhaustive list of headache disorders and their relevant criteria for diagnosis. Majority of headaches are primary headache disorders, which are discussed in other chapters. It remains important to be aware of secondary headaches, a group of disorders that often necessitate urgent care. This chapter discusses giant cell arteritis and cervicogenic headache including red flags which may indicate serious conditions such as acute haemorrhage, thrombosis or neoplasia that warrant urgent medical attention.

98.2 Headache Red Flags

Headaches are heterogeneous in nature and often present with concurrent temporomandibular disorder and/or non-specific orofacial pain. When symptoms do not fit within diagnostic criteria of primary headaches, other differential diagnosis must be considered which may require immediate medical assessment (Table 98.1).

S. Wong
Perth Children's Hospital, Nedlands, WA, Australia

R. Balasubramaniam (✉)
UWA Dental School, The University of Western Australia, Nedlands, WA, Australia
e-mail: ramesh.balasubramaniam@uwa.edu.au

R. Balasubramaniam et al. (eds.), *Oral Medicine - A Clinical Guide*,
https://doi.org/10.1007/978-3-031-36797-7_98

Table 98.1 Headache red flags and associated differential diagnosis

Headache with associated symptoms	Possible differential diagnosis
New onset where there is a history of cancer, HIV, head trauma	Metastasis, opportunistic infection or subdural haematoma
Progressive headache	Mass lesion, subdural haematoma or medication overuse
Headache beginning after 50 years of age	Cranial arteritis or mass lesion
Headache with a sudden onset	Subarachnoid haemorrhage, haemorrhage into a mass lesion or vascular malformation, cerebral parenchyma
Headache with a rash or neck stiffness	Meningococcal meningitis or Lyme disease
Headache with papilledema	Possible raised intracranial pressure due to a mass lesion or benign intracranial hypertension
Non-migraine headache in pregnancy or post-partum	Cerebral vein thrombosis
Headache triggered by postural changes	Possible low cerebrospinal fluid (CSF) pressure as a result of a CSF leak
Headache with visual disturbances	Possibly associated with optical conditions (glaucoma, optic neuritis)

98.3 Giant Cell Arteritis

Giant cell arteritis (GCA) is a granulomatous inflammatory vasculopathy affecting medium and large sized arteries.

Pathogenesis of GCA involves the activation of dendrites in the vessel adventitia-media border where T cells and monocytes are recruited into the vessel wall. Macrophages coalesce forming multinucleated giant cells that secrete metalloproteinases leading to compromised integrity of the vessel and potential occlusion.

GCA predominantly affects women over the age of 50 years, presenting with headache, scalp and temporal artery tenderness or decreased pulsation, jaw claudication with chewing and visual disturbances.

Erythrocyte sedimentation rate and C-reactive protein are typically elevated. A temporal artery biopsy shows vasculitis characterised by mononuclear cell infiltration or granulomatous inflammation with multinucleated giant cells and breakdown of the elastic lamina. Differential diagnosis of CGA includes primary headache disorders and granulomatosis involving the temporal artery.

Treatment with high dose corticosteroids and aspirin for individuals who are at risk of cardiomyopathy often leads to a rapid improvement within 5 days and dramatically reduces the risks of vision loss.

98.4 Cervicogenic Headache

Cervicogenic headache is pain referred from the cervical spine (C1–C3) to the head or face. It presents as unilateral neck pain and reduced range of motion. Pain is triggered by certain neck movements or pressure on the neck.

Symptoms of cervicogenic headache include unilateral pain without a side shift. The pain starts in the neck and spreads to the occulo-fronto-temporal areas. Its prevalence ranges from 1% to 4% without gender predilection, with an age of onset in the early 30s.

Pain can be moderate to severe and mimic primary headache disorders; however, its symptoms are usually less severe. Diagnosis is either made clinically, or interventionally, with diagnostic local anaesthetic blocks of the cervical structure provide temporary relief. Possible differential diagnosis include dissecting aneurysms of the vertebral carotid arteries, lesions of the posterior cranial fossa, meningitis of the upper cervical spine, herpes zoster infections and primary headache disorders.

Physiotherapy including cervical spine manipulation or mobilisation, cervical muscle strengthening exercises are usually the first-line treatments. Other less common treatments include corticosteroid injections, electrotherapy and botulinum toxin injections.

98.5 Conclusion

Headache relating to orofacial pain is usually attributed to primary headache disorders, however an awareness of secondary headaches and red clinical flags are essential for timely referral of potentially serious medical conditions.

Recommended Reading

Ahmed F. Headache disorders: differentiating and managing the common subtypes. Br J Pain. 2012;6(3):124–32.

Al Khalili Y, Ly N, Murphy PB. Cervicogenic headache. In: StatPearls [Internet]. Treasure Island (FL): StatPearls Publishing; 2023. 2022 Oct 3.

Benoliel R, Eliav E. Primary headache disorders. Dent Clin North Am. 2013;57(3):513–39.

Bogduk N, Govind J. Cervicogenic headache: an assessment of the evidence on clinical diagnosis, invasive tests, and treatment. Lancet Neurol. 2009;8(10):959–68.

Clinch CR. Evaluation of acute headaches in adults. Am Fam Physician. 2001;63(4):685–92.

Hoffman GS. Giant Cell Arteritis. Ann Intern Med. 2016;165(9):ITC65-ITC80.

Headache Classification Committee of the International Headache Society (IHS) The International Classification of Headache Disorders, 3rd edition. Cephalalgia. 2018;38(1):1–211.

Trigeminal Neuralgia and Other Craniofacial Neuralgias

99

Amrita Gokhale

99.1 Introduction

Craniofacial neuralgia can cause significant debilitating pain within a nerve dermatome. Its accurate diagnosis requires understanding of the anatomy of the peripheral branches of nerves, and detailed clinical history taking, examination and appropriate neuro-imaging. Craniofacial neuralgias are considered rare, although patients may visit the dentist seeking relief. The more common craniofacial neuralgias are discussed (Table 99.1).

Table 99.1 Other craniofacial neuralgias

Diagnosis	Signs and symptoms	Management
Occipital neuralgia	• Pain at the base of the head, scalp and behind the eye • Positive Tinel's sign is noted • Response to occipital nerve block is a diagnostic criterion	• Infiltration of local anaesthetics and corticosteroids • Pulsed radiofrequency treatment
Nervus intermedius neuralgia	• Pain in the auditory canal, auricle, mastoid process and soft palate	• Carbamazepine • Microvascular decompression

A. Gokhale (✉)
Royal Dental Hospital of Melbourne, Carlton, VIC, Australia
e-mail: amrita.gokhale@dhsv.org.au

R. Balasubramaniam et al. (eds.), *Oral Medicine - A Clinical Guide*,
https://doi.org/10.1007/978-3-031-36797-7_99

99.2 Trigeminal Neuralgia

Trigeminal neuralgia (TN) is characterised by excruciating, sharp, stabbing or electric shock like, short lasting unilateral pain. It is classified as classical trigeminal neuralgia (CTN) where neurovascular compression of the trigeminal nerve root is noted on magnetic resonance imaging (MRI) or during surgery. Idiopathic TN has typical features of TN without nerve compression. Secondary TN is associated with local or systemic disease such as multiple sclerosis or a space occupying lesion. Compression of the trigeminal nerve at or near the dorsal root entry zone (DREZ) is believed to be the major causative factor of TN likely related to demyelination and/or axonal dysfunction. The pathophysiology of TN without imaging evidence of nerve compression remains unclear.

The lifetime prevalence of TN is estimated to be 0.3% with median age of 48 years. The prevalence increases with age and is estimated to be 17.5/100,000 in patients under 70 years and rises to 25.9/100,000 in >80 year olds.

99.2.1 Clinical Presentation and Investigations

Paroxysmal unilateral facial pain in the distribution of one or more branches of the trigeminal nerve, lasting from a few seconds to 2 minutes is noted. The pain is precipitated by innocuous stimuli (e.g. touch, wind, shaving) within the affected trigeminal distribution. A latent period between trigger and pain may be observed. Following an attack, a refractory period occurs where pain cannot be triggered. Periods of pain remission from days to years may occur; however, TN typically recurs without apparent reason.

MRI of the brain and skull base is the gold-standard investigation to exclude secondary causes of TN. If MRI is contraindicated, a CT scan and CT cerebral angiogram may be requested.

99.2.2 Management and Prognosis

Abortive treatments are not useful in the management of TN due to the brevity of the attacks. The mainstay of management is pharmacological preventive treatments. Carbamazepine or oxcarbazepine are the first-line treatment and offer meaningful initial pain control in almost 90% of patients. Lamotrigine, gabapentin, pregabalin, botulinum toxin type A and baclofen can be used as second-line treatments. Surgical interventions (microvascular decompression or neuroablative surgery) are reserved for patients who fail to respond to or inadequately tolerate pharmacotherapeutics.

99.3 Glossopharyngeal Neuralgia

Glossopharyngeal neuralgia (GN) is characterised by sharp, stabbing, needle-like or electric shock like pain at the posterior region of the tongue, tonsils, oropharynx, larynx, auditory canal, middle ear, angle of the mandible and retro-molar region.

Approximately 10% of patients with GN experience vagal effects during an attack which can lead to bradycardia, hypotension, syncope, seizures or even cardiac arrest. Majority of GN are idiopathic. Diagnosis of GN is by clinical assessment and high-resolution MRI of the brain and skull base or CT scan.

Management of GN is similar to TN, with carbamazepine being the drug of choice. Administration of atropine prevents the associated cardiac phenomena.

Recommended Reading

Alfieri A, Strauss C. Microvascular decompression may be an effective treatment for nervus intermedius neuralgia. J Laryngol Otol. 2011;125(7):765; author reply 765.

Headache Classification Committee of the International Headache Society (IHS) The International Classification of Headache Disorders, 3rd edition. Cephalalgia. 2018;38(1):1–211.

Blumenfeld A, Nikolskaya G. Glossopharyngeal neuralgia. Curr Pain Headache Rep. 2013;17(7):343.

Cruccu G, Di Stefano G, Truini A. Trigeminal Neuralgia. N Engl J Med. 2020;383(8):754–62.

Kao CH, Lee MH, Yang JT, Tsai YH, Lin MH. Percutaneous Radiofrequency Rhizotomy Is Equally Effective for Trigeminal Neuralgia Patients with or Without Neurovascular Compression. Pain Med. 2022;23(4):807–14.

Katta-Charles SD. Craniofacial neuralgias. NeuroRehabilitation. 2020;47(3):299–314.

Lambru G, Zakrzewska J, Matharu M. Trigeminal neuralgia: a practical guide. Pract Neurol. 2021;21(5):392–402.

MacDonald BK, Cockerell OC, Sander JW, Shorvon SD. The incidence and lifetime prevalence of neurological disorders in a prospective community-based study in the UK. Brain. 2000;123(4):665–76.

Mathew PG, Cooper W. The Diagnosis and Management of Posttraumatic Headache with Associated Painful Cranial Neuralgias: a Review and Case Series. Curr Pain Headache Rep. 2021;25(8):54.

Mueller D, Obermann M, Yoon MS, Poitz F, Hansen N, Slomke MA, et al. Prevalence of trigeminal neuralgia and persistent idiopathic facial pain: a population-based study. Cephalalgia. 2011;31(15):1542–8.

100 Post-Herpetic Neuralgia

Amrita Gokhale

100.1 Introduction

Acute herpes zoster (HZ) or shingles is a reactivation of latent varicella virus infection that may occur after the primary infection. The exact mechanisms leading to viral reactivation and the subsequent appearance of acute HZ are unknown. HZ presents as a dermatomal vesicular eruption. Ramsay Hunt syndrome is a rare neurological disorder caused by reactivation of HZ in the geniculate ganglion of the facial nerve causing unilateral facial paralysis and hearing loss. Post-herpetic neuralgia (PHN) is defined as pain persisting after rash onset, occurs in 20–50% of patients, and can lead to several months of treatment and loss of quality of life. PHN can result in scarring of sensory ganglia and peripheral nerve damage with loss of large, myelinated nerve fibres.

Neurophysiological abnormalities in the A-delta and C fibres are associated with the constant burning pain in trigeminal PHN. Paroxysmal pain in trigeminal PHN patients correlates with dysfunction of the A-beta fibres.

The incidence of PHN is estimated at 3.9–42.0/100,000 person-years. The frequency of persistent pain increases with age.

100.2 Clinical Presentation and Investigations

PHN presents with unilateral facial pain in the distribution of a trigeminal nerve branch or branches. The first division of the trigeminal nerve is most commonly affected, but the second and third divisions can also be involved. The quality of pain associated with PHN has been described as burning, throbbing, stabbing, shooting, or sharp. Affected areas are usually hypoaesthetic or anaesthetic, with pale or red/

A. Gokhale (✉)
Royal Dental Hospital of Melbourne, Carlton, VIC, Australia
e-mail: amrita.gokhale@dhsv.org.au

R. Balasubramaniam et al. (eds.), *Oral Medicine - A Clinical Guide*,
https://doi.org/10.1007/978-3-031-36797-7_100

purple scars which exhibit allodynia and hyperalgesia. Between 30% and 50% of PHN patients report itching of the affected area, termed post-herpetic itch. At times this itch is graded as more severe than the pain related to PHN. Burning pain has been reported significantly more frequently in patients who did not undergo anti-viral therapy in the acute HZ stage.

100.3 Management and Prognosis

Treatment of HZ is focused on controlling pain, accelerating healing, and reducing the risk of complications such as dissemination, PHN, and local secondary infection. It is unclear if PHN incidence reduces following oral acyclovir therapy. However, early treatment with anti-virals in patients over 50 years decreases rash duration and pain severity.

Tricyclic antidepressants, gabapentin, pregabalin, opioids, and topical lidocaine or capsaicin patches are evidence-based treatment options for management of PHN pain. Lidocaine patches are useful for allodynia associated with PHN and have fewer side effects compared to systemic drugs. High concentration of topical capsaicin (8%) is efficacious but not recommended with trigeminal PHN. Generally early treatment of PHN improves prognosis. Invasive treatment options such as sympathetic and sensory nerve blocks are available for cases that do not respond to pharmacotherapeutics. Vaccination against VZV aims to reduce the incidence of HZ and therefore PHN.

Recommended Reading

Attal N, Cruccu G, Baron R, Haanpää M, Hansson P, Jensen TS, et al. EFNS guidelines on the pharmacological treatment of neuropathic pain: 2010 revision. Eur J Neurol. 2010;17(9):1113–e88.

Chen N, Li Q, Yang J, Zhou M, Zhou D, He L. Antiviral treatment for preventing postherpetic neuralgia. Cochrane Database Syst Rev. 2014;2014(2):CD006866.

Dworkin RH, Gnann JW, Oaklander AL, Raja SN, Schmader KE, Whitley RJ. Diagnosis and assessment of pain associated with herpes zoster and postherpetic neuralgia. J Pain. 2008; 9(1 Suppl 1):S37–44.

Headache Classification Committee of the International Headache Society (IHS) The International Classification of Headache Disorders, 3rd edition. Cephalalgia. 2018;38(1):1–211.

Oaklander AL. Mechanisms of pain and itch caused by herpes zoster (shingles). J Pain. 2008; 9(1 Suppl 1):S10–8.

Opstelten W, McElhaney J, Weinberger B, Oaklander AL, Johnson RW. The impact of varicella zoster virus: chronic pain. J Clin Virol. 2010;48 Suppl 1:S8–13.

Sayanlar J, Guleyupoglu N, Portenoy R, Ashina S. Trigeminal postherpetic neuralgia responsive to treatment with capsaicin 8 % topical patch: a case report. J Headache Pain. 2012;13(7):587–9.

Truini A, Galeotti F, Haanpaa M, Zucchi R, Albanesi A, Biasiotta A, et al. Pathophysiology of pain in postherpetic neuralgia: a clinical and neurophysiological study. Pain. 2008;140(3):405–10.

van Hecke O, Austin SK, Khan RA, Smith BH, Torrance N. Neuropathic pain in the general population: a systematic review of epidemiological studies. Pain. 2014;155(4):654–62.

Post-Traumatic Trigeminal Neuropathic Pain

101

Sarah Chaw

101.1 Introduction

Post-traumatic trigeminal neuropathic pain (PTTNP) is a neuropathic pain developing within 6 months after an identifiable traumatic event to the trigeminal nerve and can affect all parts of the head, including the face and oral cavity. PTTNP is defined by the International Classification of Orofacial Pain (ICOP-I) as *unilateral or bilateral facial or oral pain following and caused by trauma to the trigeminal nerve(s), with other symptoms and/or clinical signs of trigeminal nerve dysfunction, and persisting or recurring for more than 3 months*. PTTNP has previously been described as anaesthesia dolorosa and painful post-traumatic trigeminal neuropathy.

PTTNP may result from a wide variety of nerve injuries including mechanical, chemical, and thermal radiation. The traumatic event may be mild to severe, ranging from major craniofacial trauma, orthognathic surgery, dental extractions, dental implants, oral surgery, endodontic therapies, local anaesthetic injections, and minor dental procedures. Following an identifiable injury, the onset and features of PTTNP are widely variable between individuals; therefore, a combination of environmental, psychosocial, and genetic factors likely play a role. PTTNP develops via both peripheral and centrally mediated pain mechanisms.

The prevalence of PTTNP has been difficult to establish, partly due to evolving definitions and diagnostic criteria. Importantly however, traumatic injuries to the trigeminal nerve only rarely develop into a painful neuropathy. It is more common in females, with an age of onset typically around 45–50 years.

S. Chaw (✉)
Queensland Oral Medicine, Spring Hill, QLD, Australia

School of Dentistry, The University of Queensland, Herston, QLD, Australia
e-mail: s.chaw@uq.edu.au

R. Balasubramaniam et al. (eds.), *Oral Medicine - A Clinical Guide*,
https://doi.org/10.1007/978-3-031-36797-7_101

101.2 Clinical Presentation and Investigations

The ICOP-1 diagnostic criteria for PTTNP are shown in Table 101.1.

The pain symptoms can vary considerably, and some common features are summarised in Table 101.2.

PTTNP is associated with a substantial psychosocial burden, and a significant proportion of patients have other psychological comorbidities including anxiety,

Table 101.1 ICOP-1 diagnostic criteria for post-traumatic trigeminal neuropathic pain. (Adapted from: International Classification of Orofacial Pain, first edition (ICOP). Cephalalgia. 2020;40(2):129–221)

Criteria		Comments
A	Pain, in a neuroanatomically plausible area within the distribution(s) of one or both trigeminal nerve(s), persisting or recurring for >3 months and fulfilling criteria C and D	• Unilateral. Rarely crosses the midline, but over time may become more diffusely distributed
B	Both of the following: 1. History of a mechanical and thermal radiation or chemical injury to the peripheral trigeminal nerve(s) 2. Diagnostic test confirmation of a lesion of the peripheral trigeminal nerve(s) explaining the pain	The severity of nerve injuries may range from mild to severe. Tests that confirm a trigeminal nerve lesion may include: • surgical or radiological confirmation of nerve compression or lesion • nerve conduction studies • laser-evoked potentials • blink reflex • skin biopsy confirmation of reduced nerve fibre terminals Findings must be considered carefully with all clinical and diagnostic contexts of the pain
C	Onset within 6 months after the injury	
D	Associated with somatosensory symptoms and/or signs in the same neuroanatomically plausible distribution	Somatosensory symptoms or signs may be negative (e.g. hypoaesthesia and/or hypoalgesia) and/or positive (e.g. hyperalgesia and/or allodynia)
E	Not better accounted for by another ICOP or ICHD-3 diagnosis	

Table 101.2 Clinical features of post-traumatic trigeminal neuropathic pain

Quality	Often described as burning or shooting, but other pain qualities may also be reported. Patients may also describe a swollen or foreign body sensation, heat or cold
Intensity	Moderate to severe
Duration	Continuous, episodic/paroxysmal, or mixed
Other associated signs and symptoms	PTTNP can sometimes present with local redness or flushing (i.e. neurogenic inflammation)

depression, pain catastrophising, multiple somatic symptoms, and reduced quality of life and coping abilities.

Diagnosis of PTTNP requires taking a careful and thorough clinical and psychosocial history, clinical examination, and supplemented by investigations as indicated.

A cranial nerve examination (which includes sensory testing of the trigeminal nerve) is recommended. Imaging modalities such as cone beam computerised tomography, multislice computed tomography, and magnetic resonance imaging may be useful in assessing injury from a surgical procedure, detecting foreign bodies and sequestra, and to exclude other structural causes of the symptoms.

Differential diagnoses of PTTNP may include odontogenic pain, trigeminal neuralgia and other painful neuropathies, painful temporomandibular disorders, and persistent idiopathic facial pain.

101.3 Management and Prognosis

Patients with PTTNP commonly present with a history of multiple, repeated consultations with different health care practitioners, seeking and receiving various treatments aimed at eliminating pain. This is complicated by patient frustration, distress, and suffering when the condition is not diagnosed and managed in a timely manner. Therefore, patients with suspected PTTNP should be referred promptly to an appropriate specialist for formal diagnosis and management. Strategies to prevent and manage established PTTNP are summarised in Table 101.3 and often involve a multidisciplinary approach. Outcomes are variable and while a proportion of patients with PTTNP report improvement of symptoms with time, complete pain relief is difficult to achieve. Management is aimed at improving quality of life and establishing coping strategies.

Table 101.3 Strategies for preventing and managing post-traumatic trigeminal neuropathic pain

Prevention Strategies for PTTNP	Management Strategies for PTTNP
• Minimise pain and inflammation at time of surgery – Utilise procedural techniques that minimise tissue damage and nerve involvement – Provide adequate and effective peri-operative anti-inflammatory and analgesic regimens – Provide adequate local or regional anaesthesia	• Avoid irreversible dental treatment in the absence of an obvious dental cause • Pharmacologic therapy – Topical—e.g. topical anaesthetics, capsaicin – Systemic—e.g. tricyclic antidepressants, gabanoids and anticonvulsants, selective noradrenaline reuptake inhibitors • Psychological techniques (e.g. psychotherapy, mindfulness) • Cognitive behavioural therapy

Recommended Reading

Baad-Hansen L, Benoliel R. Neuropathic orofacial pain: Facts and fiction. Cephalalgia. 2017;37(7):670–9.

International Classification of Orofacial Pain, 1st edition (ICOP). Cephalalgia. 2020;40(2): 129–221.

Other Orofacial Neuropathies 102

Sarah Chaw and Anura Ariyawardana

102.1 Introduction

Painful trigeminal neuropathy is an umbrella term for several orofacial pain conditions affecting one or more branches of the trigeminal nerve, caused by a disorder that leads to neural damage. Trigeminal neuropathies may be idiopathic or caused by inflammatory or non-inflammatory processes. When the pain develops as a result of inflammation of the nerve(s), it is specifically referred to as *neuritis*.

Orofacial complex regional pain syndrome (CRPS) is a rare and chronic, painful traumatic neuropathy characterised by autonomic, trophic, and motor changes. CRPS is subtyped into CRPS I (reflex sympathetic dystrophy), CRPS II (causalgia), and CRPS-NOS (not otherwise specified), depending on severity of the inciting trauma and whether all diagnostic criteria are fulfilled.

102.2 Clinical Presentation and Investigations

Trigeminal neuropathy may be idiopathic or may develop secondary to trauma or a disorder such as systemic disease (e.g. inflammatory or autoimmune disease), demyelinating condition, infection, neoplasm or vascular disorder (Table 102.1).

S. Chaw (✉)
School of Dentistry, The University of Queensland, Herston, QLD, Australia

Queensland Oral Medicine, Spring Hill, QLD, Australia
e-mail: s.chaw@uq.edu.au

A. Ariyawardana
College of Medicine and Dentistry, James Cook University, Townsville, QLD, Australia

School of Medicine and Dentistry, Griffith University, Southport, QLD, Australia

Metro South Oral Health, Queensland Health, Brisbane, QLD, Australia

R. Balasubramaniam et al. (eds.), *Oral Medicine - A Clinical Guide*,
https://doi.org/10.1007/978-3-031-36797-7_102

Table 102.1: Causes of trigeminal neuropathy

Trauma	• Mechanical, chemical and thermal radiation
Inflammatory/autoimmune	• Connective tissue disease • Sarcoidosis • Multiple sclerosis
Infection	• Bacterial (e.g. syphilis, leprosy) • Viral (e.g. varicella zoster, herpes simplex) • Fungal (e.g. actinomycosis)
Vascular	• Cerebrovascular accident
Neoplasm	• Intra- or extra-cranial compression • Perineural spread • Metastasis
Toxic/metabolic	• Chemotherapy • Diabetes mellitus
Degenerative	• Huntington's disease
Congenital	• Skull base anomalies
Other	• Idiopathic trigeminal neuropathy • Amyloidosis

In CRPS, the traumatic event may be minor (e.g. injections, local infection) to major (e.g. surgery, high-impact trauma). Following neural injury, the pain may be "sympathetically maintained" or augmented via sympathetic efferent neural activity (e.g. increased densities of axonal adrenergic receptors, sympathetic fibre sprouting).

The primary pain is often continuous, with qualities commonly described as burning, squeezing, or "pins and needles". There may be superimposed paroxysmal pain attacks, but these are not the predominant pain type. Clinically detectable somatosensory changes within the trigeminal distribution are prevalent in painful trigeminal neuropathies, and these may be negative (e.g. hypaesthesia and/or hypalgesia) and/or positive (e.g. hyperalgesia and/or allodynia).

Trigeminal neuropathic pain follows the distribution of one or more branches of the trigeminal nerve (with the exception of CRPS which may spread in a non-dermatomal pattern over time). The specific distribution of pain is dependent on the cause of neural injury. Secondary trigeminal neuropathy caused by connective tissue disease or hereditary disorder usually presents as bilateral continuous pain, but may initially present asymmetrically and occasionally present with paroxysmal pain superimposed on the background pain.

While secondary trigeminal neuropathy may share causes with secondary trigeminal neuralgia (TN), it is distinct from TN due to the differences in clinical characteristics (e.g. quality of spontaneous and evoked pain, and the presence of clinically detectable somatosensory changes). The allodynic areas are different to the trigger zones present in TN (they are generally much larger than the TN trigger zones, with absence of "latency" and "refractory periods" as seen in TN).

CRPS has distinct features that differentiate from non-CRPS neuropathic pain, including regional changes in skin colour, temperature, oedema, sweating and motor function. The severity of the signs/symptoms is disproportionate to the injury.

Diagnosis of trigeminal neuropathy is based on careful history and thorough clinical examination to identify potential causes of trauma/injury to the trigeminal nerve(s). Additional investigations can help to confirm or exclude possible causes.

These may include blood tests (e.g. haematinics, glucose testing, autoimmune serology) to assess for systemic causes, cranial nerve examination, quantitative sensory testing, diagnostic block injections, and imaging modalities.

Intra- and extra-oral plain radiography and cone beam computerised tomography may assist in assessing/excluding odontogenic and paranasal sinus pain. Multislice computed tomography and magnetic resonance imaging may be useful in identifying space-occupying lesions, inflammation, and other structural causes along the course of the trigeminal nerve(s).

Due to the autonomic changes (e.g. redness, swelling, heat), CRPS often mimics common inflammatory causes of orofacial pain, such as odontogenic or paranasal sinus pain. Avoid irreversible dental treatment if an obvious dental cause cannot be identified.

Differential diagnoses of trigeminal neuropathy include odontogenic pain, trigeminal neuralgia, painful temporomandibular disorders, paranasal sinus pain, oral dysaesthesia (burning mouth syndrome), and persistent idiopathic facial pain.

102.3 Management and Prognosis

Management of painful trigeminal neuropathies is often multidisciplinary and is dependent on the specific diagnosis and cause (Table 102.2). Some conditions may be self-limiting (e.g. postherpetic neuralgia); however given the wide range of

Table 102.2 Management strategies for painful orofacial neuropathies

Condition	Management
Idiopathic/secondary trigeminal neuropathy	• Manage the underlying cause (if present) • Symptoms may improve or resolve following treatment of the underlying cause • Pharmacologic therapy may be used for residual pain – Topical—e.g. topical anaesthetics, capsaicin – Systemic—e.g. tricyclic antidepressants, gabanoids and anticonvulsants, selective noradrenaline reuptake inhibitors
Neuritis	• Remove/manage any inciting cause of neural inflammation • Treatment options depend on site involved, and severity of pain and inflammation • Supportive care • Pharmacologic therapy (e.g. non-steroidal anti-inflammatories, systemic corticosteroids)
Complex regional pain syndrome	• Pharmacologic therapy (e.g. topical anaesthetics, capsaicin, systemic corticosteroids, tricyclic antidepressants, gabapentin) • Physiotherapy • Regional anaesthesia techniques (e.g. sympathetic blockade, sympathectomy, somatic blockade) • Neuromodulation • Psychological techniques (e.g. mindfulness, psychotherapy) • Cognitive behavioural therapy • Comorbidities (e.g. anxiety, depression) should also be managed concurrently

causes (many of which are chronic systemic diseases or possibility of a serious/ sinister condition), patients would be best assessed and managed under specialist dental and medical care.

Recommended Reading

Borchers AT, Gershwin ME. Complex regional pain syndrome: a comprehensive and critical review. Autoimmun Rev. 2014;13(3):242–65.

International Classification of Orofacial Pain, 1st edition (ICOP). Cephalalgia. 2020;40(2):129–221.

Melis M, Zawawi K, al-Badawi E, Lobo Lobo S, Mehta N. Complex regional pain syndrome in the head and neck: a review of the literature. J Orofac Pain. 2002;16(2):93–104.

Okeson JP. Neuropathic pains. In: Okeson JP, editor. Bell's Oral and Facial Pain. 7th ed. Illinois: Quintessence; 2014. p. 456–67.

Smith JH, Cutrer FM. Numbness matters: a clinical review of trigeminal neuropathy. Cephalalgia. 2011;31(10):1131–44.

Burning Mouth Syndrome

103

Amanda Phoon Nguyen

103.1 Introduction

Burning mouth syndrome (BMS) presents as altered sensations in the mouth. BMS commonly presents with burning, with or without somatosensory changes. Other described symptoms include xerostomia, unpleasant sensations, and an altered or abhorrent taste. BMS may be referred to as oral dysaesthesia or complex oral sensitivity disorder. Previously used terms include stomatodynia, glossodynia (when confined to the tongue), and primary burning mouth syndrome.

Whilst largely unknown, the aetiology and pathophysiology of BMS are thought to be neuropathic or nociplastic in origin. Both peripheral and central neuropathies appear to play a role, with intraoral small-fibre changes, somatosensory changes, and abnormal brainstem reflex responses demonstrated.

BMS affects between 0.1% and 3.9% of the general population, associated with older age and female gender. It has been reported to occur more frequently in women in the seventh decade of life.

103.2 Clinical Presentation and Investigations

Current diagnostic criteria of the International Classification of Orofacial Pain include daily intraoral burning or dysaesthetic sensation for more than 2 h per day for more than 3 months without causative lesions on clinical examination and investigation.

A. P. Nguyen (✉)
Perth Children's Hospital, Nedlands, WA, Australia

School of Nursing, Curtin University, Bentley, WA, Australia
e-mail: Amanda.PhoonNguyen@health.wa.gov.au

R. Balasubramaniam et al. (eds.), *Oral Medicine - A Clinical Guide*,
https://doi.org/10.1007/978-3-031-36797-7_103

Dysaesthetic sensations may include taste changes, foreign body sensations, and oral dryness. These sensations are usually bilateral. The most common site is the tip of the tongue, although any part of the mouth may be affected including the palate and lips. BMS can appear suddenly or develop gradually over time and may last for months to several years. The intensity may fluctuate varying from mild to severe. The pain may follow a diurnal pattern, most commonly worsening as the day progresses.

Psychosocial factors and distress are frequently identified in BMS patients. These include somatisation, high harm avoidance, low self-directedness tendencies, obsessive-compulsive behaviour, anxiety, depression, and personality disorders.

BMS is a diagnosis by exclusion. Local causes with similar symptoms should be first excluded. These may include oral infections such as candidosis and other oral mucosal disease such as oral lichen planus. Systemic causes that may present with similar symptoms include vitamin B, folic acid, iron or other haematinic deficiency, medication-related adverse effects, metabolic or endocrine-related disorders, allergies, and hormonal disturbances such as thyroid dysfunction.

Diagnostic tests to be considered may include blood tests, cranial nerve testing, and MRI of the brain to exclude CNS pathology or a demyelinating disease.

103.3 Management and Prognosis

Symptoms may respond to topical or systemic medications. Pharmacological management includes topical use of clonazepam or capsaicin. Systemic clonazepam, gabapentinoids, tricyclic antidepressants, and alpha-lipoic acid have been trialled with some success. Non-pharmacological management includes cognitive behavioural therapy, multisensory illusions, and protective barriers (oral stents). Thorough patient education is important, particularly levelling the expectation regarding a cure for BMS. Management needs to consider the psychosocial contribution and may require multidisciplinary treatment involving a psychologist and/or psychiatrist.

Recommended Reading

International Classification of Orofacial Pain, 1st edition (ICOP). Cephalalgia. 2020;40(2):129–221.

Jääskeläinen SK, Woda A. Burning mouth syndrome. Cephalalgia. 2017;37(7):627–47.

McMillan R, Forssell H, Buchanan JA, Glenny AM, Weldon JC, Zakrzewska JM. Interventions for treating burning mouth syndrome. Cochrane Database Syst Rev. 2016;11(11):CD002779.

Persistent Idiopathic Facial Pain

104

Amanda Phoon Nguyen

104.1 Introduction

According to the International Classification of Orofacial Pain 1, persistent idiopathic facial pain (PIFP) is defined as a persistent facial pain, with variable features, recurring daily for more than 2 h per day for more than 3 months, in the absence of clinical neurological deficit or preceding causative event. Other terms for this condition have included 'atypical facial pain' and 'phantom toothache' although these terms may encompass other diagnoses.

The cause of PIFP is unknown, and its pathogenesis remains poorly understood. It is generally accepted to be a nociplastic pain with centrally mediated mechanisms. In a research setting, some patients with PIFP demonstrated neurophysiological findings that may be indicative of peripheral nerve dysfunction including abnormal nerve responses, increased excitability, and deficient habituation. Positron emission tomography (PET) studies reveal blood flow and dopamine receptor changes in areas of the brain and brain stem of PIFP patients, suggesting altered central pain processing.

There is significant association with psychiatric comorbidities such as anxiety disorders and affective disorders. PIFP patients often present with associated central sensitivity syndromes such as fibromyalgia, chronic fatigue syndrome, pelvic floor pain, interstitial cystitis, and irritable bowel syndrome.

The lack of a previous definition of PIFP has contributed to the difficulty in understanding the epidemiology of PIFP. Whilst the exact prevalence is unknown, PIFP is thought to be uncommon, with onset in middle age and with a female predominance.

A. P. Nguyen (✉)
Perth Children's Hospital, Nedlands, WA, Australia

School of Nursing, Curtin University, Bentley, WA, Australia
e-mail: Amanda.PhoonNguyen@health.wa.gov.au

R. Balasubramaniam et al. (eds.), *Oral Medicine - A Clinical Guide*,
https://doi.org/10.1007/978-3-031-36797-7_104

104.2 Clinical Presentation and Investigations

PIFP is often continuous with exacerbations. The distribution is typically poorly localised and not within a peripheral nerve distribution. Involvement of the maxilla is more often described with extension to the eyes, nose, cheeks, and temples. The pain may be described as deep or superficial, mild to severe, dull or aching, and may radiate from face to mouth or vice versa. With time, there may be involvement of a wider area of the craniocervical region and in a non-dermatomal pattern. A variety of words can be used to describe the character of the pain such as 'tedious' and 'boring'. Some patients may report exacerbation with stress. There may be somatosensory changes such as dysaesthesia, a subjective sensation of numbness and/or swelling, but with no objective sensory deficits or swelling.

The history, clinical (including cranial nerve), and radiographic examinations should exclude any potential head and neck, or dental pathology, and may include volumetric imaging such as orofacial cone beam computed tomography and MRI of the brain and skull base where applicable. Unlike neuropathic orofacial pains, PIFP is not usually associated with severe allodynia, hyperalgesia, or reproducible negative neurosensory signs.

Differential diagnoses may include post-traumatic trigeminal neuropathic pain, post-herpetic neuralgia, regional myofascial pain, and neurovascular orofacial pains.

104.3 Management and Prognosis

PIFP can be difficult to treat and commonly associated with significant psychosocial distress. Dissatisfaction with treatment is high, and prognosis is poor. PIFP patients typically require multidisciplinary management. Pharmacological therapy may include tricyclic antidepressants, anxiolytics, anticonvulsants, and selective noradrenaline reuptake inhibitors. Non-pharmacological therapies include cognitive behavioural therapy, hypnosis, biofeedback, and psychotherapy.

Recommended Reading

Benoliel R, Gaul C. Persistent idiopathic facial pain. Cephalalgia. 2017;37(7):680–91.
International Classification of Orofacial Pain, 1st edition (ICOP). Cephalalgia. 2020;40(2):129–221.

Persistent Idiopathic Dentoalveolar Pain 105

Amanda Phoon Nguyen

105.1 Introduction

As the term suggests, persistent idiopathic dentoalveolar pain (PIDAP) is a persistent, unilateral intra-oral pain in the dentoalveolar region. It rarely occurs in multiple sites. According to the International Classification of Orofacial Pain-1, it has variable features but recurs daily for more than 2 h per day for more than 3 months, in the absence of any preceding causative event. The pain is described as deep and dull. Previous terms that have been used to describe this entity and other overlapping diagnoses include atypical odontalgia and phantom tooth pain.

The aetiology of PIDAP remains poorly understood. It was postulated to be a neuropathic pain; however, given the centrally mediated mechanisms, it is now considered a nociplastic pain. There is a significant association with central sensitivity syndromes such as fibromyalgia, chronic back pain, irritable bowel syndrome, and depression.

Due to differences in nomenclature and previous non-standardised diagnostic criteria, epidemiological data is difficult to obtain. There is a female predominance, and the condition is believed to be rare. The most common age of first diagnosis is between 30 and 60 years of age.

105.2 Clinical Presentation and Investigations

The pain described in PIDAP is deep, dull, and pressure-like with descriptors such as "pulling/dragging", "pulsating", "throbbing", "pressing", and "burning" used. The pain is localised to the dentoalveolar region and rarely involves multiple sites.

A. P. Nguyen (✉)
Perth Children's Hospital, Nedlands, WA, Australia

School of Nursing, Curtin University, Bentley, WA, Australia
e-mail: Amanda.PhoonNguyen@health.wa.gov.au

R. Balasubramaniam et al. (eds.), *Oral Medicine - A Clinical Guide*,
https://doi.org/10.1007/978-3-031-36797-7_105

With time, and central sensitisation, it may spread to a wider area of the craniocervical region. Unlike odontogenic pain, PIDAP is rarely associated with nocturnal awakening. Stress may cause exacerbation. Hyperalgesia, hypoaesthesia, and/or hypoalgesia are rare, although allodynia in the gingival sulcus of the affected teeth appears to be a more common reported feature.

The clinical examination and radiological imaging should thoroughly exclude other orofacial or dental causes. This may include volumetric imaging such as cone beam computed tomography of the orofacial region and dentition, and magnetic resonance imaging of brain and skull base where applicable.

If the pain onset is within 6 months of a causative event, a diagnosis of post-traumatic trigeminal neuropathic pain may present similarly. Other differential diagnoses include primary headache such as migraine and trigeminal autonomic orofacial pain, idiopathic trigeminal neuralgia, and persistent idiopathic facial pain. In trigeminal neuralgia, pain is more likely described as short, sharp paroxysms of electric shock-like episodic pain.

105.3 Management and Prognosis

Misdiagnosis of PIDAP as odontogenic pain may lead to unnecessary dental treatment. Patients may seek or insist on dental treatment and require education on their diagnosis. Management can be challenging, and different agents and strategies have limited predictable therapeutic effect. Pharmacological therapy may include systemic medications such as tricyclic antidepressants, anxiolytics, gabapentinoids, anticonvulsants, and selective noradrenaline reuptake inhibitors. Topical forms of tricyclic antidepressants, gabapentinoids, anticonvulsants, and ketamine may be compounded and applied to the painful dentoalveolar site with the aid of a neurosensory stent. Other management options include acupuncture, hypnosis, and cognitive behavioural therapy. Complete resolution of pain is uncommon and hence management is aimed at reducing symptoms and improving quality of life.

Recommended Reading

International Classification of Orofacial Pain, 1st edition (ICOP). Cephalalgia. 2020;40(2):129–221.
Sanner F, Sonntag D, Hambrock N, Zehnder M. Patients with persistent idiopathic dentoalveolar pain in dental practice. Int Endod J. 2022;55(3):231–9.

Constant Unilateral Facial Pain with Additional Attacks

106

Amanda Phoon Nguyen

106.1 Introduction

According to the International Classification of Orofacial Pain-1, constant unilateral facial pain with additional attacks (CUFPA) is defined by constant, unremitting dull unilateral facial pain of mild to moderate intensity present for longer than 3 months, which is accompanied by distinct attacks of moderate to severe pain in the same location lasting 10–30 min. There are no typical autonomic and/or migrainous features accompanying either the constant pain or the additional pain attacks. The attacks can occur up to six times daily.

The cause of CUFPA is unknown, and its pathogenesis remains poorly understood. There are no available studies on the aetiopathogenesis of CUFPA.

CUFPA is thought to be rare, but the epidemiology is largely unknown.

106.2 Clinical Presentation and Investigations

The predominant feature of CUFPA is well-defined and distinct paroxysmal attacks on a background of unilateral dull pain along the same distribution. The attacks last at least 10 min and at least some are spontaneous. The background pain is unilateral, side locked, of mild to moderate intensity and typically in the maxilla or mandible.

A. P. Nguyen (✉)
Perth Children's Hospital, Nedlands, WA, Australia

School of Nursing, Curtin University, Bentley, WA, Australia
e-mail: Amanda.PhoonNguyen@health.wa.gov.au

R. Balasubramaniam et al. (eds.), *Oral Medicine - A Clinical Guide*,
https://doi.org/10.1007/978-3-031-36797-7_106

The pain is constantly present for more than 3 months, with distinct attacks up to six times day.

The clinical examination and diagnostic imaging including volumetric imaging such as cone beam computed tomography of the orofacial region, and magnetic resonance imaging of brain and skull base where applicable. Any potential orofacial or dental cause is ruled out. If cranial autonomic features are present, consideration should be given to other primary headache disorders including trigeminal autonomic orofacial pain.

Paroxysmal pain is also a feature of other facial pain conditions such as trigeminal neuralgia and post-traumatic trigeminal neuropathic pain. In select cases, an indomethacin trial excludes paroxysmal hemicrania and hemicrania continua. In CUFPA, the pain attacks are well-defined, distinct, and spontaneous, without associated symptoms such as photophobia or phonophobia. Restlessness or agitation is more commonly seen in trigeminal autonomic cephalgias.

106.3 Management and Prognosis

Misdiagnosis may lead to unnecessary dental treatment, and caution should be exercised to avoid this. Pharmacological therapy may include systemic and topical medications such as tricyclic antidepressants, anxiolytics, gabapentinoids, anticonvulsants, and selective noradrenaline reuptake inhibitors. Some patients remain unresponsive to medication, and the prognosis of this condition is unknown.

Recommended Reading

International Classification of Orofacial Pain, 1st edition (ICOP). Cephalalgia. 2020;40(2):129–221.

Moore RJ, Dave M, Stocker J, Aggarwal VR. Simplifying differential diagnoses of orofacial conditions - a guide to surgical sieves and red flags. Br Dent J. 2021;230(5):289–93.

Ziegeler C, May A. Facial presentations of migraine, TACs, and other paroxysmal facial pain syndromes. Neurology. 2019;93(12):e1138–e1147.

Orofacial Movement Disorders

107

Lalima Tiwari, Jacinta Vu, and Alissa Jacobs

Orofacial movement disorders include dystonia, dyskinesia, and drug-induced extrapyramidal reactions.

107.1 Orofacial Dystonia

Orofacial dystonia, also known as oromandibular dystonia, is a movement disorder characterised by intermittent, involuntary, and momentary sustained contraction of orofacial muscles.

It is postulated that basal ganglia dysfunction, hyperexcitability of motor signalling interneurons, reduced inhibition of spinal cord and brainstem signals, and dysfunction of neurochemical systems involving dopamine, serotonin, and noradrenaline contribute towards the pathophysiology of dystonia.

L. Tiwari (✉)
UWA Dental School, The University of Western Australia, Nedlands, WA, Australia
e-mail: lalima.tiwari@uwa.edu.au

J. Vu
Perth Children's Hospital, Nedlands, WA, Australia
e-mail: jacinta.vu@health.wa.gov.au

A. Jacobs
Fiona Stanley Hospital, Murdoch, WA, Australia
e-mail: alissa.jacobs@health.wa.gov.au

R. Balasubramaniam et al. (eds.), *Oral Medicine - A Clinical Guide*,
https://doi.org/10.1007/978-3-031-36797-7_107

107.1.1 Clinical Presentation and Investigations

Prevalence of orofacial dystonia ranges between 3 and 30 per 100,000 and mostly occurs in a focal form rather than part of a generalised systemic form.

The patient demonstrates repeated, intermittent, involuntary spasms of one or more orofacial muscles during waking periods but not sleep. The ability to partially control or suppress the movement with the use of tactile stimulation (e.g. touching the chin) known as a "sensory trick" or "geste antagoniste" is pathognomonic for dystonia in the orofacial region.

Investigations and imaging, including CT and MRI, are targeted towards ruling out central nervous system disease and local pathological conditions.

107.2 Orofacial Dyskinesia

Orofacial dyskinesia is defined as involuntary, repetitive, continuous movement of the face, tongue, and jaw that may be painful and can be either drug-induced (tardive) or spontaneous.

The presence of spontaneous orofacial dyskinesia can occur in central nervous system conditions, whilst drug-induced dyskinesia is a response to chronic blocking of dopamine receptors in the basal ganglia by neuroleptic medications (Table 107.1).

107.2.1 Clinical Presentation and Investigations

Prevalence rates for spontaneous dyskinesias range from 1.5% to 38%. Prevalence rates of drug-induced dyskinesia ranges from 15% to 30%.

Orofacial dyskinesias can present as repetitive, non-random, continuous, stereotypic movements involving the tongue, lips, and mandible. Spontaneous dyskinesia is milder in intensity compared to drug induced. (Table 107.2).

The diagnosis of drug-induced dyskinesia requires minimum of 3 months of cumulative exposure to the offending drug and persistence of symptoms for more than 3 months after withdrawal of the offending medication.

Table 107.1 Medications associated with drug-induced dyskinesia and extrapyramidal reactions

Conventional antipsychotics	Haloperidol Chlorpromazine Perphenazine Pimozide Trifluoperazine
Atypical antipsychotics	Clozapine Olanzapine Risperidone
Antiemetics	Prochlorperazine Promethazine Metoclopramide
Antiparkinsonian agents	Levodopa Benztropine Trihexyphenidyl
Anticonvulsants	Phenytoin Carbamazepine
Antihistamines	Diphenhydramine Ranitidine
Tricyclic antidepressants	Amitriptyline Doxepin
Selective serotonin reuptake inhibitors	Fluoxetine Fluvoxamine Paroxetine Sertraline Citalopram Escitalopram
Prescription stimulants	Methylphenidate Phentermine Pemoline Dextroamphetamine Amphetamines Diethylpropion
Illicit drugs	Methamphetamine Cocaine 3,4-methylenedixoymethamphetamine (Ecstasy)

Table 107.2 Clinical presentations of orofacial dyskinesia

• Rapid movements
• Tongue twisting and protrusion
• Lip smacking and puckering
• Chewing movements
• Can affect extremities and trunk
• Oral pain
• Dysphagia
• Dysarthria
• Can be associated with ill-fitting removeable prosthesis
• Perception of inadequate oral hygiene

107.3 Drug-Induced Extrapyramidal Reaction

Drug-induced extrapyramidal reactions (EPR) refer to patients who develop a medication-induced orofacial motor hyperactivity. It often involves the muscles of the head, face, and mandible, in addition to other parts of the body including back and extremities, abdominal and pelvic muscles can also be affected.

Medications and illicit drugs that typically block dopamine receptors such as haloperidol, chlorpromazine, methamphetamine, and cocaine are most associated with EPR (Table 107.1).

Less than 5% of patients exposed to antipsychotics develop EPR, which manifest after exposure to the medication. The craniocervical region is the most affected site.

107.3.1 Clinical Presentation and Investigations

EPR usually present as dystonia, akathisia, and parkinsonism (Table 107.3). The onset of EPR can be classified as acute (occurring within hours to days after exposure), subacute (occurring weeks after drug exposure), and tardive (occurring months after exposure).

A diagnosis of EPR compared to other orofacial movement disorders is apparent from an accurate history and physical examination, especially noting a history of medication exposure.

Table 107.3 Clinical presentation of drug-induced extrapyramidal reaction

Dystonia	Involuntary, sustained, patterned, and often repetitive muscle contractions causing twisting movements or abnormal postures
Akathisia	A subjective report and objective manifestations of restlessness in the form of movement of the limb
Parkinsonism	Bradykinesia (slowness of movement) associated with at least one of rigidity, tremor, or postural instability

107.4 Management and Prognosis

Management for orofacial movement disorders is based on pharmacotherapy including cholinergic antagonists, GABA agonists and antihistamines, and botulinum toxin injections or, if not adequate, neurosurgical therapy or implanted medication pumps. The general rule for drug-induced dyskinesia or EPR is to withdraw the offending medication or drug in anticipation that the dyskinesia or reaction will resolve. Usually, the movement disorder resolves within few days to months following discontinuation of the medication, although in some cases it is irreversible.

Recommended Reading

Balasubramaniam R, Ram S. Orofacial movement disorders. Oral Maxillofac Surg Clin North Am. 2008;20(2):273–85, vii.

Clark GT, Ram S. Orofacial Movement Disorders. Oral Maxillofac Surg Clin North Am. 2016;28(3):397–407.

Cornett EM, Novitch M, Kaye AD, Kata V, Kaye AM. Medication-Induced Tardive Dyskinesia: A Review and Update. Ochsner J. 2017;17(2):162–74.

Dresel C, Bayer F, Castrop F, Rimpau C, Zimmer C, Haslinger B. Botulinum toxin modulates basal ganglia but not deficient somatosensory activation in orofacial dystonia. Mov Disord. 2011;26(8):1496–502.

D'Souza RS, Hooten WM. Extrapyramidal Symptoms. 2021. In: StatPearls [Internet]. Treasure Island (FL):StatPearls Publishing.

Bruxism

108

Lalima Tiwari, Jacinta Vu, and Alissa Jacobs

108.1 Introduction

Bruxism is an umbrella term used to describe a repetitive jaw-muscle activity characterised by clenching or grinding of the teeth and/or by bracing or thrusting of the mandible. It has two distinct circadian manifestations: during wakefulness (awake bruxism) or sleep (sleep bruxism). In healthy individuals, bruxism can be considered behavioural, rather than a pathological disorder of various aetiologies, with harmless, harmful or protective clinical consequences.

108.2 Clinical Presentation and Investigations

Awake bruxism is often observed as a transitory reaction to daily stress, in periods of concentration, or as a manifestation of anxiety. It can also be associated with Huntington's disease, orofacial dystonia, and illicit drug use.

The aetiology of sleep bruxism remains largely unknown but is considered to be multifactorial and centrally regulated in association with an excessive sleep arousal response and autonomic nervous system activity.

L. Tiwari (✉)
UWA Dental School, The University of Western Australia, Nedlands, WA, Australia
e-mail: lalima.tiwari@uwa.edu.au

J. Vu
Perth Children's Hospital, Nedlands, WA, Australia
e-mail: jacinta.vu@health.wa.gov.au

A. Jacobs
Fiona Stanley Hospital, Murdoch, WA, Australia
e-mail: alissa.jacobs@health.wa.gov.au

R. Balasubramaniam et al. (eds.), *Oral Medicine - A Clinical Guide*,
https://doi.org/10.1007/978-3-031-36797-7_108

Potential risk factors for sleep bruxism include excessive alcohol and caffeine use, cigarette smoking, medications including selective serotonin reuptake inhibitors, illicit drugs such as cocaine and methylenedioxymethamphetamine, and sleep disordered breathing. It currently remains unknown whether these risk factors increase the probability of sleep bruxism or aggravate pre-existing sleep bruxism.

Prevalence of awake bruxism is 5%. Reported prevalence of sleep bruxism ranges from 8% to 15% among adults, and up to 40–50% of children and adolescents.

Consequences of bruxism can manifest as varying clinical signs and symptoms (Table 108.1).

Diagnosis of awake bruxism is typically based on patient self-report and clinical examination. A clinical diagnostic criteria of sleep bruxism has been proposed by American Academy of Sleep Medicine (AASM) (Table 108.2).

Table 108.1 Signs and symptoms of bruxism

Signs	Symptoms
Tooth wear	Temporomandibular disorders
Masseter hypertrophy	Headache
Tongue indentation	Tooth pain and mobility
Linea alba	
Fractures	
Torus and exostosis	
Dental pulp alterations	

Table 108.2 American Academy of Sleep Medicine criteria for the diagnosis of sleep bruxism

Criteria A and B must be met: A. The presence of regular or frequent tooth grinding sounds occurring during sleep B. The presence of one or more of the following clinical signs: 1. Abnormal tooth wear consistent with above reports of tooth grinding during sleep 2. Transient morning jaw-muscle pain or fatigue; and/or temporal headache; and/or jaw locking upon awakening consistent with above reports of tooth grinding during sleep

Note: Although polysomnography is not required for the diagnosis, sleep bruxism, as described in the most recent version of the AASM Manual for the Scoring of Sleep and Associated Events, is ideally recorded with masseter muscle activity with audio-video signal to increase diagnostic reliability

108.3 Management and Prognosis

Management of bruxism is targeted towards addressing the consequences of bruxism rather than the muscle activity. Management of awake bruxism is focussed on habit awareness and behavioural therapies including habit reversal training, relaxation, hypnosis, and biofeedback therapy to reduce underlying psychosocial stressors. Avoidance of known risk factors should also be discussed with the patient. The primary management for sleep bruxism includes an oral appliance during sleep to protect dental structures. Additional therapies also include cognitive behavioural therapy and pharmacological treatment including muscle relaxants and botulinum toxin.

Recommended Reading

Berry R, Brooks R, Gamaldo C, Harding S, Lloyd R, Marcus C, et al. The AASM manual for the scoring of sleep and associated events: rules, terminology and technical specifications. Version 2.3. 2016. Available from: www.aasmnet.org

Khoury S, Carra MC, Huynh N, Montplaisir J, Lavigne GJ. Sleep Bruxism-Tooth Grinding Prevalence, Characteristics and Familial Aggregation: A Large Cross-Sectional Survey and Polysomnographic Validation. Sleep. 2016;39(11):2049–56.

Lobbezoo F, Naeije M. Bruxism is mainly regulated centrally, not peripherally. J Oral Rehabil. 2001;28(12):1085–91.

Lobbezoo F, Ahlberg J, Glaros AG, Kato T, Koyano K, Lavigne GJ, et al. Bruxism defined and graded: an international consensus. J Oral Rehabil. 2013;40(1):2–4.

Manfredini D, Colonna A, Bracci A, Lobbezoo F. Bruxism: A summary of current knowledge on aetiology, assessment and management. Oral Surg. 2020;13(4):358–70.

Soto-Goñi XA, Alen F, Buiza-González L, Marcolino-Cruz D, Sánchez-Sánchez T, Ardizone-García I, et al. Adaptive Stress Coping in Awake Bruxism. Front Neurol. 2020;11:564431.

Yap AU, Chua AP. Sleep bruxism: Current knowledge and contemporary management. J Conserv Dent. 2016;19(5):383–9.

Psychosocial Considerations for Orofacial Pain

109

Lalima Tiwari, Alissa Jacobs, and Jacinta Vu

Psychosocial factors such as depression, anxiety, somatoform disorders, catastrophising and fear avoidance require evaluation and inclusion in the treatment plan in patients suffering with orofacial pain.

109.1 Depression and Anxiety

Depression is a common and disabling psychiatric condition that is characterised by persistent sadness and a lack of interest or pleasure in previously rewarding or enjoyable activities. Anxiety refers to multiple mental and physiological phenomena, including a person's conscious state of worry over a future unwanted event, or fear of an actual situation. Quite often, anxiety and depression coexist with chronic pain.

Although depression and anxiety are not the cause of the patient's orofacial pain complaint, they are both known to aggravate and perpetuate the pain and require evaluation as part of care.

Prevalence of depression and generalised anxiety is higher in females (8.8% and 12.8%) compared to males (2.3% and 4.4%). These disorders are more prevalent in

L. Tiwari (✉)
UWA Dental School, The University of Western Australia, Nedlands, WA, Australia
e-mail: lalima.tiwari@uwa.edu.au

A. Jacobs
Fiona Stanley Hospital, Murdoch, WA, Australia
e-mail: alissa.jacobs@health.wa.gov.au

J. Vu
Perth Children's Hospital, Nedlands, WA, Australia
e-mail: jacinta.vu@health.wa.gov.au

R. Balasubramaniam et al. (eds.), *Oral Medicine - A Clinical Guide*,
https://doi.org/10.1007/978-3-031-36797-7_109

patients with temporomandibular disorders (TMD), estimated at 16–40%. Depression and anxiety may amplify pain perception with resultant hypervigilance and are considered predictors of chronic TMD progression in patients with acute TMD. Depression and anxiety have also been found to occur more commonly in patients with persistent oral burning with reported prevalence of 27.2–56.7%.

109.2 Somatic Symptom Disorder

Somatic symptom disorder was introduced in Diagnostic and Statistical Manual of Mental Disorders (DSM-5) to replace the diagnoses somatoform disorder. It is characterised by one or more somatic symptoms that are accompanied by excessive thoughts, feelings and/or behaviours related to the somatic symptoms. In addition, the symptoms cause significant distress and/or dysfunction. They may or may not be explained by a recognised medical condition. These include functional somatic syndromes, now known as central sensitivity syndromes, with comorbid fibromyalgia, interstitial cystitis, irritable bowel syndrome, back pain, chronic fatigue syndrome and vulvodynia. Patients suffering with central sensitivity syndrome often present with painful TMD. This process has been attributed to alterations in peripheral and central nervous system processes, that have the net effect of amplifying the perceptual response to nociceptive stimuli. The presence of underlying somatic symptoms in patients with painful TMD and dysfunction is likely to be associated with poorer treatment response and continued perpetuation of pain.

109.3 Catastrophising and Fear Avoidance

Pain catastrophising is a maladaptive cognitive-affective response to pain that involves negative thinking regarding the pain experience. It is a multidimensional construct consisting of rumination, magnification, and helplessness. Baseline catastrophising is strongly associated with increased pain intensity and may increase the risk for transition from acute to chronic TMD pain. Studies have also shown that pain catastrophising was closely associated with the disruption on daily activities and affective dysfunction noted in patients with burning mouth disorder.

Fear avoidance is the avoidance of activity and subsequent disuse due to fear of pain, resulting in a downward cycle of pain and disability. Research relating to fear avoidance in orofacial pain is sparse. The concept however can be applicable in chronic orofacial pain conditions such as persistent TMD and orofacial neuropathic pain.

Table 109.1 Checklist of psychological and behavioural factors for orofacial pain (adapted from the American Academy of Orofacial Pain Guidelines for assessment, diagnosis and management of orofacial pain, sixth Edition)

- Inconsistent, inappropriate and/or vague reports of pain
- Symptoms incompatible with the innervation and function of anatomical structures
- Overdramatisation of symptoms
- Symptoms that vary with life events
- Significant pain of greater than 6 months' duration
- Repeated failures with conventional therapies
- Inconsistent response to medications
- History of other stress-related disorders
- Major life events (e.g. new job, marriage, divorce, death of a family member or friend)
- Evidence of alcohol and drug abuse
- Clinically significant anxiety, depression or suicidal or homicidal ideation
- Evidence of secondary gain

Table 109.2 Instruments available for screening and comprehensive assessment of psychosocial aspects for temporomandibular disorders based on Diagnostic Criteria for Temporomandibular Disorders

Domain	Instrument	Screening	Comprehensive
Pain location	• Pain drawing	✓	✓
Pain intensity	• Graded chronic pain scale	✓	✓
Pain disability	• Graded chronic pain scale	✓	✓
Distress	• Patient Health Questionnaire-4 (PHQ-4)	✓	✓
	• Patient Health Questionnaire-9 (PHQ-9) - depression • General Anxiety Disorder-7 (GAD-7) - anxiety		✓
Physical symptoms	• Patient Health Questionnaire-15 (PHQ-15)		✓
Limitation	• Jaw functional limitations scale – 8- item • Jaw functional limitations scale – 20-item	✓	✓
Parafunction	• Oral behaviours checklist	✓	✓

109.4 Management and Prognosis

Early consideration and identification of psychosocial factors, which impact prognosis, are important in treating orofacial pain conditions. Behavioural components of the history may be suggestive of underlying psychosocial factors (Table 109.1), and validated instruments may be utilised for the assessment of psychosocial characteristics of patient's with TMD (Table 109.2). Behavioural self-regulation is key to management and often involves a multidisciplinary team approach including mental health professionals such as psychiatrists and psychologists utilising therapies such as pharmacotherapeutics, mindfulness, relaxation techniques, cognitive behavioural therapy, biofeedback, and psychotherapy.

Recommended Reading

Balasubramaniam R, Tiwari L, Phoon-Nguyen A. Psychosocial considerations for temporomandibular disorders. Ann R Australas Coll Dent Surg. 2018;24:33–7.

Bartley EJ, Schmidt JE, Carlson CR, Fillingim RB. Psychosocial considerations in TMD. In: Gremillion HA, Klasser GD, editors. Temporomandibular Disorders: A Translational Approach From Basic Science to Clinical Applicability. Cham: Springer International Publishing; 2018. p. 193–217.

Chana P, Smith JG, Karamat A, Simpson A, Renton T. Catastrophising, pain self-efficacy and acceptance in patients with Burning Mouth Syndrome. J Oral Rehabil. 2021;48(4):458–68.

De Leeuw R, Klasser GD, editors. Orofacial pain: guidelines for assessment, diagnosis, and management. Hanover Park, IL, USA: Quintessence Publishing Company, Incorporated; 2018.

Edmond SL, Enriquez CS, Millner MH, Nasri-Heir C, Heir GM. Is there an association between fear avoidance beliefs, and pain and disability in patients with orofacial pain. J Oral Rehabil. 2017;44(6):426–33.

Fillingim RB, Ohrbach R, Greenspan JD, Knott C, Diatchenko L, Dubner R, et al. Psychological factors associated with development of TMD: the OPPERA prospective cohort study. J Pain. 2013;14(12 Suppl):T75–90.

Häggman-Henrikson B, Bechara C, Pishdari B, Visscher CM, Ekberg E. Impact of Catastrophizing in Patients with Temporomandibular Disorders-A Systematic Review. J Oral Facial Pain Headache. 2020;34(4):379–97.

Löwe B, Levenson J, Depping M, Hüsing P, Kohlmann S, Lehmann M, et al. Somatic symptom disorder: a scoping review on the empirical evidence of a new diagnosis. Psychol Med. 2022;52(4):632–48.

Tu TTH, Watanabe M, Suga T, Hong C, Takao C, Takenoshita M, et al. Personality Traits in Burning Mouth Syndrome Patients With and Without a History of Depression. Front Psychiatry. 2021;12:659245.

Yeung E, Abou-Foul A, Matcham F, Poate T, Fan K. Integration of mental health screening in the management of patients with temporomandibular disorders. Br J Oral Maxillofac Surg. 2017;55(6):594–9.

Halitosis

110

Jeremy Lau

110.1 Introduction

Halitosis can be defined as malodour of the breath. Halitosis is an extremely common condition and is estimated to affect 32% of the global population. It can be subcategorized into genuine halitosis, pseudo-halitosis, and halitophobia. Some individuals may present with pseudo-halitosis where they may perceive a malodour in their breath despite not being objectively detected. Halitophobia is the pathological fear of having bad breath despite objective evidence that it does not exist.

110.2 Clinical Presentation and Investigations

The cause of halitosis is multifactorial and can be subcategorised into oral factors and extra-oral factors (Table 110.1). Intra-oral malodour primarily results from microbial biofilm forming interdentally but also within the tongue crypts, fissures, and papillae. Volatile organic compound including sulphur compounds can be generated leading to halitosis. Metabolic disorders including diabetic ketoacidosis may also cause halitosis. Furthermore, halitosis can also be induced or exacerbated by medications and substances. Substance-induced halitosis includes smoking, alcohol, as well as food items such as garlic and onion.

Psychogenic factors do not have clinical correlation to pathology, and psychological aspects play a major role. Individuals with significant stress and anxiety more commonly experience halitosis.

J. Lau (✉)
UWA Dental School, The University of Western Australia, Nedlands, WA, Australia
e-mail: jeremy.lau@uwa.edu.au

R. Balasubramaniam et al. (eds.), *Oral Medicine - A Clinical Guide*,
https://doi.org/10.1007/978-3-031-36797-7_110

Table 110.1 Common associations of halitosis

Oral factors- local associations	Extra-oral factors- systemic associations	Extra-oral factors- psychogenic associations

• Intra-oral bacteria
  • Periodontal disease
  • Caries
• Poor oral hygiene
• Acute infection
• Salivary gland hypofunction
• Oral cancer
• Tonsillar pathology
	• Acute pharyngeal infection
	• Sinusitis
	• Febrile illness
	• Respiratory conditions
	• Gastrointestial disease
	• Kidney disease
	• Liver disease
	• Ketoacidosis
		• Psychosocial factors
		• Stress and anxiety
		• Economic status
		• Underlying psychiatric condition

There is a varied clinical presentation of halitosis. Where local factors are present, patients frequently present with poor oral hygiene, heavy bacterial load on the tongue, significant dental plaque, dental caries, and periodontitis. Other clinical signs and symptoms can include hyposalivation, xerostomia, and altered taste.

Clinical investigations of halitosis include discriminating between genuine halitosis, pseudo-halitosis or halitophobia. It is important to take a thorough history of the presenting condition and a thorough medical history.

Organoleptic: Organoleptic scale measures the intensity of malodour between 0 (no odour) and 5 (intense). This test is considered the gold standard. In the test, the examiner performs a sniff test of the patient's breath.

Gas chromatography: This test measures the concentration of volatile sulphur containing compounds in samples including in saliva, tongue coating, and expired breath. Gas chromatograph is highly accurate but can be expensive.

Microbiologic tests: Microbes from biofilms causing malodour can be detected and analysed quantitatively or qualitatively. Examples of such microbes include Treponema denticola, Porphyromonas gingivalis, and Tannerella forsythia.

Blood tests: Due to the systemic association with halitosis, blood tests may be considered. Blood tests may be conducted to screen for conditions including diabetes, kidney or liver disorders, gastrointestinal disorders, and Sjogren's syndrome.

110.3 Management and Prognosis

Treatment and management of halitosis include identifying and addressing its cause. Some common strategies involve:

- Improving oral hygiene.
- Comprehensive dental care including treatment of dental pathology and management of oral lesions.

- Management of dry mouth.
- Masking agents including mints and chewing gum.
- Antimicrobials that target volatile sulphur compound producing bacteria in the oral microbiota.
- Medical management of any systemic conditions associated with halitosis.
- Psychosocial interventions.

Management of patients with pseudo-halitosis involves careful education, counselling, and reassurance. Many patients with pseudo-halitosis may have underlying psychosocial issues. Overall, management of patients with halitosis requires a multidisciplinary approach. Many patients may have overlapping aetiologies, and this may lead to challenges in its management.

Recommended Reading

Aylıkcı BU, Colak H. Halitosis: From diagnosis to management. J Nat Sci Biol Med. 2013;4(1):14–23.

Mokeem SA. Halitosis: a review of the etiologic factors and association with systemic conditions and its management. J Contemp Dent Pract. 2014;15(6):806–11.

Silva MF, Leite FRM, Ferreira LB, Pola NM, Scannapieco FA, Demarco FF, et al. Estimated prevalence of halitosis: a systematic review and meta-regression analysis. Clin Oral Investig. 2018;22(1):47–55.

Facial Palsy

111

Guru O and Mathew Lim

111.1 Introduction

The facial nerve (cranial nerve VII) has five branches that provide motor supply to the facial muscles as well as taste sensation to the anterior two-thirds of the tongue (chorda tympani), and secretomotor function to the submandibular and sublingual glands. Facial palsy is a term used to describe conditions with varying aetiology that present as paralysis or weakness of the facial muscles. Facial palsies are most commonly unilateral.

The aetiology of facial palsy is numerous and includes swelling, inflammation or direct trauma to the nerve (Table 111.1). The most common cause is Bell's palsy, which is a type of facial palsy for which the cause is largely unknown. Postulated theories include anatomical (e.g. decreased size of the facial canal and/or increased cross sectional size of the nerve), viral reactivation, ischaemia (resulting in compression of the nerve) and inflammation.

Bell's palsy is the most common cause of facial paralysis, accounting for 50–75% of all cases. The incidence of Bell's palsy ranges from 7 to 40 cases per 100,000 with no significant age or gender predilection. Causal incidences include idiopathic (49.5%), infectious (15.3%), neoplastic (13.5%), neurologic (13.5%) and traumatic (8.2%).

Guru O (✉)
UWA Dental School, The University of Western Australia, Nedlands, WA, Australia
e-mail: guru.o@uwa.edu.au

M. Lim
Melbourne Dental School, The University of Melbourne, Carlton, VIC, Australia

Dental Services, Alfred Health, Melbourne, VIC, Australia

Royal Melbourne Hospital, Parkville, VIC, Australia

R. Balasubramaniam et al. (eds.), *Oral Medicine - A Clinical Guide*,
https://doi.org/10.1007/978-3-031-36797-7_111

Table 111.1 Causes of facial nerve palsy

Trauma	Base of skull fractures, facial injuries, otitic barotrauma
Congenital	Birth trauma, geniculate ganglion vascular malformation, pontine vascular malformation, pontine developmental venous anomaly, Mobius syndrome, Goldenhar syndrome
Neurologic	Infarction of brainstem, multiple sclerosis, myasthenia gravis, Melkersson-Rosenthal syndrome, cerebrovascular accidents
Infection	*Bacterial*: Necrotising otitis externa, Borreliosis (Lyme disease), syphilis, leprosy, cat scratch disease, skull base osteomyelitis, rickettsia, Ehrlichiosis *Viral*: Ramsay hunt syndrome, herpes zoster oticus, mumps, HIV, cytomegalovirus, rubella virus, influenza B virus, Coxsackie virus *Others*: Otitis media, mastoiditis, Guillain-Barré syndrome, cat scratch
Metabolic	Diabetes mellitus, pre-eclampsia, acute porphyria, hypertension, hypothyroidism, pregnancy, vitamin A deficiency
Neoplastic	Lymphoma, tumour of parotid gland, Tumours of the middle ear (including cholesteatoma and schwannomas) Tumours of the petrosal bone, Tumours of the cerebellopontine angle (including acoustic neuroma) Other tumours—Haemangioma, meningioma, glomus tumour, leptomeningeal carcinomatosis, acute lymphatic leukaemia
Idiopathic	Bell's palsy
Iatrogenic	Brain, middle ear, mastoid, parotid or facial surgery, carotid artery dissection in the neck, dental local anaesthesia
Autoimmune	Temporal arteritis, systemic lupus erythematosus, sarcoidosis, amyloidosis, Thrombotic thrombocytopenic purpura, Wegener disease, Sjogren syndrome

111.2 Clinical Presentation and Investigations

Facial palsies may be classified into two primary groups depending on the motor neurones affected. Upper motor neurons supply the muscles of the lower face and arise from the contralateral motor cortex, thereby causing a unilateral palsy with some sparing of the frontalis and orbicularis oris muscles. Lower motor neuron lesions are characterised by unilateral paralysis of all facial muscles. The typical presentation includes an unfurrowed forehead, Bell's sign, diminished lacrimation, loss of the nasolabial fold, and drooping of the corner of the mouth. This muscle weakness may be preceded by pain in the ear region, hyperacusis and loss of taste.

Despite being the most common presentation, Bell's palsy should be a diagnosis of exclusion. The most important differential is palsy associated with an acute stroke. This can be differentiated by the movement of the frontalis, as it receives input from both hemispheres and hence retains its function in a unilateral central lesion compared to a peripheral facial palsy whereby all facial muscles are affected. A thorough history and examination should cover:

- Patient details: age, ethnicity, geographic location (including recent travels) and medical history.
- History of presenting complaint, including any changes in the time course of paralysis.

- Complete head and neck examination:
  - Facial evaluation in horizontal thirds—instruct to smile, close eyes and raise the eyebrows (House-Brackmann scale or Sunnybrook Facial grading system).
  - Otologic examination, including tuning fork evaluation.
  - Skin examination-noting rashes, lymph nodes (systemic, autoimmune, neoplastic causes), vesicles (zoster).
  - Palpation of parotid glands to rule out mass.
  - Intra-oral examination—rashes and tongue presentation (Melkersson-Rosenthal syndrome, zoster).

Further tests such as imaging (CT temporal bone and/or MRI brain and skull base) and serology may be ordered based on the examination findings (e.g., full blood count, Lyme serology, serum ACE for sarcoidosis).

111.3 Management and Prognosis

Management of a facial palsy is determined by the attributed causation. Supportive therapy includes corneal protection and use of ocular lubricant to prevent irritation and dryness. Bell's palsy medical management often involves a short course of oral steroids (prednisolone 0.5 mg/kg. Consider 60 mg for 5 days, followed by 10 mg of tapering for 5 days). Antiviral therapy is controversial unless a viral cause is supported. Initiation of steroid therapy within 72 hours of symptom onset usually results in recovery within 2 weeks. The prognosis varies between causes. Better long-term outcomes are usually associated with those who recover within 2 weeks, decreased severity of paralysis and younger age.

Recommended Reading

Bleicher JN, Hamiel S, Gengler JS, Antimarino J. A survey of facial paralysis: etiology and incidence. Ear Nose Throat J. 1996;75(6):355–8.

Heckmann JG, Urban PP, Pitz S, Guntinas-Lichius O, Gágyor I. The Diagnosis and Treatment of Idiopathic Facial Paresis (Bell's Palsy). Dtsch Arztebl Int. 2019;116(41):692–702.

House JW, Brackmann DE. Facial nerve grading system. Otolaryngol Head Neck Surg. 1985;93(2):146–7.

Ross BG, Fradet G, Nedzelski JM. Development of a Sensitive Clinical Facial Grading System. Otolaryngol—head neck surg. 1996;114(3):380–6.

Spencer CR, Irving RM. Causes and management of facial nerve palsy. Br J Hosp Med (Lond). 2016;77(12):686–91.

O TM. Medical Management of Acute Facial Paralysis. Otolaryngol Clin North Am. 2018;51(6):1051–75.

Zhang W, Xu L, Luo T, Wu F, Zhao B, Li X. The etiology of Bell's palsy: a review. J Neurol. 2020;267(7):1896–905.